# *America Is Me*

# America Is Me

170 Fresh
Questions and Answers
on Black American History

Kennell Jackson

HarperPerennial
*A Division of* HarperCollins*Publishers*

*To my mother and father–Lottie F. Jackson and Kennell A. Jackson Sr.–who, as my first real teachers, taught me to love knowing*

A web page for this book is available at: http://www–leland.stanford.edu/~kennell/

A hardcover edition of this book was published in 1996 by HarperCollins Publishers.



HarperCollins books may be purchased for educational, business, or sales promotional use. For information please write: Special Markets Department, HarperCollins Publishers, Inc., 10 East 53rd Street, New York, NY 10022.

First HarperPerennial edition published 1997.

*Designed by Nancy Singer*

---

The Library of Congress has catalogued the hardcover edition as follows:

Jackson, Kennell A.
America is me : 170 fresh questions and answers on Black American history / by Kennell Jackson.
p. cm.
Includes bibliographical references (p.) and index.
ISBN 0-06-017036-0
1. Afro-Americans—History—Miscellanea. I. Title.
E185.J323 1996
973'.0496073—dc20 95-48927

---

ISBN 0-06-092785-2 (pbk.)
97 98 99 00 01 ❖ /RRD 10 9 8 7 6 5 4 3 2 1

# Acknowledgments

My debt is immense. I owe many thanks to David Cornell, Bart Kilpatrick, Rene Romano, Leslie Harris, Todd Benson, Steve Pitti, Wilmetta Tolliver, Christy Kim, Chi-hui Yang—Stanford undergraduates and graduate students who were true researcher-friends throughout this work. Special thanks go to the Reference Services librarians at Stanford University's Cecil H. Green Library, who answered literally hundreds of my questions: Joanne Hoffman, Eric Heath, Elisabeth Green, John Rawlings, Richard Fitchen, Rose Adams, and Kathryn Kerns. Sonia H. Moss, of Interlibrary Loan, helped with all my special manuscript requests. Brian Keizer—music critic extraordinaire—taught me nearly everything I know about Black music.

To the agent and editors who "coached" me through this process, I owe an African salute—especially to Janet Goldstein and Betsy Thorpe, and also to Hugh Van Dusen and Kate Ekrem of HarperCollins; and my agents Barbara Lowenstein and Madeleine Morel.

# Contents

# Introduction

In the year 2019, Blacks will have been in America for four hundred years. In less than twenty-five years, we—as a nation—will reach this landmark. Being so focused on the end of the millennium, hardly anyone has mentioned that we are fast approaching this watershed in American history.

Four hundred years confers a permanence, an aura of true and lasting quality, on the Black American past. Many nations in today's world are younger than four hundred years. Empires have been swept away during this time. Revolutionary movements have come and gone. Whole systems of thought have risen and disintegrated during these years. Through all of this, Black Americans have continued, winding their way toward the future. They are an enduring part of American and world history. The year 2019 will simply highlight this fact.

Blacks in America have created a major historical story. Actually, it is more on the order of an epic. A few months into writing this book, the true scope of Black American history—in time, places, and people—descended on me. One spring morning, a book collector friend of mine arrived at my door with a small cache for me to see of early nineteenth-century documents written by Blacks.

In a slim pine box, she had a few sermons and addresses, a small book, and a few parts of letters. Many were almost two centuries old. They had to be handled with the care given to radioisotopes in a lab. As we spread these writings out on a big table, read, and discussed them, we began to realize—in a much deeper sense than we had before—how vast the Black past was. It was something about those almost-two-centuries-old fading documents—the jumbled printing, the scribbling, the old English usage—that got to us. We talked about what a tiny portion in time and deed our own generation of Black American history represented. We felt like latecomers to the story.

After that day, I would occasionally steal away from writing this book and visit places that would let me absorb more of the epic quality of the Black American past. My hasty trips gave me even a broader context for my friend's documents. They took me further back in time and out to a greater geography. My first trek, in the summer of 1993, was traveling along coastal West Africa. The soft moist winds of Africa greeted me. I lived in villages; acquainted myself with local manners; listened to the euphonious languages; and, alas, touched the earth on which the old slave cargoes had been assembled. A few months later, I was in the Caribbean looking at old plantations and still-existing runaway slave villages—suitable places for imagining the early Africans, who were brought from those islands to America rather than directly from Africa.

On a magnificent rainy day in November 1994, I found myself at Jamestown, Virginia, the lowland ocean-facing settlement to which Africans were first brought in 1619. Leaving Jamestown, my next stop was South Carolina. In the eighteenth century, thousands of Africans were enslaved there. Later, I devised my own tour of Black New York City, revisiting Black history sites that reached back to the 1710s. One such site was Maiden Lane—just a few blocks from Wall Street—where in 1712 thirty slaves made a desperate bid for freedom. Chicago, Omaha, Oklahoma were soon added to my travels—places to which Blacks have fled as they sought to use their freedom. On Labor Day 1995, my wanderings came to an end. I was in Los Angeles's Black communities, on the edge of the continent. Places that are so different and distant from one another conveyed to me the epic sweep of Black American history.

Like any epic, Black American history has had a succession of great narrators. In this, it has been blessed. These narrators have kept it alive. Each teller has told their part—African village storytellers; captives whispering in the holds of slave ships; plantation women with their tales of heroes and heroines; plantation men who spoke of remote lands left behind. And then, there were the narrators who began to write Black histories in the nineteenth century. Lastly, there are those narrators that we now call historians, those systematic students of the Black past, who have come from every background and who have labored so diligently to build a modern Black American history.

*America Is Me* belongs to this procession of narratives. It attempts to capture the epic that is the Black American past. This book spans an enormous time. It goes deeper into the African past than many books on Black American history, hopefully painting a richer picture of the societies from which Black Americans have come. And it runs into the 1990s, coming as close to recent events as possible.

As the narrator, I have chosen a slightly different style for telling of this past—a question-and-answer format. I pose a question or series of questions and then give an answer that is moderate in length. No answer is much over fifteen hundred words. Many are under a thousand words. In the answers, I have tried to blend a trio of features: facts, analysis, and storytelling. The latter, I think, is essential to interesting historical writing and hopefully it will be restored to its proper place in the future. My passion is for finding the revealing anecdote or the illuminating fact. Whenever I succeeded, I included them in the answers. In sum, I have tried to make the answers into an essay rather than a mere recitation of facts. My greatest desire is to draw the reader in.

Why have I chosen the question/answer format? It is an interesting story about the education of a historian. For almost the whole of the 1980s, I was director of Stanford University's Undergraduate Program in African and Afro-American Studies. Nearly every week, groups—churches, sororities, schools, senior citizens homes, youth clubs, shelters—would call the program's office requesting a speaker. Requests came from a variety of groups. Often, they wanted someone to speak briefly on Black history and issues. Desiring to get beyond the university, to hear what was on people's minds, I jumped at the opportunities. I spoke to far more than a hundred Black groups. And, I have also been honored by invitations from many others—for example, an Asian-American veterans association, a Muslim day school, exclusive prep schools, Jewish groups, business alliances, even a collection of organic farmers.

To me, these occasions were greatly instructive. Indeed, they transformed my vision of my future work in Black history. They reminded me of my responsibility to what historian Benjamin Quarles has called "Black public history."

Usually when I finished with my talk of about twenty minutes, I would ask for questions. Often, this was where the challenge and fun began. Stand before an audience for a Black history talk and you can expect a wealth of questions. In fact, questions came from everywhere and on an amazing array of historical issues. At first, I must confess the range of the questions and the seriousness of the interrogators made me anxious. It was a little intimidating. In particular, Black audiences always want to probe their past, and they are often sharp and want substantive answers. I met people who had lived through experiences I was talking about. I encountered many people who had read widely. Some were writing their own works. Trepidation was in order: I was no Black history reference text.

As I began to prepare more for these occasions, my command over the myriad facts of Black history was strengthened, and I began to relish the give-and-take of these robust sessions. They tested my knowledge and wits. I even enjoyed the odd questions, the embarrassing ones and the ones with charged racial implications. After a few years "on the circuit," I kept a record of these questions.

Into the 1990s, this has continued. In February 1995, at an excellent Black history month presentation, I had what I considered the ultimate opportunity, speaking to an ethnically mixed group of young men and women in a juvenile detention center. That evening was one of real sharing—giving rise to a generous, surprisingly informed dialogue on Black history. On top of this, I get letters with questions from individuals, especially from high school students. Last year, a letter from students in San Jose, California, asked, "What kind of clothing did the slaves and their masters wear?" and "Were there any specific slaves that are historically famous for what they've gone through?"

Primarily, this book grows out of my traveling question-and-answer road show of the last fifteen years. But, the specific catalyst was my reading of Kenneth C. Davis's *Don't Know Much About History* (1990), a popular question-and-answer American history text. His format paralleled my own experience. Moreover, I thought he had hit upon a great 1990s idea: these days, people often want shorter introductions to history's events, personalities and processes.

Three types of questions are presented here. First, there are those basic questions that are essential to Black American history. One should know the answer to these questions to be Black history literate. For example, to know Black history up to the Civil War, there were certain fundamental questions. Generally, they pertain to African societies' values, the creativity of Africa, the slave trade, plantation economies, the retention of African culture by slaves, Black rebellions against slavery, Black abolitionists such as Frederick Douglass, Sojourner Truth, and Harriet Tubman, and the Black Civil War soldiers. Basic questions are the perennials of Black American history.

Second, this book contains questions that are newer in origin. Beginning with the mid-1980s, these types of questions were increasingly asked of me in my talks—questions about the role of women in Black history; questions about new controversies (pertaining to ancient Egypt, the Black Panther Party's work, Black male issues, the underclass); questions about ideologies (Afrocentrism, Black conservatism, Jesse Jackson's rainbow coalition); and questions about sports, music, and literary history. Third, the book contains a number of essays of a somewhat philosophical nature, such as what American parents should tell their children about slavery. Many of these questions have been suggested—in some version—by my public encounters.

Roughly, the questions are positioned in chronological order. When a person achieves prominence, a question about them will appear. When an event occurs, the question has a place in time reflecting that fact. Sometimes, when dealing with a movement covering several years, such as the Harlem Renaissance of the 1920s, I have had to be a little arbitrary in placing the question. In most instances, I try to write about events confined to the timeframe of each chapter. Sometimes, however, I have had to bring in events from earlier times or later to make sense in the answer.

From photographer Gordon Parks's photoessay *Born Black* (1963) comes the title for this book. When Parks (1912– ) wrote those three words, "America is Me," he asserted that his experience and the American experience were woven together. In a real sense, this is true not only for a single Black, like Parks, but for the whole Black people in America.

Blacks and America were joined at birth. In Arthur Ashe's posthumous *Days of Grace* (1993), he writes his daughter ("My Dear Camera"), telling her "You are part of a tree" and "You are a tenth-generation African-American." Furthermore, Blacks have contributed immensely to America—building it, helping to produce one of the most unique cultures in the world, expanding its legal definition of itself, lending its geniuses to the improvement of American life. This is almost a mantra with me.

But, this book demonstrates more than that. There is more to the meaning of *America Is Me.* It demonstrates how Blacks have epitomized and even stimulated the formation of core American values. How they accomplished this while being enslaved and, when freed, treated badly has never been told fully. For example, as I wrote this book, it was astonishing to discover the prevalence in Black history of what we call today "risk-taking." Gordon Parks—trying to break into high-class photography—was such a risk-taker.

As the book took shape, it grew more and more obvious that Blacks have helped inspire that deep American feeling for challenging imposed limits, for going all out, for adventure, pioneering, exploration, and daring inquiry. The past Blacks I encountered while writing this book were always moving ahead. Often, they were setting the pace—not just for themselves, but for their country. As Black Americans have made their spirited assault on limits, they have taught other Americans lessons about forging ahead. They have helped lift the American spirit. Often, I felt humble in retracing the lives of these past Blacks: they had such a profound sense of the bounty and the promise of themselves as humans and of their country. Hence, the title *America Is Me* is intended to suggest many links between Blacks and this land.

It is my intention to emphasize what historian Nathan Huggins advocated in his last volume *Revelations* (1995): "We might . . . see as if for the first time the elemental truth in the black American experience; rather than being an anomaly, it is central to the story."

A word about a name: after much consultation and discussion, I decided to use "Black" as the main designation in the text. The other major designation is "Black American." When "African-American" was introduced in the mid–1980s, I supported its adoption. But, while writing this book, I found it hard to use "African-American" for the very earliest periods in this story, the seventeenth and eighteenth centuries. It appeared anachronistic when inserted into the text for these early periods. Equally so, for the period from 1910 to 1950, African-American was also jarring in the text. Quotes from 1910–1950 heavily used the word "Negro." From 1965, "Black" was also extensively used. Finally, it was important to choose one designation and stick with it for consistency. Using "Black" and "Black American" implies no distaste for "African-American." Historical issues were the deciding factors here.

These days, my morning papers regularly bring me editorials and op-ed pieces that speak of America as pulling apart along racial lines, with words like "the gulf between the races," "racial chasm," "racial impasse." The other day, I heard a local talk show host refer to our racial issues as a "trembling faultline"—serious words in earthquake-conscious California. No one really knows whether our current national racial unease is just an episode, part of a cycle, or something truly long term.

Yet, one thing is sure. A proper understanding of the Black American past can be a spiritual resource for the country as it passes through these troubling times. It stands ready to teach us as we work through our present quandary. The history that this question/answer book brings forth is a grand story, generated by a future-facing people, who believed deeply, as Gordon Parks wrote, "there is yet a chance for us to live in peace beneath these restless skies."

For readers wanting to continue the dialogue of this book, this e-mail address for the author is provided: AmericaIsMe.@ leland.stanford.edu

*Stanford*
*October 13, 1995*

# 1

# Living Gifts from Old Africa

## To the 1400s–1500s

4–3.5 Million Years B.C.: Earliest hominids are in East Africa

150,000 B.C.: Homo erectus—our direct ancestor—in East Africa

100,000 B.C.: First complex toolmaking appears (in Africa—Zaire)

8000–6000 B.C.: First African settlements form and agriculture emerges

3100 B.C.: The rise of ancient Egypt

600–500 B.C.: Early African ironworking (in Sudan) advances

800 A.D.–1500 A.D.: West Africa's great kingdoms—Ghana, Mali, Songhai—flourish

1375: Earliest detailed European map of western Africa is made

1400–1600: Africans on western coast—make great progress

1440s: First Africans taken as slaves by Portuguese

1450: Portuguese ships reach Senegal in West Africa

1482: Portuguese build Elmina Castle (on modern Ghana coast)

1498: Vasco da Gama sails to India by circling the Cape of Good Hope

1450–1600: Slave trade alliances between Europeans and Africans are formed

From five hundred miles above the earth's surface, a satellite's cameras click hundreds of times per minute, capturing the smallest features of a whole continent. The year is 1985. The continent is Africa.

Satellite photos are technological wonders. Everything can be seen so clearly: coastal forests, broad deserts, tiny Saharan oases, long rivers, slivers of valleys, minute lakes, artful cloud formations, even cattle trails. These pictures capture today's Africa, its unique V form a familiar shape on the world's maps. This Africa is also a very old place, imposing in its antiquity.

"Venerable Africa—ancient place, one of the oldest of this old earth" is how an African popular song puts it. Black American history began in this land. Africa's human seed has been scattered throughout the world for centuries. Photojournalist Chester Higgins Jr. has spent his career "searching the world for the people of Africa"—photographing them in Belize, Jamaica, Mexico, Grenada, Alabama, and New York City. Today's Black Americans are one flowering of Africa's human seed.

Africa partly molded Black Americans into the diverse people they are today. In 1619, when Africans were first introduced into the North American English settlements, they came as trade commodities. But in their minds and hearts, they carried the living cultures of Old Africa. The English in these settlements had little or no knowledge of Africans, but they too were embarking on an historic journey involving the ancient, vital cultures of Africa. They too would be touched by the African continent.

This chapter goes beyond the story of Africans as traders' merchandise. Many Americans have questions about Old Africa, about the times before Africans were sold as commodities. In answering these questions, we begin to create a picture of

Africa before the contact with Europeans was made; we begin to understand the cultural gifts that the Africans who were brought to America carried with them to this civilization in the making.

### What are some basic facts about Africa?

Africa is the second largest continent in the world; only Asia is larger. In land area, it is four times the size of the United States, three times larger than Europe. The equator runs through it. A substantial part of the continent lies in the tropical zones. Thus, temperatures are usually high for most of the year. Not many islands surround Africa, but Madagascar, the world's fourth largest island, is associated with the continent. Africa has five big regions: North Africa, West Africa, Central Africa, East Africa, and southern Africa.

At present, 650 million people live in Africa, one-eighth of the world's total. The world's largest desert—the Sahara, in West Africa—almost divides the continent. The majority of Africa's peoples live in sub-Saharan Africa, the area south of the Sahara, "Black Africa." North of the Sahara is Arab Africa.

Most of Africa's land is plateau. The continent has few major mountains. The highest is Mount Kilimanjaro (19,341 feet) in Tanzania, East Africa. Africa is bound on the west by the Atlantic Ocean, on the east by the Indian Ocean and the Red Sea, and on the north by the Mediterranean Sea. Two of the world's longest rivers are in Africa: the Nile, running 4,145 miles, and the Zaire (formerly the Congo), traveling 2,900 miles.

Currently, some 850 ethnic groups live in Africa, each with a distinct language. The groups range from those with millions of members (such as the Hausa in West Africa, the Swahili peoples in East Africa, and the Zulu in South Africa) to those with a few thousand members (such as the Okiek of Kenya, East Africa). All of these people speak languages that belong to four large language families.

Three religious systems exist in Africa. There are the powerful, traditional, older local religions. There is Islam, which has had followers in Africa for more than 1,200 years. And there is Christianity, mostly disseminated by missionaries during the last century. Ethiopia, however, was one of the first nations in the world to become Christian (around 300 A.D.), developing its own Coptic church.

Today, all this diversity is contained within fifty-four nations. Eritrea is the newest one, voting for its independence in May 1993. The three largest nations in land area are Sudan (970,000 square miles), Algeria (920,000 square miles), and Zaire (905,000 square miles). The three largest states in population are Nigeria (109 million), Egypt (52 million), and Ethiopia (47 million). Overwhelmingly, African nations were created by colonial rule, which began in the nineteenth century. Even their present boundaries were drawn by Europeans. With colonial rule, the French, English, and Portuguese languages became more widely used.

Only the nation of Liberia, founded in 1821 by Blacks from the United States, managed to escape colonial rule. Ethiopia was independent for thousands of years. However, in 1935–41, fascist Italy took it by force.

Africa produces half the world's diamond and gold supply. Large quantities of oil, natural gas, cobalt, wood, coffee, rubber, iron, and coal are also produced. But today, Africa struggles with increasing poverty. It is also struggling with ethnic conflicts, oppressive rulers, and disease. Africa's greatest resource has always been its people, with their intelligence and persistence. In many African countries today, citizens are working passionately to build democracies and stable economies.

The year 1994 was one of extremes for Africa, with the nadir of genocide in Rwanda and the apex of South Africa's liberation. Several events at the end of the year pointed to a more positive future for Africans: Nelson Mandela's pragmatic leadership; an end to Mozambique's prolonged civil war and the establishment of free elections; the rising free-market economies in Ghana and Botswana; the vitality of campaigns for women's rights, human rights, and environmental causes. Africa has the ability to rise again. As Mozambican poet Jose Craveirinha writes, "the sunflower of hope grows in the world."

## Why is Africa called the "original home of humankind"?

Over a century ago, the English naturalist Charles Darwin claimed that Africa was "the cradle of mankind." Darwin, who had spent a lifetime studying human origins, developed the theory of evolution. His pronouncement came in 1871, long before any major human fossils had been discovered in Africa.

Today, we know that his words were prophetic. Scientists have discovered astonishing bone remains from both early males and females. These bones tell us of the very first ancestors of each of us—regardless of our race, color, or nationality. Just as Darwin forecast, "it is somewhat more probable that our early progenitors lived on the African continent than elsewhere."

These discoveries have been made in one of the great geological wonders of the world—the awe-inspiring Great Rift Valley of East Africa. The Rift is a valley but on the grandest scale. It is a deep gash in the earth that runs almost the whole length of East Africa, some three thousand miles from Eritrea in the northeast down through the southern nation of Tanzania. It is many miles wide. From a small plane, you can follow this vast trough as it wanders southward, changing from canyons to plains to lakes and rivers. The herding Maasai and Afar peoples still live in this huge crevice.

Millions of years ago, the earliest humans and near-humans found this area to be a good place for settlement. Different edible plants grew there: seeds, nuts, berries, roots, flowers. Antelope, elephants with elaborate tusks, and birds could be hunted. Some spots were warm all year. Water was not hard to find, especially after the rainy seasons. Movement from place to place was easy. Yet the area could also be dangerous, with its erupting volcanoes and prowling big cats.

Like persistent detectives, archaeologists had tried since the 1920s to solve the mystery of early human life in the Rift Valley. Their clues to the rich possibilities in this area were human remains; sometimes they found only bone fragments—a jaw, the top of a skull, a leg bone—or a trail of petrified footprints. But their art was to turn these bits and pieces into a story that could bring our earliest ancestors to life.

Finally, starting in the late 1950s, five discoveries in the Rift Valley changed the way we view ourselves.

- In 1959 in Tanzania, at a place called Olduvai Gorge, the skull of an older man was found. His cranium had both human and ape features. He had an enormous lower jaw, helpful for chewing rough, raw food. His front teeth were like tiny stones, but his back molars were extremely large. He had been a biped—that is, he walked upright on two feet. "Zinjanthropus" was the first name given this human, meaning "man from the land of Zinj," an ancient Arab name for East Africa. His discoverers, Mary and Louis Leakey, calculated that he lived 1.75 million years ago.

- The second discovery was made in 1974 by Don Johanson in Ethiopia, far north of Olduvai. He found the partial skeletal remains of a female. Johanson called her "Lucy," after the Beatles' song "Lucy in the Sky with Diamonds"—the record the archaeologists played to celebrate their find. Later, she was given the Ethiopian name "Dinquinesh," meaning "thou art beautiful." Because her skeleton was nearly complete (but unfortunately, headless), many things could be learned about her. She was only three and a half feet tall, weighed about seventy-five pounds, and had a touch of arthritis. She was twenty-five to thirty years old when she died peacefully. She lived 3.5 million years ago.

- In 1984, Richard Leakey—the son of Mary and Louis Leakey—added another chapter to the dramatic Rift Valley story. Near Kenya's Lake Turkana, his team—led by Kamoya, a talented African fossil finder—found a skeleton of a young man. He was called "Turkana boy" and was different from the preceding discoveries. He had been a tall, strapping youth, almost five feet eight inches tall. His skeleton was 150,000 years old. Most significantly, he was a direct ancestor of ours (Homo erectus). With his big jaw and molars, "Turkana boy" was more like us than Lucy or "Zinjanthropus." With him and his contemporaries, our present human line began.

- In 1994, newspapers' front pages announced a fourth major discovery. A male skull—with an apelike heavy brow, jutting jaw, and small brain case—was found near Ethiopia's Awash River, about three miles from where Lucy's skeleton had been found. This bone was three million years old and was referred to as the "Son of Lucy." In his prime, he stood five feet tall and weighed about 110 pounds. This skull's discovery completed the picture of Lucy's species.

- In the summer of 1995, another Rift Valley find commanded front page coverage—this time, fossil jawbones that suggested a new species of early human ancestor dating to about four million years ago. The newly discovered species was a composite of primitive and and modern features: apelike facial features, like big jaws, ears, and large canine teeth, with more modern human characteristics such as upright posture. The discoverer was paleoanthropologist Maeve Leakey.

### What are the main myths about Africa? What is the reality?

An East African tale tells of a woman who is about to wed in an arranged marriage and who is told a hundred little stories about her intended husband. Eventually, she becomes puzzled, wondering what kind of man could fit all these stories. When she finally meets him, standing tall and draped in a splendid toga at the wedding ceremony, she breaks down and sobs. Startled, he asks, "Why are you crying so?" She answers, "I am so glad to see that you are not the person in the stories people told me. You are *real.*"

Many stories have been told about Africa, and many are false. Africa has been transformed by these myths. But if we take a closer look at them, then we can come to know Africa as a real place.

*Myth: Africa's peoples were the least evolved of the world's humans; their contributions to the world have been few.*

Reality: There is no scientific proof to support the notion that Africans or Blacks in general have ever ranked in the lower reaches of human societies or human intellect. Neither continents nor races have monopolies on abilities. Peaks of achievement in societies frequently result from special combinations of circumstances. Many African ideas have contributed to the world's inventory of major achievements.

A well-known example is African music and dance. From the first plantation drummers to the Florida A & M band percussionists, from the African call-and-response singing to the spirituals, from the African village dances to the Charleston and shimmy, Africa has truly enriched the world's musical and dance cultures. And this continues today. *Africa: Never Stand Still*, a 1994 music anthology, highlights fifty African musical styles, all gaining new worldwide audiences.

Less well known is Africa's contribution to modern art. Some of the greatest artists of the twentieth century—Europeans like Picasso, Modigliani, and Matisse and Black Americans like Aaron Douglas and Lois Mailou Jones—have borrowed heavily from African art, especially from African wooden masks and sculpture. Artists have loved the Africans' concise, abstract representations of the human form, which use few details to create a powerful impression.

### *Myth: Africa does not have a history.*

Reality: In 1830, the German philosopher Georg Wilhelm Friedrich Hegel claimed that Africa was a great stagnant mass. "Africa," he wrote, "is no historical part of the world; it has no movement or development to exhibit." A century later, the British historian Reginald Coupland said Africans were "dumb actors" in history, "doing nothing that seems important." Many took these attitudes farther. They claimed that because most African societies did not possess written records—like the leather-bound chronicles of Europe—Africans had no history. Writing equaled history.

Over the past half century, historians have refuted these ideas. Africans have an intense interest in the past. This passion for history is revealed in oral traditions—in songs, ceremonial recitations, proverbs, and epics—that are passed down from generation to generation by word of mouth. Jan Vansina, collecting traditions in the 1950s Congo, was told by locals, "We too know the past, because we carry our newspapers in our head." They told him poems (called *shoosh*) that contained intriguing references to past events. These oral poems were documents. Using oral traditions for making history demands the utmost care: like written records, they cannot be accepted at face value. Today, African history is no longer denied but is a thriving field.

### *Myth: Africa was one big jungle.*

Reality: Tropical rain forest is the right name for what used to be called jungle, and this accounts for only 8 to 10 percent of Africa's total territory. Rain forest runs mainly along the West African coast down to Central Africa where it pushes deep into the interior. Africa, like most of the world's tropical places, had more rain forest in past centuries than it does today.

Believing that Africa was uniformly covered with jungle created a thicket of misconceptions. Most people thought Africans lived like animals in the jungle, and it was assumed that they followed "the law of the jungle"—that is, the rule of survival of the fittest—in a human-eat-human world. The picture was not of a place of lawgiving bodies and organized ways of settling disputes. Probably, Edgar Rice Burroughs's Tarzan stories and movies—starting as early as 1914 with *Tarzan of the Apes*—planted far more jungle in Africa than nature ever did.

### *Myth: African societies are primitive.*

Reality: Before the turn of the century, it was hard to find many who understood the true sophistication of African societies. Even the brilliant, insightful British explorer and linguist Richard Burton, who traveled a great deal in Africa in the 1800s, could not bring himself to see much good in them. In his travel books he repeatedly alleged that African communities were "savage," "anarchic," even "gorilla republics." A few people here and there knew differently, but they were a distinct minority.

In fact, African societies have always been elaborate and productive human entities. Kingdoms in Africa—of which there were many, with great-sounding names like Buganda (Uganda), Asante (Ghana), Bornu (Nigeria), Shambaa (Tanzania)—had kings, queens, royal families, and elaborate courts. They were run with efficiency and boasted treasuries, militaries, foreign policies, and coherent religious ideologies. Many other African societies were not kingdoms; they were more loosely structured. These societies—such as the Ibo (Nigeria), Akamba (Kenya), and Tonga (Zambia)—were nevertheless highly organized, with governments, laws, rituals, and marketplaces.

Even the communities most often presented as simple—such as the Pygmies of Central Africa—were actually complex enterprises, particularly skillful in using their ecosystems and sustaining a spiritual vision of the world.

### *Myth: Africans lived in huts.*

Reality: The *Oxford English Dictionary* defines hut as "a dwelling of ruder and meaner construction . . . as is inhabited in primitive societies . . . or constructed for temporary use." In other words, little planning or building skill goes into making a hut. Once, "hut" described many dwellings, such as housing for the poor and for armies, as well as ancient houses. But when colonialism took over Africa in the late nineteenth century, the word was applied increasingly to African household dwellings.

African shelter covers a wide variety of styles, from the hill enclosures of West Africa's Dogon, to Ethiopia's high oval houses, to the Zulu domed dwelling. Often, a single household's compound consists of many small buildings for different family branches. Always, shelter is made with great care. It has to work: for family life, as a social place, for protection. Flimsy it cannot be. Many homes are decorated. In Muslim West Africa, the entrance is embellished with circular designs. In other places, African women paint the walls in bright colors or decorate them with murals. They use colors made from natural substances and the painting is a group project executed by the women.

African house ideas have influenced American architecture. The front porch is one example. Nowhere in England or northern Europe were porches found. Verandahs were common in West African houses, owing to the tropical heat. In the New World, European carpenters borrowed this design feature from African slave builders. Later, the front porch became fashionable for American houses. When Americans sit on their front porches, resting their bones and swapping stories, they are sitting on a bit of inheritance from West Africa.

### *Myth: Africa was a continent of ceaseless wars, where life was always in jeopardy.*

Reality: Africa has had its share of wars. Like any other place, it has had warmongers—individuals hell-bent on conquering and subjugating others. Ferocious military machines existed in Africa. The Zulu armies of southern Africa offer a premier example; they were led by the megalomaniac military genius Shaka from around 1815–28. In many African societies, both men and women earned fame through military exploits—men like Mirambo of nineteenth-century Tanzania and women like the "Amazon" fighters of Dahomey.

Yet Africa was not engulfed by wars. Great periods of stability existed. The idea of warring Africa was propagated by Africa's colonizers. They claimed to bring peace, "a new Pax," to Africa. This justified the often brutal wars that *they* waged in order to take control.

### *Myth: African languages are unsystematic, simple, and limited in what they can express.*

Reality: In those old adventure films about Africa—the ones on late-night television—the "natives" often speak in nonsense language. Except for a few Swahili phrases in *King Kong* (1933), *African Queen* (1951), and a few other films, most Africans in old Hollywood productions spoke mumbo jumbo, a gibberish uttered in guttural tones. This was Hollywood's definition of African speech.

Long before Hollywood' depictions, though, travelers in Africa claimed that African languages were too simple to convey complex thoughts. "One hundred to two hundred words" was the limit once put on an entire African language's vocabulary. Concepts like love and compassion were supposedly absent. The popular 1980 film *The Gods Must Be Crazy* asserted that the Kalahari Bushmen's language lacked words for "property" and personal possessions—a romantic idea, but wrong.

The truth is that African languages have precise structures and extensive vocabularies that can achieve great depth of meaning. Using an expressive tonal language like Yoruba, spoken by eighteen million Nigerians, writers have produced village operas, poetry, novels, plays. African languages are also famous for their everyday proverbs. Language is one of the richest cultural resources of Africa. Africans savor the subtleties of their languages, often commenting that European languages are limited.

*Myth: Africa has had a long history of despotic rule. Therefore, democracy cannot grow in Africa.*

Reality: Tyrannical rule was not standard in the old states of Africa. For hundreds of years, elaborate processes influenced the selection of monarchs. Checks against a king's abuse of power existed. Kings occasionally encouraged dissent; they could be criticized and even removed. When a man became king over the Asante (in what is now Ghana), he was advised, "We do not wish him to be greedy. . . . We do not wish that he should call people 'fools.'. . . We do not wish that he should act without advice."

Idi Amin, the Ugandan tyrant of the 1970s, claimed, "Parliaments are a waste here." Contemporary African tyrants, like Zaire's Mobutu, claim, "Democracy is not for Africa." But this is propaganda to strengthen their hold, not historical practice.

## When and why was Africa called "the dark continent"?

Before 1700, Africa and Africans were seen in a more positive light. Foreigners had a gentler view of the continent.

Ancient Greek stories depicted Africa—they called it all Ethiopia—as a good, valuable, creative place. The poet Homer depicted his Olympian gods climbing down from their heavens to visit "the blameless Ethiopians." *The Iliad*'s gods were fond of their visits to Africa. Happy to abandon their weighty divine affairs, they lost themselves in twelve days of feasting with the "Ethiopians."

During later centuries, Africa and Africans continued to be lauded. The great Arab traveler Ibn Batuta journeyed widely in West Africa (1351–53) and found much to praise. He reported on "the small amount of injustice," "the prevalence of peace," and the sophisticated mining technology. By 1240, Saint Maurice—the patron saint of the German city of Magdeburg—was painted as a Black.

The great Dutch mapmaker William Blaeu also depicted a congenial continent. Borders to his maps showed pretty, strapping Africans, with burnished skins and colorful clothing, placed against a background of pleasant countryside. When Shakespeare wrote *Othello* (1603), he created an admired Moor who commanded the Venetian republic's military. Othello was a man of high accomplishment. The great masters of the Renaissance, Rembrandt and Velázquez, painted Black persons similarly to their depiction of whites.

Gradually, however, these kinder, gentler views of Africa and Africans were disappearing. Europeans increasingly saw the African's dark skin as symbolic of evil—of the dark forces mentioned in the Bible. The seventeenth and eighteenth centuries saw the advent of even harsher European images of Africa. As the slave trade grew, the "two words *Negro* and *slave* . . . by custom [grew] Homogenous and Convertible," and the image of the African was denigrated. By the nineteenth century, popular books on Africa bore lurid titles: *Central Africa: Naked Truth of a Naked People* and *The Cannibal's Rage.* Missionaries, sent to convert Africans, wrote home of "poor lost Africa" and "this sinful place, Africa."

In 1878, Henry Morton Stanley delivered one of the meanest blows to the image of Africa. His *Through the Dark Continent,* a thousand-page, two-volume book, was a major publishing event.

(It contained fabulous illustrations of African crafts and houses; later, the Black American intellectual W. E. B. Du Bois actually used them in his books!) The book traced Stanley's journey—his "perilous labors"—across East and Central Africa. From the day the book was published, the appellation "the dark continent" stuck. Way into the 1950s, it was outrageously popular and synonymous with Africa.

Had the book been written by anyone other than Stanley, the phrase would not have been fixed in the public mind. He was the most famous explorer of his time. Part of his reputation was earned: he was indomitable and ruthlessly pushed his caravans to cover enormous territory in record time. In addition, he found the famous missionary David Livingstone—missing for two years—in East Africa in 1872. (On spotting Livingstone, Stanley uttered the now-famous line, "Dr. Livingstone, I presume.") Part of his fame, however, came from self-promotion. Still, whether his stature was real or manufactured, the public believed Stanley about Africa.

"Dark continent" was not just a clever marketing device cooked up by a London publisher. It also stood for a set of ideas. Africa, to Stanley, was an unknown and mysterious land. The world had been so long "in the dark" about it. Africa had a "dark interior . . . still unknown to the world." Africa was cut off, isolated from the outside world. Its coasts were known, but its heartland was not. Stanley went "to pierce" and "to penetrate" Africa's insides. Most of all, Stanley wanted to get missionaries into Africa's interior, to save Africa for a Christian future. In his thinking, Christian light would dispel Africa's darkness.

Stanley always exaggerated Africa's isolation. Trans-Saharan trading routes linking Africa to Europe were at least a thousand years old. The florin of Florence and the ducats of Genoa and Venice had been made from Black Africa's gold. For centuries in northeast Africa, trading had gone on across the Red Sea. Within Africa itself, peoples had many contacts with each other. Even news of happenings far away traveled quickly.

When British novelist Joseph Conrad named his famous short novel *Heart of Darkness* in 1902, he was echoing Stanley. By that year, colonialism had overwhelmed Africa. Marlow, Conrad's leading character, complained that in his lifetime, Africa "had become a place of darkness." In Europe's eyes, it had. Stanley had seen to that.

### What is all the fuss today about ancient Egypt?

Ancient Egypt made a big news splash when British archaeologist Howard Carter opened Tutankhamen's tomb in 1923. Newspapers ate it up: "Ancient Egyptian Child-King Found Amid Gold and Rare Stones."

Now ancient Egypt is back. But this time it is raising disputes left and right. How can a society that passed from the world's stage almost 2,500 years ago be the center of an American debate today, from Alice Walker's novel *Temple of My Familiar* (1990) to *Newsweek's* "Is Cleopatra Black?" (1991) and from Cornell University to the intellectuals who frequent Harlem's Pan-Pan restaurant?

Understandably, Egypt has always provoked interest. It was one of the two oldest civilizations in world history. (The other was Mesopotamia, in what is now Iraq.) Egypt flourished first around 3100 B.C. After that, it survived a full three thousand years. Even in its early days, Egypt had a system of writing and record keeping, a strong government, many outstanding kings (called "pharaohs") and queens, bustling towns and markets. Egypt was the home of powerful ideas and philosophies. And of course Egypt had impressive monuments, perhaps the most spectacular buildings in human history.

Because of its splendor, Egypt always attracted travelers. Herodotus (circa 485–425 B.C.), the father of Western history and the roving reporter of the ancient world, was a thrilled visitor. "Egypt . . . contains more wonders than any other country. . . . It is admirable beyond the powers of description," he wrote. Many centuries later, Egypt was still enchanting foreigners. Listen to Frenchman Jean Savary in 1779: "We saw the summit of two great pyramids. The appearance of these ancient monuments that have survived the destruction of nations . . . inspires a kind of reverence."

Accolades like these have been used by many thinkers who

argue that Egypt is a symbol of African greatness. Furthermore, they contend that ancient Egypt's monuments refute the idea of a technologically inferior Africa. From the Step Pyramid built in 2700 B.C. to Cheops's Great Pyramid to the Great Sphinx, Egyptian builders mastered exceedingly complex conceptual problems, as well as amazing feats of labor and engineering. For example, Cheops's gargantuan pyramid—451 feet tall—was made of 2.6 million blocks. Its sides were almost precisely the same length, within eight inches of each other. Egypt's temple art was another stunning accomplishment. Serene, lifelike statues of elite Egyptians reveal a remarkable understanding of the human form.

Today, the fuss about ancient Egypt is not about big buildings. Instead, the argument is that Egypt's real achievement lies not in its stone monuments but in its vital thought, which has shaped Western ideas—in particular, ancient Greek ideas. People argue that knowledge of this contribution has been suppressed. Can some of the West's basic notions have derived from Egypt? Most of the world has been taught that Greece is the birthplace of Western thought, that Plato and Aristotle were the apogee of the West's ideas, and that Greek thought grew mostly from Greek roots.

But new evidence suggests that ancient Greek thought owes a debt to Egypt and, hence, to Africa. How so? One argument is that many of the most important Greek philosophers (Thales, Pythagoras, Solon, Plato, Eudoxus) went to Egypt and studied there. Just as twentieth-century thinkers went to London and Paris, ancient ones went to the banks of the Nile. They brought home major ideas and reworked them into new forms. It is now clear, for instance, that Greek gods derived from Egyptian deities. And the Greek language owes some of its vocabulary to Egyptian.

It has also been argued that the basic Western concepts—the notions of ideal form, the afterlife, resurrection, the idea of one high god—were born in Egypt. These ideas became crucial to Western philosophy and religion. The great sixteenth-century Italian philosopher and mathematician Giordano Bruno wrote, "We . . . owe Egypt, the grand monarchy of letters and nobility, to be the parent of our fables, metaphors, and doctrines." Herodotus, "the father of history," admitted this. In fact, the Greeks never downplayed the critical role that ancient Egypt had in molding their civilization.

But this is not the whole story. Over the centuries, it has been hard to make the world see ancient Egypt as belonging to Africa, even though Egypt is on African soil, fed by Africa's Nile, and tied to other African societies. Here is where Black Americans come in. Already in the early nineteenth century, African-Americans were waving the banner of "Egypt is African."

In 1833, Maria Stewart, a prominent Black leader, challenged a Boston audience by asserting that "history informs us that we sprung from one of the most learned nations [Egypt]. Despised Africa . . . was the esteemed school for learning." In 1854, Frederick Douglass argued, "Europe and America have received their civilization from the ancient Egyptians. . . . Egypt is in Africa. Pity that it had not been in Europe . . . or better still in America. . . . The ancient Egyptians were not white people." Other early Black writers had the same passion for Egypt, and it is easy to see why: surrounded by American slavery, they found in Egyptian grandeur and ingenuity good arguments against the debasement of Blacks. These arguments also began the habit of imposing American notions of race on Egyptian culture, a habit that Black classicists, like Frank Snowden, have criticized.

In this century, Black thinkers—notably George G. M. James, Cheikh Anta Diop, and Yosef ben-Jochannan—have continued to uphold Egypt as a great contributor to world civilization and a crucial part of Africa. Their ideas have swept through Black America. Joining them is Martin Bernal, whose two-volume *Black Athena: The Afro-Asiatic Roots of Classical Civilization* rocked universities beginning in the mid 1980s.

Sharp clashes of opinion have erupted over these points, prompting a rethinking of Africa's place in world history. And there have been downsides. The rest of ancient Africa has been neglected in favor of Egyptomania. Seldom mentioned are the six thousand years of African farming innovation. Rarely cited are the artistic traditions of Nigeria's ancient Nok culture. Even ancient Nubia, south of Egypt, has been overlooked. Yet exciting new evidence shows that Nubia was a rival powerhouse to Egypt. Nubia might even have supplied models for the Egyptian kingship. Nubia is becoming a new hot spot of research for ancient Africa.

### What do Africans treasure about their societies and their world?

In the novel *Fragments* by Ghana's A. K. Armah, an aged blind woman, Naana, is keeping faith with the familiar African truths as the world around her is changing. She believes that "each thing that goes away returns and nothing in the end is lost." Daily she reflects on the treasures of her life.

Naana is not alone. For centuries, African writers, philosophers, artists, and musicians have tried to define what is most valuable to them. Edward Wilmot Blyden (1832–1912), a prodigious West Indian thinker who went to live in West Africa's Liberia, spent his life identifying the basics of African culture. From all this speculation have come a few fundamentals:

- Africans are committed to kin and family. Different from our nuclear families of father, mother, and children, Africans live within extended families—often large collections of kin. A single person has many influential relatives, who are sources of authority, instruction, and joy. Beyond their families, Africans are deeply involved in a web of kinship. A Nigerian anthropologist tells us that his great-grandmother knew how tens of thousands of people in her town were related to one another. She was a walking encyclopedia of kinship. Africans conceive of themselves as born into a world of kin, of passing their days among kinfolk, and when they die, of having their lives commemorated in song and sayings by the kin they have left behind.

- Africans treasure their exploration of the spiritual world. What is a person's place in the universe? Are you born on a specific path? How do past generations relate to us? Major questions like these have preoccupied Africans. Films such as

*Mammy Water: In Search of the Water Spirits of Nigeria* (1991) superbly illustrate just how seriously—and joyously—Africans approach religious matters. Mammy Water is a female river spirit, drawing all types of people to praise her and present offerings. She responds with help, encouragement, protection, and the righting of wrongs.

For centuries, Africans have been learning their gods' ways, especially those of ancestral gods. An East African tradition asks, "Can I be far from myself if I know my gods? When I pour my honey beer to my ancestors, can I be wasting it?" Even today, Africans demonstrate their genius for spiritual things by reinventing the Christianity that Europeans spread; here, Christianity has been Africanized.

- Trading is another beloved pursuit in African life. Have you ever wondered why there are so many African street traders in places like New York, Washington, and Atlanta? What explains their affection for the capitalist marketplace and wealth? The answer is simple: Africans—both men and women—are some of the world's most dynamic trading peoples. The street traders are the latest version of the centuries-old proclivity of Africans for trade.

  Centuries ago, African traders established long-distance trade routes and built major markets. In these markets the necessities of life were traded: grains, meat, iron hoes, cloth. But luxuries were also there: snuff, rare skins, carnelian stones, exotic meats, unusual peppers, cooling jugs, leather toys, magical colored powders, even European crossbows, lances, and drinking urns brought across the Sahara. The great towns of Africa—Memphis, Tombouctou, Gao, Jenne—rose at the crossroads of trading routes. During recent decades, African women have become such successful traders that they are affectionately known as "Mama Benz" because of their trademark chauffeured Mercedes Benzes.

- Africans are passionate investigators of the mysteries of nature. Each year brings more exciting news about African discoveries in industry, science, and mathematics. A few years ago, J. Worth Estes, a pharmacologist, created a stir by demonstrating that ancient Egyptian healers had discovered a number of creams and ointments useful in fighting disease-causing bacteria. The healers had also discovered means of contraception and of fixing fractures and removing skin tumors. Recently, archaeologist Peter Schmidt announced that African ironworking was thousands of years old and that African ironworkers were ahead of Europe by five hundred years in certain methods of making steel.

Yale art historian James Faris Thompson has been holding audiences spellbound with his presentations on complex African sign-writing and memory systems. Ivan Van Sertima, author of paradigm-busting *They Came Before Columbus* (1976), excites readers with his argument that pre-Columbus African voyagers sailed westward into the Caribbean basin. He has also been spreading the news of African calculating systems, numbers games, geometric codes, time reckoning, and astronomical observation. His point is this: "Africa has to be considered a place where sciences and mathematics and serious thought had their fullest expression."

- One last treasure on which Africans agree is their earth, their soil, their land. Travel to a place like Senufo country on the Ivory Coast in West Africa and you can see neatly plowed fields that stretch for miles, circling clay-walled homesteads—a picture of symmetry. African songs tell of love for the continent's ocher-colored soils and the land's sweet smells after a rain. Annually, Africans have given thanks to the earth at harvesttime or during "first fruits" ceremonies. It is this earth that African farmers—often women—have patiently tilled. Moreover, African farmers have been innovators. They tamed wild grain crops thousands of years ago, as early as did farmers in China and central Asia.

Other aspects of African life are also treasured: the love of leisure and a willingness to work to get it; the beautiful crafts, especially woven cloth; personal power and discipline. Today these values are being tested as never before, as cities grow to bursting, women seek more independence, and educated people run faster and faster in their careers. But this will not prevent Africans like Naana from treasuring the traditional values.

### What is a griot?

In Baltimore's Great Blacks in Wax Museum, there was once an exhibit that included a griot—a male figure named Mamoueou Kouyate. The exhibit showed a group of children at his feet, listening closely to his story.

For centuries, African griots like Kouyate have been famous as "people of words," shaping crucial cultural information into songs and spreading this information to a variety of audiences. Griots have lived in the Sudanic region of West Africa, since the kingdom of Mali, in the thirteenth century.

Griots are keepers and tellers of history. As a griot's song says, "we sing of the past, and from its ashes comes forth the dust of the future." Their singing does much to keep the knowledge of the past alive. As they say, "the written word dies with the drying ink. Warrior, scholar, and magician, peasant, tradesman and slave, we offer them the same water for all their thirst. He who is wise shall heed what we say, what we sing. . . . We are the memory of man."

During earlier centuries in West Africa, the griot was a bard or minstrel employed at royal courts, preserving in memory the line of descent of all the ruling families and keeping the record of battles, victories, events. A griot was expected to offer advice on matters of state. He knew a specific monarchy's history and could extract lessons for the rulers from that knowledge.

There was an even deeper side to this role. Interpreting dreams was a griot specialty. Knowing the secrets of healing—spells, charms, and herbal remedies—was another art. Some griots knew how to manipulate sound to produce certain mental effects, such as relaxation, meditation, or trances. Most important, griots represented the common people; they were the voice of the people beyond the monarchs' court. They told monarchs what the people were saying and thinking. They could be outspoken and critical, even to a ruler's face. Their criticism was held in high regard, and they could not be punished. They helped democracy live in their societies.

Griots are trained from childhood. Often, they come from particular families whose members, for generations, have been griots. Djimo Koyate, a griot who has performed in America, says he was taken everywhere with his father, who was his teacher. "I was born a griot," he says. Interestingly, different griot families often use different accompanying instruments, such as the *kora*, a large plucked harp-lute, and the *balafon*, a wood-frame xylophone.

Griots have always traveled the countryside, carrying the news of the day. As one wag has said, "they were the CNN of the medieval Sudan." The West African griot is a crucial figure in Africa's and the world's oral heritage. The griot is Africa's analog to the tellers of the *Iliad* and *Odyssey*, to the strolling Celtic bards, to reciters of *Beowulf*.

Today, the griot tradition continues in Africa, and it has jumped across the Atlantic to America. Rappers style themselves after griots. In 1991, Black choreographer Garth Fagan created dance stories, accompanied by trumpeter Wynton Marsalis's music, under the title "Griot New York." Brooklyn is now the home to prominent griot Al-Haji Papa Bunka Susso from Gambia, who gives performances around the country.

### Was Richard Nixon right? Was there never "an adequate Black nation" in history?

While working as White House chief of staff for Richard Nixon, H. R. Haldeman kept a diary from 1969 until 1973. When it was published in 1994, readers learned of Nixon's opinion of the place of Blacks in world history.

On Monday, April 28, 1969, Nixon was discussing welfare reform. Suddenly, he veered from the subject. Nixon, Haldeman reports, "pointed out that there has never in history been an adequate Black nation, and they [Blacks] are the only race of which this is true. . . . Africa is hopeless."

This was Nixon's way of saying that American Blacks had done poorly, that they were the welfare wards of America. "You have to face the fact," he said, "that the *whole* problem [welfare] is really the Blacks."

Nixon's few words were volatile ones. First, by claiming that Blacks had "never in history" produced a decent nation, he was harshly judging the whole of Black history. Second, he singled out Blacks as the only deficient "race" in the production of nations. Saying that Blacks—or any "race"—are historically backward guarantees an uproar. Had Nixon let these casual "bon mots" slip publicly in 1969, Black Power intellectuals—still an alert strike force in America at that time—would have rushed into battle. Think of the political harvest they would have reaped in attacking the president's theory of Black history.

Nixon's remarks are offensive. But in their candor, they raise an important issue. They make us confront the question that many people of all groups have in their minds—but were too afraid to ask—about Black history, about Black achievement, and especially about African history.

The president was wrong in his appraisal of Black state building in world history. Africa produced a plethora of major states. After ancient Egypt, from 500 A.D. to the 1800s, Africa generated numerous massive kingdoms and monarchies, many with well-engineered techniques of governing. For more than a thousand years, the continent was one of the leaders worldwide in state formation.

Ghana, Mali, and Songhai—unquestionably, great world civilizations—spread across West Africa's Sudan from 800 to 1500. Camel caravans carried news of their achievements across the Sahara to Europe. Simultaneously, Great Zimbabwe in southern Africa, with its precision-built stone monuments, was being formed. Following the Sudanic explosion came more West African kingdoms. (Nixon should have known this; as vice president, he had been to West Africa.) Central Africa had states in the fifteenth to seventeenth centuries. And out on the tropical East African coast, Swahili trading city-states were successes.

H. R. Haldeman could have gone over to the Howard University library in Washington, D.C., and checked out an historical atlas of Africa for his boss, like the atlas produced by expert J. D. Fage in 1958. All these states are there. Nixon would have been enlightened.

Or maybe not. We usually fail to see what we aren't looking for. Nixon was interested in what he called "an adequate Black nation." He looked at history in terms of nations, the modern organizations that appeared in Europe in the seventeenth to nineteenth centuries. To him, these were the real motors of history. Forget those old states and kingdoms; for Nixon, they were just quaint artifacts in a world history museum!

Answering Nixon's challenge about Black nations is not easy. While Europe was developing nations, Africa was fighting for its life against the draining slave trade. After the slave trade was prohibited in the 1800s and before Africa could catch its breath, European colonialism ravaged the continent. By 1890, Africa was well on its way to foreign control. In other words, Africa did not have a historical chance to develop nations. Not until the 1960s did Africans become free again.

Yet it is possible to refute Nixon's condemnation. Look at Ethiopia in northeast Africa. Over four thousand years old, it is the world's oldest country, prominent in biblical times. Sequestered in hills, Ethiopia was spared the slave trade's ravages. By the nineteenth century, Ethiopia's expanding territory was being unified. After the charismatic Menelik II (1844–1913) became emperor in 1889, he modernized the country, introducing railways, roads, education, and telecommunications.

In 1896, Menelik's army had soundly defeated a major Italian invasion at the Battle of Adowa, a crushing setback for colonialism. European nations had to recognize Ethiopia as a nation—even "an adequate Black nation."

Italy struck back in 1935, this time taking Ethiopia and sending Emperor Haile Selassie into exile. Ethiopian patriots fought the Italians, and the British helped free the country in 1941. When World War II was over, the idea of Ethiopia was still intact—a clear sign of nationhood.

### Who were the great male movers and shakers of Old Africa?

*Narmer (fl.*[1] *3150* B.C.*)* King Narmer fused northern and southern Egypt, becoming the first ruler over the united kingdom. His title was "Lord of the Two Lands." There were many innovations during his reign: hieroglyphic writing; the emergence of the king's majesty as an idea; and the king's renewal ritual, the *heb-sed* festival. With these legacies, Menes, Narmer's successor, inherited a greater country.

*Imhotep (fl. twenty-seventh century* B.C.*)* Even after all these centuries, Imhotep is still one of Africa's most fascinating individuals. Amazingly talented, he was a scholar, sage, astronomer, counselor to the pharaoh Zoser, the kingdom's chief administrator, and architect of the first pyramid, the Step Pyramid at Saqqara. He was extremely knowledgeable about drugs, surgery, and healing lore. After his death, he was deified; commoners wrote odes to him, and thousands visited his tomb. To the Greeks, he was the "father of medicine."

*Tuthmosis III (1504–1450* B.C.*)* Tuthmosis was the most famous member of the most famous Egyptian dynasty, the eighteenth. Later, he was called "the Napoleon of Ancient Egypt." Like Napoleon, he was short in stature (less than five feet tall) and a successful general. Tuthmosis led eighteen massive campaigns into western Asia, created an empire, and made Egypt invincible. On his death, his soldiers' eulogy said, "Lo, the King completed his lifetime of many years, splendid in might, in valor, and triumph."

*Taharqa (710–664* B.C.*)* As a king of Nubia, Taharqa used his army to help ancient Israel and earned mention in the Bible as a friend of the Israelites (in Isaiah 37:9 and 2 Kings 19:9). He became a pharaoh, one of the Nubian kings of Egypt. He was a

[1] "Fl." stands for flourished. In this case, it means that the person was definitely alive in this year, but more precise dates are not known.

great builder of public complexes, in both Egypt and Nubia. He was unseated by Assyrian invaders and returned to Nubia, where he died.

*Hannibal (247–182 B.C.)* A North African from the royal house of Carthage, Hannibal became a great general who fought many campaigns in Spain and Italy, using large armies. In the Second Punic War (218–202 B.C.), he surprised Romans by invading Italy from the north with elephants, and he defeated the Romans in a series of battles. During World War II, allied generals studied his Italian campaigns. After his warring life, he was an unsuccessful statesman. Later, he committed suicide to avoid Roman capture.

*Saint Augustine (354–430 A.D.)* One of the world's greatest thinkers, Saint Augustine was born in the North African town of Tagaste (in what is now Algeria). Around 383, he left Africa to teach in Rome. A profound spiritual upheaval caused him to convert to Christianity. He returned to North Africa to become bishop of Hippo in 396. He wrote two classics of world literature: the *Confessions* (400 A.D.) were the story of his restless journey to God, and *The City of God* (412–27) was a philosophy of history from the Christian viewpoint. "The true philosopher," he wrote, "is a lover of God."

*Tenkameinin (fl. eleventh century A.D.)* King of ancient Ghana in West Africa, Tenkameinin promoted the gold trade. But his main reputation was as a just and good king: he listened to his subjects, traveled and lived among them, and cared about average citizens. Legend pictured him as a democratic monarch. Commoners said, "A just man, with the heart of a friend, kind and loving, is his mark."

*Mansa Musa (?–1337 A.D.)* The ninth king of Mali, Mansa Musa headed a powerful, wealthy state. He lived life on a grand scale: in 1324, he captured the imagination of Europe and the Middle East with a lavish pilgrimage to Mecca, involving sixty thousand people and eighty camels laden with gold. The first European maps of West Africa, drawn by Abraham Cresques in 1375, presented Mansa Musa as the king of gold. He built Tombouctou into a prosperous city that was also the center of Muslim culture and scholarship, and he created the university at Sankore. Ironically, he died a poor man.

*Nzinga Mvemba (?–1545 A.D.)* For thirty years, he was king of the Kongo, the largest state in Central Africa. After meeting early Portuguese travelers, he asked Portugal for teachers and technology to help him transform his kingdom. He converted to Christianity, taking the name of "Afonso I." Portuguese greed turned his modernizing dream into a nightmare: the visitors captured people for the slave trade. He complained to the Portuguese king: "We cannot reckon how great the damage is. . . . Our country is being completely depopulated." As his country fell apart, he died disillusioned.

*Osei Tutu and Anokye (fl. 1670s)* If there ever was a dynamic duo, it was these two. Osei Tutu, starting out as a mere chief, quickly built his chiefdom through trade, conquered other chiefs, and established the Asante kingdom. His position was helped by Anokye, a brilliant shrine priest. Anokye came up with the idea that the Asante's soul was contained in the Golden Stool (Sika Dwa), a sacred seat he claimed to have brought down from heaven. Using the stool as a symbol of unity, Osei Tutu started the Asante on the road to becoming a megastate.

## Who were the great female movers and shakers of Old Africa?

*Hatsheput (circa 1540–1483 B.C.)* This formidable Egyptian let nothing stand in her way. She started as a coregent, ruling with child-king Tuthmosis III. But later she claimed her place as the queen. Her achievements were many: trading expeditions, a surge in building, and excellent government. Murals portrayed her as a male pharaoh, wearing a man's crown and a beard. She built her splendid tomb beneath the cliffs of Deir el-Bahari. This is how she assessed her reign: "I have restored that which had been ruined. I have raised up that which had gone to pieces before."

*Tiy (1415–1340 B.C.)* Originally from Nubia, Tiy came to Egypt from an important noble family. Her intellect, style, and character inspired Pharaoh Amenhotep III to marry her. She was the chief wife among his many wives. Amenhotep said she was "the body of grace, sweet in her love, who fills the palace with her beauty." Together, they presided over a prosperous, stable kingdom. Statues of Tiy and Amenhotep showed them united, confident, proud, and serene.

*Nefertiti (fl. 1365 B.C.)* Her name meant "the beautiful one has come." A famous limestone bust of her, found in 1912 and now at the Berlin Museum, captured her elegant regal face. More than a beauty, though, Nefertiti was a power in royal politics. As Pharaoh Akhenaton's wife, she also promoted his radical idea that the king was the channel to a god, Aten.

*The Queen of Sheba (960–930 B.C.)* Her original name was Makeda; Sheba was her birthplace. She forged the early identity of the kingdom of Ethiopia and linked the kingdom to other countries. Most famous was her journey to Jerusalem to visit King Solomon, whose wealth and wisdom she had heard of. The Bible (2 Chronicles 9) reported that she came "with a very great company . . . camels that bare spices, and gold in abundance and precious stones." The revered Ethiopian book *Glory of Kings* tells of her romance with Solomon, to which a son was born.

*Candace of Meroe (third to second centuries B.C.)* This was not the name of a single person; rather, it was a title for women of royal blood, queens and queen mothers. Meroe was the capital of Nubia, a kingdom south of Egypt. Over two centuries, many queens—*candace*—reigned. They actively promoted trade, the building of public works, and diplomatic ties. Because of its many women monarchs, Nubia was called "the land of queens."

*Dahya of the North African Berbers (575–702 A.D.)* Another of Dahya's names was "Kahina," meaning prophetess. Her skills in predicting the future helped the Berber warriors to resist invaders. Later, she was made queen. While queen, she fought alongside her troops and was killed in battle. So feared was she that the enemy demanded to see her severed head as proof of her death. Legends claimed she lived nearly 120 years.

*Amina of Hausaland (fl. fifteenth to sixteenth centuries A.D.)* Amina was a queen of the Hausa kingdom of Zaria (in modern northern Nigeria). An extremely successful military ruler, she camped among her warriors and planned and led military campaigns. Under her, Zaria dominated the region. Hausa stories called her "a woman as capable as a man." In 1973, Nigeria issued a stamp in her honor, showing her seated on a charging stallion, lifting a tasseled spear in the air.

*Nzinga (circa 1581–1663)* Nzinga was queen of the Ndongo-Matamba kingdoms (in what is now Angola), succeeding her brother. At the time, the Portuguese were active near her kingdom. Nzinga defied their schemes to control her country and used many strategies to stop them, including diplomacy, warfare, and even converting to Christianity. Her court exhibited unusual splendor, and the highest posts went to women. A Dutch observer wrote that she was "a crafty, proud, and stubborn woman, fascinated by weapons." At the time of her death, the country was still free.

*Macarico (fl. 1600)* A noblewoman among West Africa's Mane people, she was exiled from her homeland by the emperor. In response, she organized an exodus of her followers. Under her leadership, they became a conquering army, overrunning large territories out of which she created her own country.

*The* Signares *(sixteenth to eighteenth centuries)* These successful businesswomen lived in West Africa's Senegal-Gambia area. Trading with early Europeans, they became wealthy, owned land, built major houses, and bought trading vessels. Fashionable public figures, they wore lavish dresses, fabulous gold and silver jewelry, and smart headdresses. Often, griots walked in front of them, praising their deeds. The most important *signare* was Bibiana Vaz, who created a short-lived "republic" over which she ruled.

**What was the quality of African life at the moment of earliest European contact—prosperous and developing, or poor and regressing?**

Portuguese sailors were the first Europeans to reach the western coast of Africa in the fifteenth century. Rising along the West African coastline in the fifteenth century was a dynamic world with its own direction and momentum. A land of many peoples, it was a prosperous territory four thousand miles long.

During the next decades, the Portuguese became a familiar presence in the region. At the time, Portugal was not a European superpower; it was small, largely rural, and only moderately developed. But it had a future-oriented elite, epitomized by Prince Henry ("the Navigator"), who was eager to explore sea travel. By 1500, the Portuguese had sailed thousands of miles along the shoreline of Africa, from what is now Senegal to the Cape of Good Hope.

The Portuguese were startled by the Africa they saw. It was not unlike their home country. In some respects, Africa and Europe were mirror images of each other: both were centered on large families; both had flourishing religions; both had trading towns; the lifeline of both was farming. The average Portuguese sailors would not have thought western Africa inferior to their homeland. They would have seen it as different, interesting, curious, perhaps, but not inferior and certainly not in decline.

It was a time of experimentation. New styles of life were developing. Trading was breaking new ground. New cultivation and irrigation methods were being devised. Agriculture was productive. Many herbal cures were being refined. The mining of gold was being advanced. Big states were taking shape in the interior. Along the coast, new states also were appearing. These new kingdoms gave a sense of pulsing life throughout the region.

Soon, other Europeans came to western Africa. They too discovered its spectacle. Towns and cities were especially lively, offering interesting houses and street scenes, colorful festivals, music, religious rites, markets stuffed with goods, ostentatious displays, strange visitors from other African lands, busy messengers, thievery, intrigues, and, sometimes grisly public punishments. More than twenty different hairstyles were seen on the women of one town. A European ambassador visiting a city had "great feasts . . . held in his honor, and he was shown . . . many good things."

Ife and Benin—now in Nigeria—were two such cities. Like the powerful European cities of the time, they were centers for the arts. In Benin, artists produced unsurpassed realistic bronze likenesses of humans and animals, known today throughout the world as the Benin brasses. In Ife, artists made exceptional terra-cotta heads.

On seeing Benin, a Dutchman wrote, "The town seems very great. . . . You go into a great broad street . . . seven or eight times Warmoes Street in Amsterdam. The king's palace . . . [occupies] as much as the town of Haarlem [in Holland]." Then he paid the ultimate Dutch compliment: "The people are in no way inferior to the Dutch as regards cleanliness." Another Dutchman wrote, "Everyone wants to see the king, just as in Rome when the pope celebrates his 'Jubilee.' Six thousand people press toward him, and when he comes they all go down on their knees and clap their hands in greeting. For his part, he waves his hand back and forth, just as does the pope in Rome when he blesses the people."

As Europeans grew more acquainted with West Africa, they confronted another stunning reality, one hardly known today. Local industries produced all of the basic goods and all the necessities of life for the region. West Africa was a bustling manufacturing zone and had an independent economy. Hundreds of types of textiles were produced, from the simplest to the most elegant. Household objects—pottery, carved dishes, wooden drinking cups, decorated gourd containers, wooden stools, bowls—were all made locally. So were flutes, drums, harps, and xylophones. Some craft objects were also fabulous art: velvety raffia cloth, costumes, carved doors, bracelets, ivory combs, headrests, special spoons, drinking cups, beaded caps. Today, these objects are in museums all over the world.

Metal knives, spears, blades, arrow tips, hoes, axes, and weapons were also made. Africa had been one of the world's great iron-producing regions since ancient times. Africans had already engineered ironworking furnaces using heating techniques that wouldn't be discovered in Europe for another four hundred years. As a result, Africans produced perhaps the best steel in the world at this time.

Men were dominant in many of these societies, as they were in Europe. But the region's women were also a force, using many strategies to advance themselves. They were traders, merchants of specialized commodities (pottery, foods, cloths), pivotal figures in farming, prophetesses and diviners, powerful wives of elite men, and queen mothers in royal courts, playing the political game, advising, and often acting as confidantes. In some places, kinship rules favored the women's line, thus empowering them further.

West Africa was no paradise. It had its wars and petty tyrants, and it exploited its poor. But it was a vital, complex place when Europeans first put their feet on its shores.

**Who started the Atlantic slave trade? When did it start?**

Although many disputes surround the Atlantic slave trade, it is certain who started it: the Portuguese initiated this awful commerce. For centuries, there had been a huge trade in Africans conducted by Arabs from Muslim North Africa. Most of this ran across the Sahara desert. But it was the Portuguese who "pioneered" the trade along the Atlantic coast of West Africa.

Still, it is unfair to blame only them. In time nearly every European nation lusted after the profits of the Atlantic slave trade. Actually, the British did the most business in this human commerce. The French too managed a tidy profit. So did Spanish, Dutch, German, Danish, and Swedish traders. All these merchants made the Atlantic slave trade into the biggest emporium of humans in world history. It is this trade that channeled Africans into the New World.

When the Portuguese started sailing along the African coastline, they were looking for any opportunity to establish trade and allies. At first, Africans were not interested in their goods. Undeterred, the Portuguese bought African goods to resell to other Africans—woven mats, bark cloth, big sharp knives, peppers, dried fish, hides. Eventually, Africans began to buy

Portuguese copper pots, linen, colored cloth, clothes, beads, and bracelets.

The dream of luminous gold kept the Portuguese coming back to Africa. Columbus, who went to West Africa on a Portuguese ship before sailing to find the Indies, had the fantasy of getting rich through African gold. "Anyone who has gold," he wrote boldly, "can do whatever he likes in the world." He can "even bring souls into paradise." Decades later, Portuguese persistence paid off. At one point, they were taking back to Portugal nearly one-tenth of the world's gold supply from Africa. Some Portuguese were getting rich. They could see now the potential of Africa.

At first, the Portuguese tried to advance their interest in slave trading by attacking and raiding local areas. After a few attacks, Africans began to offer vigorous resistance. Using small vessels suited for the coastal waters, Africans repelled the intrusions. Several Portuguese crews were killed. Clearly, another strategy had to be adopted. From the 1450s on, the Portuguese relied more on diplomacy in contacts with Africans and remained near the coastline.

Two small groups were now coming together: the Portuguese traders and African allies along the Atlantic coast. Portuguese traders became friends with some rulers, giving them gifts. Diplomatic contacts were established. The language Crioulo, a blend of Portuguese and various West African languages, sprang up. The coming together of these two groups marks a momentous turning point, and you might think that historians would know the exact day, hour, and circumstances under which these transactions took place. Yet we can only assert that the alliance was formed sometime from the 1440s to the 1490s.

Later, a special class of middlemen helped the Portuguese penetrate the area. Sometimes these were Portuguese living in African villages, speaking local languages, intermarrying with local women. Other times they were biracial offspring of Portuguese-African couples, who spoke Portuguese, wore European dress, and built homes in the European style.

What really put the Portuguese into the slave-trading business was the building of forts along the West African coast. Forts gave them permanence, protection, presence, and a place from which to manipulate the trade. Negotiations with African rulers gave them the land to build on. Largest among the forts was São Jorge Mina, built in 1482, named after the patron saint of mining. But the Portuguese did not rely on a saint's blessings for the fort's survival; they sent from Portugal every stone, door, and iron bar for the fort's construction. In fact, no saint would have blessed what El Mina was to become: a depot for the slave trade. Humans bound for slavery were to be collected in its dank chambers.

Beginning in the 1440s, Africans were taken to Europe as slaves, primarily going to Portugal and Spain. The trade began slowly, but within twenty years, five hundred people were being taken annually, and by the 1490s, the number of African slaves in Europe had climbed to more than fifteen thousand. They worked mostly as domestic servants, but many were farm laborers, crews on ferries, street vendors, food hawkers, and waterfront stevedores. Slaves were another kind of gold—Black gold. In a way, they were better than gold, because they gave their owners a lifetime of work, whereas gold could be a fleeting thing.

By the late 1480s, the Portuguese and trading in humans were synonymous. A West African man told an English traveler, "The Portugales were bad men, . . . they made [the Blacks] slaves if they could take them, and would put yrons [irons] upon their legges." In 1502, the Portuguese landed the first Africans in the New World, on the island of Hispaniola (today's Haiti)—just ten years after Columbus's voyage. Everything was now in place for the transatlantic slave trade in Africans.

### As the slave trade grew, who were its victims?

Africans did not volunteer for the slave trade. People who were taken were victims, tragic cases whose lives were changed forever against their will.

Victims of the trade came from several sources: human booty—that is, prisoners resulting from major wars; people kidnapped by small raiding parties; and people accused of witchcraft, vicious magic, violating community rules, and antisocial behavior. People who fell into indebtedness were often victims too, as were alleged criminals who were sentenced to death. The charges against these people might be false; a corrupt ruler could accuse people unfairly of crimes and send them off to the trade. For the most part, West African societies on the coast did not sell their own citizens.

Of all these sources, raids, kidnapping, and wars were the most productive in generating people to be sold. This was capture by force. Many young girls and boys were taken for the trade by this method. A traveler later recorded, "An abundance of little Blacks of both sexes are also stolen away . . . in the cornfields."

As the demand for slaves increased, Africans organized raids solely for the purpose of capturing bodies. Small, unruly independent parties roamed the countryside. Often these raiders went deep into the interior, miles from the coast, to acquire captives, sending shock waves through the land. When Europeans began to sell firearms to African raiders, they demanded lots of them and lots of ammunition so that they could carry out even more devastating raids. Big wars, when they erupted, guaranteed even greater numbers of slaves.

Africans who have managed to tell their story about being made a slave always talked about the moment of their capture. It was a shattering experience. "Snatching people from the paths of the country is how the poor captives are gotten," one African captive recounted. Another said, "We were innocent until we knew that they would hold us."

Once captured, Africans were shackled, tied together, and marched to the coast. By the 1780s, tethered slave gangs were called coffles. Upon reaching the coast, they were held in cages, prisons, or booths—often called barracoons—waiting to be sold. Trader Frances Moore wrote in his journal, "Their way of bringing them in is tying by the neck with leather thongs, at about a yard distance from each other, thirty or forty in a string." These Africans had reached the edge of their continent. This ended the first phase of their enslavement.

### Why did Africans trade Africans into the Atlantic slave commerce?

No question is asked more frequently or is more vexing, and no question starts bigger arguments than this one. A famous African historian, Adu Boahen, puts the question another way: "Why then did African rulers cooperate with the Europeans in this hideous trade?" This question touches on so many issues—moral, economic, and interpretive. It raises another big, unresolvable question: Who was to blame for the trade?

Recently, Black critic Stanley Crouch took a provocative swing at this question. According to him, Africans who sold slaves did not worry over the enormity of what they were doing. "There was no great African debate," he wrote, "over the moral meanings of slavery itself." Africa did not produce any movement against the slave trade—"no African William Lloyd Garrisons, no African abolition movement, no African Underground Railroad." Africans betrayed the region's citizens for profits. For Crouch, this is not a scandalous admission, just a historical fact that needs to be faced.

Today, it is widely recognized that many groups and individuals in western Africa collaborated with the Europeans in the Atlantic slave trade. It is also recognized that European traders were the principal driving force behind the trade. They had uses for Africans as labor, at first in Europe and later in the establishment of settlements in the New World. Their demands set the agenda for the trade, and they understood the place of African slaves in the new global order. Had there been no European scheme, Africans could not have invented and carried out this trade.

Still, there is no getting around the fact that Africans of various stripes assisted Europeans in the slave trade. In fact, the Europeans working in West Africa were never numerous enough to manage single-handedly the raiding aspect of the trade. Nor could they have survived searching in the interior for slaves; they would have died of diseases (malaria, for one) for which they had no immunity. Africans, on the other hand, were powerful enough to help European traders, and they became stronger as the trade matured.

History teaches us to understand the causes of great inhumane actions, such as enslaving others. It is important to understand—but not to excuse—what motivated African collaborators, what caused them to join in selling humans. There are five leading factors:

- The elites and ruling classes of West Africa desired the goods that European traders offered in exchange for African bodies: iron bars, copper basins, cloths, medicines, muskets, ammunition, gunpowder, brandy and rum, other African goods. Europeans also offered gold. Africa's elites liked the power that the slave trade gave them.

- As the slave trade grew, European merchants used guns and gunpowder to acquire slaves. Very slowly, an African arms race developed. Power in West Africa began to come from the barrel of a gun. Many rulers were forced to enter the slave trade in order to survive in a turbulent world.

- Slaves and servile people already existed in West Africa before the European demand arrived. Not treated as outcasts, not set apart, not barred socially as barely human, they were instead a part of extended family structures. They could rise in society, do skilled work (as blacksmiths, fishermen, weavers), enter the military, and even rule. An Asante proverb said, "A slave who knows how to serve succeeds to his master's property." But these slaves and servile people were also exchanged, given away, and accepted as gifts. Trading Africans to Europeans, therefore, was not such a radical departure.

- When the Atlantic slave trade first started in the fifteenth century, it was small. Thus, it could be accommodated without being threatening. It began in the Upper Guinea area, between today's Senegal and Ghana, and it crept down the coastline, to Lower Guinea and Central Africa, insinuating itself. By the time the radical demand for slaves started—with the settlement of the New World—the slave trade with Europeans was already woven into African life.

- There was no idea of a unified Africa. Overwhelmingly, the people captured for slavery were taken by one group from other groups, people with slightly different customs and languages living in other, often distant places. The sense of difference made it easier to enslave people and to trade them. Mostly, they were not relatives, neighbors, or friends.

Africans did resist the early trade. In 1526, Nzinga, king of the Kongo, complained by letter to Portugal's king, "It is our will that in these kingdoms there should not be any trade of slaves nor outlet for them." Other kings interrupted the trade. Whole villages fled the raiding grounds. Even ordinary citizens tried to stop early slave auctions on the coast.

But to no avail. The trading in Africans galloped ahead, its growth dependent on a tight partnership between European traders and African rulers. The French trader Jean Barbot, who spent a long time on the coast, captured the truth in 1680: "The trade in slaves is the business of kings, rich men, and prime merchants." Together, they pursued profits and power.

From this evil partnership came the spread of Old Africa's peoples into the new settlements across the Atlantic.

## RESOURCES

Adler, Peter, and Nicholas Barnard. *African Majesty: The Textile Art of the Ashanti and Ewe.* London: Thames and Hudson, 1992. Text with textile artwork.

Clayton, Peter A. *Chronicle of the Pharaohs: The Reign-by-Reign Record of the Rulers and Dynasties of Ancient Egypt.* London: Thames and Hudson, 1994. Text with illustrations.

Dassow, Eva, ed. *Egyptian Book of the Dead: The Book of Going Forth by Day.* San Francisco: Chronicle Books, 1994. Featuring integrated text and full-color images.

Davidson, Basil. *Africa in History.* New York: Macmillan, 1991.

Davidson, Basil. *The African Slave Trade.* Boston: Little, Brown, 1980.

Davidson, Basil. *The Search for Africa: History, Culture, Politics.* New York: Times/Random House, 1994.

Hamdun, Said, and Noel King. *Ibn Battuta in Black Africa.* Princeton: Markus Wiener Publishers, 1994.

Hickey, Dennis, and Kenneth C. Wylie. *An Enchanting Darkness: The American Vision of Africa in the Twentieth Century.* East Lansing: Michigan State University, 1993.

Holloway, Joseph E., ed. *Africanisms in American Culture.* Bloomington: Indiana University Press, 1990.

Lipschutz, Mark R., and R. Kent Rasmussen. *Dictionary of African Historical Biography.* 2nd ed. Berkeley: University of California Press, 1986.

Mullin, Michael. *Africa in America: Slave Acculturation and Resistance in the American South and the British Caribbean, 1736–1831.* Urbana: Illinois University Press, 1994.

Murray, Jocelyn, ed. *A Cultural Atlas of Africa.* New York: Equinox Books, 1989.

Northrup, David, ed. *The Atlantic Slave Trade.* Lexington, Mass.: D. C. Heath, 1994.

Phillipson, David W. *African Archaeology.* 2nd ed. New York: Cambridge University Press, 1993.

Shillington, Kevin. *History of Africa.* New York: St. Martin's Press, 1989.

Sweetman, David. *Women Leaders in African History.* Oxford: Heinemann International, 1984.

Thompson, Robert Farris. *Flash of the Spirit: African And Afro-American Art and Philosophy.* New York: Vintage Books, 1983. Text with photos.

Thornton, John. *African and Africans: In the Making of the Atlantic World, 1400–1680.* New York: Cambridge University Press, 1992.

Van Sertima, Ivan. *They Came Before Columbus.* New York: Random House, 1976.

Vlach, John Michael. *The Afro-American Tradition in Decorative Arts.* Athens: University of Georgia Press, 1990. Text with photos.

TEXTS FOR YOUTH

Roehrig, Catherine. *Fun with Hieroglyphs: Guidebook, Rubber Stamps, and Ink Pad.* New York: Metropolitan Museum of Art, 1990.

Hyman, Mark. *Blacks Before America.* Trenton, N.J.: Africa World Press, 1994.

# 2

# Africans Enter American Slavery

## Before 1619 to the 1720s

1100–1500s: Highly developed Indian civilizations populate the Western Hemisphere

1492: Columbus's first voyage opens the New World to European settlement

1502: First African slaves introduced in the New World by the Spanish

1519: Spanish Blacks—descendants of African slaves in Spain—accompany Cortés in the conquest of Mexico

1522: First slave revolt on island of Hispaniola (today's Haiti/Santo Domingo)

1527: Spanish Black (Estevanico) explores Florida and Mississippi

1562: John Hawkins raids African coast for slaves: first English role in slave trade

1560–1605: Spanish and French explore and settle North American colonies

1606: English king James I grants colonizing charters to Virginia and Plymouth Companies

1607: Jamestown colony is established in Virginia

1619: At Jamestown, first twenty Africans arrive

1638: African slaves brought to New England

1641: Massachusetts legally recognizes slavery

1661: Virginia recognizes slavery, establishes strict laws in favor of slaveholders

1664: Maryland prohibits marriage of English women to Black men

1680–1808: Height of slave trade to North American colonies

1712: Slave revolt breaks out in New York City

1720: Many slaves come to American colonies directly from Africa

1730: Slave population in British North America is over 90,000

In 1789, the British public was introduced to a most unusual book, *The Interesting Narrative of the Life of Olaudah Equiano, the African.* It was one of the first books by a Black person to be published in English. Copies of it were sent to members of Parliament. Within a few years, the two-volume book was an international best-seller. On opening the book, its readers undoubtedly paused to study the author's likeness—a full-page cameo showing a handsome man, of soft Black complexion and with curly hair, bedecked in a fine jacket, vest, and ruffled shirt, and holding an open Bible. On his deathbed in 1791, Methodism's founder John Wesley had someone read this man's book to him.

Olaudah Equiano (1745–97) had penned an eighteenth-century classic that would become as influential as Voltaire's *Candide* (published in 1759) and Benjamin Franklin's *Autobiography* (1791). Equiano's goal was to write "a genuine Narrative; the chief design of which is to excite . . . a sense of compassion for the miseries which the Slave-Trade has entailed on my unfortunate countrymen." Together with his sister, he had been kidnapped at eleven, in 1756, in the West African interior, near the modern Iseke in Nigeria. Soon he was delivered to a ship headed across the Atlantic. "I was now persuaded that I had gotten into a world of bad spirits," he wrote, "and that they were going to kill me." Equiano was enslaved in many places, including Virginia. Eventually, he bought his freedom.

When Equiano entered the slave trade, it was a big ferocious business. Already more than 250 years old, it still had several decades to go before it was prohibited. Out of the slave trade came a new way of life—a tough, varied, successful system of production based on the labor of African men and women. In 1789, when Equiano's book appeared, this system was integral to American life, particularly in the South.

Millions of Africans passed from their homelands into American slavery without leaving behind a book, but if they had, their story no doubt would have read something like Equiano's. They would have told of despair, brutality, and the meanest toil. They would also have told of their exceptional intelligence, quick adaptations, and strong communities. Their stories would show us how they endured the long night of slavery.

### Was the enslaving of Africans a unique historical event?

Humans have been sold into slavery in all the regions of the world. In Europe and Africa, in the Middle East and Asia, this practice has existed from time immemorial—from Egyptian times, to ancient Rome and Greece, and throughout the Middle Ages.

Many peoples and individuals have been through the hell of slavery. The Israelites were slaves for the Egyptians. Saint Patrick (circa 385–461 A.D.), at age sixteen, was captured by pirates, taken to Ireland, and made a slave for six years. The word *slave* comes from the Slavs, who were enslaved repeatedly and for centuries. Unfortunately, enslaving humans was a time-honored tradition.

Enslaving Africans through the Atlantic slave trade was, therefore, not unique. But this slave trade had its own special features. First, the scale of it was truly extraordinary: millions were displaced, vast distances were covered, immense territories were affected on both sides of the Atlantic. More important, the transatlantic slave trade was part of the very genesis of capitalism and helped to transform the modern world economy.

What was also special about the enslavement of Africans was that by the midseventeenth century, Black skin and slavery were considered synonymous. Overwhelmingly, slaves were Africans. Historian Basil Davidson has concluded, "Africans came to be seen as mere savages whom it was legitimate and even laudable to enslave. . . . With this new attitude, we have entered the period of modern racism."

The Atlantic slave trade in Africans was the latest instance of a huge human commerce that took place throughout the course of world history. But this is another reason that we think of it as unique: it occurred so recently, remarkably close to our own time. Fewer than two centuries ago, fewer than seven generations ago, Africans could be legally traded into the United States. Fewer than 150 years ago, until 1865, the system of slavery was alive and well in this country. Had American slavery lasted only a mere thirty-five more years, it would have greeted the twentieth century. Black American ex-slaves (and white ex-slaveholders) were still living into the 1960s.

## What was the basic purpose of the African slave trade to the Americas?

Planting and harvesting crops; working in mines; building farm homes, stables, granaries; doing household jobs; constructing canals and bridges—these were some of the jobs performed by Africans in the Americas. African slaves, first and foremost, were laborers.

The basic purpose of the trade was to capture people by force in West Africa and to transport them across the Atlantic Ocean to the new European settlements. Once there, they became the primary labor not only on farms but also in cities and towns and in specialized occupations.

Laborers were needed to open up the new territories claimed by the Spanish, Portuguese, Dutch, French, and English. These territories were called the "Americas"; on early maps, this was the general designation for the areas Europeans were settling. Actually, the "Americas" were not one place but three distinct places: South and Central America, the West Indies, and North America.

Africans were crucial to the new world being born. Without their labor, the settlements would not have been as stable, would not have developed as quickly, and would not have generated such wealth. African work jump-started the European settlements in the New World and sent them off on their spectacular historical trajectory.

**How many Africans were brought to the Americas during the slave trade?**

Around ten million Africans were brought to the Americas during the nearly four centuries of the slave trade.

Ten million is a figure derived from nearly three decades of intensive work conducted since the 1960s by historians of the slave trade. In 1969, Philip Curtin, then a professor of African history at the University of Wisconsin, was the first historian to advance a rigorous global census for the Atlantic trade. Ten million covers all the Africans brought to South and Central America, the islands of the Caribbean, and North America, including the American colonies that were to become the United States.

Ten million! And this number doesn't include their children, their children's children. . . . Every new generation born to slaves increased the slave population until it becomes impossible to know the total.

This moving of Africans was one of the largest forced migrations of people in world history. To get an idea of how large this trade was, think of moving—person by person—the population of two American cities like New York and Chicago from Africa all the way across the Atlantic. Or imagine transporting the combined populations of Michigan and Montana across thousands of miles of ocean. To deliver this large population to the Americas, the trade had to become a massive, efficient, organized enterprise.

There is another figure of devastating significance. Added to the ten million Africans landed in the Americas, there were another one and a half to two million people who were captured or shipped but never made it to the Americas. Some died from the trauma of being captured. As Scottish traveler Mungo Park wrote in 1799, some died because they were "frequently transferred from one dealer to another." An ex-slave recalled, over the course of half a year, "I was sold six times over, sometimes for money, sometimes for a gun, sometimes for cloth." Some slaves succumbed on the long, dreary marches. More died while held on the African coast before being shipped to the Americas. Finally, many more perished during the long, stressful journey across the Atlantic.

Beyond these figures, thousands of Africans were maimed and killed as the slave trade's relentless search for bodies spread throughout the interior. Wars and raids wreaked havoc. Africans fleeing the wars became impoverished refugees. Mungo Park wrote, "The desolations of war often . . . produce . . . famine." Thus, people who were never captured also died as a result of the trade. In Central Africa, a premier source of slaves, thousands upon thousands died as a by-product of the struggle for captives. For these people, there are only estimates.

Since Curtin's first calculations of the slave trade's numbers, there has been some dispute. Joseph Inikori, a Nigerian historian now teaching in America, claimed that the ten million figure is too low. Using other historians' revised estimates, he has upped the figure to around 13.4 million Africans delivered to the Americas. Also according to him, nearly 15.4 million were shipped from Africa. These numbers are not too far-fetched to imagine. As more is known about the trade's various segments, the earlier lower totals might inch upward.

Why fix a number on this ghastly trade? If we knew all the

names and could carve them on a wall stretching from coast to coast, we wouldn't need a number. But a number is the only way we can "see" the magnitude of the slave trade. The number can tell us how many people Africa lost. The number also allows us to measure the contribution of Africa and Africans to the history of the Western Hemisphere.

**Should Africa be repaid—that is, given reparations—for its losses during the slave trade?**

"Nations in Slave Trade Urged to Repay Africa." So read a *New York Times* headline in 1992. A year later, several full-page articles on repayment in Kenya's *Sunday Nation* sparked a nationwide debate in that country. Today, many African intellectuals are seriously discussing these proposals.

Chief Moshood Kashimawo Abiola, a Nigerian multimillionaire businessman, is the latest proposer of repayment. He wants "white people in Europe, America, and [people of] the Middle East to pay the African continent for the damage done during the slave trade." Chief Abiola explains, "I've been almost everywhere in the world, the Black man is at the bottom of the ladder. . . . I went up and down the continent . . . and realized that there is nothing we have in Africa that is not controlled by people in Europe and the United States—absolutely nothing." From the slave trade forward, Africa slid downhill, while the nations that profited from the trade rose in the world.

Chief Abiola was able to get the Organization of African Unity (OAU)—a United Nations–type organization, with representatives from all African nations—to put this on its agenda. The OAU charged its Group of Eminent Persons with the task of designing a campaign to obtain reparations for enslavement, focusing on Western nations. The idea is that "the most evil and most extensive slavery was the transatlantic one. It is therefore right and proper that the West should be the first to be called upon to pay."

African intellectual Ali Mazrui says there is a "new moral and political breed" among Africans, "the reparationists." "They are committed to the proposition that the injustices of enslavement and bondage could not have ended with formal emancipation. They can only truly end with the atonement of reparations." Today's reparationists propose the Middle-Passage Plan (named after the transatlantic voyage from Africa to the Americas), which would involve capital and skills transfers from the Western nations to be used in the reconstruction of Africa.

To pay reparations is to acknowledge and give compensation to someone who has suffered injury, harm, or other wrong at one's hands. Nations have paid reparations to people and countries. After World War II, for instance, Japan paid war reparations to the Philippines, Indonesia, Burma, and Vietnam, all countries that had suffered under Japanese invasions. In 1952, West Germany agreed to make financial reparations to Israel and to organizations serving Jewish people for Nazi atrocities against the Jews during World War II. In 1988, the United States Congress voted an apology and a payment of twenty thousand dollars to some of the oldest survivors of the Japanese-American internment camps during World War II. In 1990 and 1991, United Nations Security Council resolutions urged Iraq to pay reparations to Kuwait for the damage done by the Iraqi army during its occupation.

Both pros and cons exist on the issue of paying reparations to Africa:

- Some pros: Africa lost more than people to the trade. The continent lost all the labor, intelligence, and skills of these people—that is, their productivity and creativity—for a lifetime. And the talents of their offspring were lost forever. The captured people were human capital that would have been invested in building Africa. Instead, slave masters in the Americas used that human capital for building their own mines, farms, craft and fishing industries, and new cities, such as Havana, Salvador, New Orleans, Richmond, Charleston, Philadelphia, and the cities of Mexico. Admittedly, masters fed, clothed, and housed slaves. But these slave provisions were notoriously deficient. In sum, repaying Africa would begin to compensate for an enormous brain and labor drain.

- Some cons: although Europeans (and Middle Easterners) spurred the slave trade's development, they were aided by Africans. This is an unpleasant historical fact. If the blame for Africa's slave trade woes cannot be exclusively assigned to foreigners, then foreigners cannot be asked to pay by themselves. Chief Abiola dismisses the fact that many Africans captured and sold slaves. He sees this as a diversionary argument.

  Another point against reparations is that contemporary nations are not likely to ante up for crimes committed against Africa centuries ago. Few twentieth-century reparations have applied to wrongs more than fifty years old. Recent injustices seem to get more attention and have a greater chance of remedy. Usually courts respect most direct victimization and a contract or treaty that has been broken. The slave trade's primary victims are not alive, nor are the primary culprits. The slave-trading companies that worked the West African coastline cannot be hauled before a tribunal because they no longer exist.

Informed of these hurdles to reparations, Chief Abiola responded with optimism: "All we want is for the world to listen. If they do, they will understand, and something dramatic will occur." In 1993, the OAU's First Pan-African Conference on Reparations was held in Abuja, Nigeria. The "Abuja Declaration" demanded capital transfer and debt cancellations. This issue is gathering support, but it will be a hard sell.

### What do the terms "African diaspora" or "the Black diaspora" mean?

*Diaspora* comes from ancient Greek, meaning "scattering or dispersion." When seeds are sown, they are scattered, thrown by hand across a plowed field. For the Greeks, this sowing action was a diaspora.

The concept of a human diaspora originates in Jewish history. One of its first expressions is found in the Bible: "The Lord shall cause thee to be smitten before thine enemies: thou shalt go out one way against them, and flee seven ways before them; and shall be removed into all the kingdoms of the earth" (Deuteronomy 28:25). Within Jewish thought, this quotation is believed to refer to the future forced dispersion of the Jews, in flight from their enemies and persecutors. Also present in Jewish thought is the notion of voluntary dispersion.

"African diaspora" and "Black diaspora" are fairly recent terms. They were coined around the 1930s among Pan-Africanists, Black thinkers from all over the world whose goal was to empower a political movement embracing all Blacks. The terms seemed to fit: like a gigantic historical centrifuge, the slave trade had flung Africans to the far corners of the globe.

Stressing the historic links among all Blacks, these thinkers sought to overcome this tragic fragmentation. American poet Langston Hughes expressed this sentiment:

*We are related—you and I,*
*You from the West Indies, I from Kentucky,*
*We are related—you and I,*

*You from Africa, I from these States,*
*We are brothers—you and I.*

Senegalese writer Leopold Senghor has imagined that African blood flowed in the veins of citizens of Harlem and that Black soldiers in Europe of the Second World War were African warriors.

The concept of an African diaspora highlights the existence of a global Africa: the continent of Africa *plus* the scattering of Blacks due to enslavement. Historian Joseph Harris has mapped global Africa and identified four major areas of scattering: Europe, Arabia and Asia, North Africa, and the Americas. Of course, the largest numbers of dispersed Africans anywhere were in the Americas. Today, thirty-four countries in the Western Hemisphere have Black populations.

Over the last two centuries, several return-to-Africa movements have flourished. Liberia, in West Africa, was created in the

1820s for the settlement of Blacks—mostly free, but also slaves—from America. Using the United States as a model, they began a new country, unfortunately colonizing the indigenous peoples. (Many Blacks saw Liberia as a scheme to rid the United States of them.) In this century, the greatest movement advocating a return involved the United Negro Improvement Association, founded in 1914 by Jamaican Marcus Garvey (1887–1940) with the slogan "Back to Africa." Garvey's message was still compelling during the 1950s and 1960s, when Blacks returned to the newly independent African states to work, study, and celebrate these new nations.

Today, the African diaspora is defined not only by a world geography of color but also by a cultural geography. Elements of African cultures have survived through slavery and its racist legacies. So global Africa today consists of cultures embodying living African elements. The Black Atlantic world is a vibrant cultural entity, pulsating with powerful African ideas, particularly religious ones.

One of the great museum exhibits of 1994–95 was *Faces of the Gods,* a traveling show recreating religious altars in the Americas inspired by West African beliefs. According to the exhibit's catalogue, it is stunning proof that "Africans arriving on the shores of North and South America did not forget their ancestors or their gods."

## What was the middle passage?

The middle passage was the name given the long ocean journey between Africa and the Americas that took Africans into slavery. At its end, they had passed from their familiar homelands into strange territories.

Once transported across the Atlantic, Africans almost never returned home. This was "the voyage of no return." The middle passage was, therefore, a destructive experience—an experience of profound distancing—for Africans. Alexander Falconbridge, an English ship doctor, wrote of the psychological stress: "Frequently, it happens that the negroes . . . become raving mad;

and many of them die in that state, particularly the women."

Ironically, the ships taking Africans to slavery often had names like the *Blossom, Rainbow, Morning Star, Recovery, Young Hero, Saint Paul, Elizabeth, Mercury, Temperance,* even *Lucky, Good Spirit, Charity,* and *Mercy*. One ship was named *Hannibal,* after the great African general. Regardless of the noble names etched on the ships' bows, beneath the main decks were row upon row of despondent captured Africans.

Shackled by locks and chains specially designed for slave ships, the captives were made to lie "spoonlike" in every available space. Male captives were chained ankle to ankle. Sometimes the women were left unshackled. A sailor described this scene: "The slaves lie in two rows, one above the other, on each side of the ship, close to each other, like books upon a shelf. I have known them so close that the shelf could not, easily, contain one more." An adult male was given a tiny space—five and a half feet long, sixteen inches wide, two to three feet high. No one could sit up; no one could stand.

Stuffing slave ships like this was called "tight-packing." The result was horrifying: "The poor creatures, thus cramped, are likewise in irons for the most part, which makes it difficult for them to turn or move. . . . Every morning . . . more instances are found of the living and dead fastened together." The dead and near dead were thrown overboard.

British antislavery clubs used slave-ship deck plans to horrify the public. The floor plan of a British ship, the *Brookes,* was made famous; it had carried over 450 slaves. Compassion supposedly moved the British Parliament in 1788 to set limits on packing. Other European countries did the same. But as profits from the trade rose, companies and some ship captains broke the rules. They figured, the more shipped, the more might arrive healthy and thus the greater the number there would be for sale.

Overcrowding was not the only assaulting experience. Taken below deck, Olaudah Equiano remembered vividly his first impression: "There I received such a salutation in my nostrils . . . so that with the loathsomeness of the stench, and crying together, I became so sick and low that I was not able to eat." John Newton, a ship captain, recalled that the heat melted his candles whenever he went below deck to inspect the human cargo. Another report said, "The floor of their rooms was so covered with the blood and mucus which had proceeded from them in consequence of flux [dysentery] that it resembled a slaughterhouse."

Captives had to endure these conditions for forty to sixty days. From the tip of West Africa to the Middle Atlantic American colonies was a trip of four thousand miles. From Central Africa to Georgia, the distance was six thousand miles. A ship's speed, the winds and weather on the high seas, and a captain's navigating skills affected the time the voyage took.

Naturally, captives fell ill from infections, intestinal worms, scurvy from lack of vitamin C, and more seriously, from dysentery and smallpox. As prevention, crews scrubbed the underdecks with brooms and then would "perfume betwixt decks with a quantity of good vinegar." Food was available twice a day. Often, this consisted of horse beans—a cheap vegetable—boiled to a coarse pulp and sprinkled with flour, red pepper, and palm oil. Occasionally, rice, plantains, yams, and manioc were provided. For a few minutes each day, the captives were fed and "exercised" on the upper deck in the fresh air. Exercising—called "dancing the slaves"—actually consisted of the captives being forced to dance and sing.

Although the majority of captives survived, tens of thousands died. A look at a few French ships' records tells the story. On the voyage from West Africa (Senegal) to Louisiana in 1726, *L'Aurore* left with 350 captives and arrived with 290—a loss of 60. *La Diane* departed Africa in 1728 with 516 and landed 464 in Louisiana—a loss of 52. *Le Saint Louis* in 1731 left with 350 and arrived with 291—a loss of 59.

Resistance was a constant threat throughout the middle passage. Bringing the captives to the upper deck was a moment that crews especially dreaded. As trader Richard Snelgrave wrote, "we were obliged to secure them very well in Irons, and watch them narrowly." Unchained, captives jumped overboard. These were freedom suicides. Jumpers sent up a great yell upon hitting the water. Sailors believed that Africans aimed to swim home.

Crews feared revolts; more than fifty substantial revolts have been documented, and references to many smaller incidents exist. "These Slaves contrived to release the others," a ship log reported, "and whole rose upon the People, and endeavored to get Possession of the Vessel." In 1730, a Massachusetts ship's crew was overwhelmed as "the Negroes got to powder and arms and about three o'clock in the morning, rose upon the whites."

In 1770, the *Don Carlos*'s first mate witnessed a serious revolt: "Most of them were yet above deck, many of them provided with knives. . . . Others had pieces of iron. . . . They had also broken off the shackles from several of their companions' feet. . . . Thus armed, they fell in crowds on our men." They killed fifteen men, cut one sailor's leg "so round the bone," and "the cook's throat to the pipe." But the revolt was put down.

Ship captains reacted swiftly and severely to revolts. After a revolt on a French ship, the captain decided to "make a necessary example for such cargoes, and to hoist him [the leader] to the top of the main mast and fire upon him with rifles until death ensued."

A memorable rebellion occurred in 1721, while the English ship *Robert* was still anchored near the Sierra Leone coast. An unshackled woman, together with two men—one called Captain Tomba—organized an uprising to get the captives back to shore. The woman stole the weapons with which the rebels killed several sailors and wounded others. Her punishment was great: "The woman [was] hoisted up by the Thumbs, whipp'd, and slashed . . . with Knives before the other Slaves till she died." Obviously, the dream of freedom existed from the moment that slavery began.

### Why do Black Americans visit the slave castles of the West African coast?

El Mina Castle. Christianbourg Castle. Fort San Anthony. Fort Saint Jorge. Fort Saint Sebastian. These are some of the slave castles and forts along the West African coast that were built by the European powers between the fifteenth and eighteenth centuries. The Portuguese were the most prolific builders of forts, but the French, Dutch, Danes, and English were all in the fort-owning business. Coastal forts were the sole anchorage for white men on the coast.

The term "slave castles" is not the best name for these places. Within them, the Europeans who ran the forts had luxurious apartments. But trading was the main purpose of these forts. Mainly, they were depots for gold exchanges and warehouses for captured Africans. The biggest of the forts were like busy miniature cities, with a military and trade commander, a few soldiers, several multistory buildings, and up to fifty cannons. Ships stopped there continually to pick up gold, slaves, and ivory and to replenish their supplies of fresh vegetables, fruit, and water.

Often painted white, set on promontories above beautiful little bays and coves, with views of the blue seas, these forts are some of the world's historic monuments. (UNESCO has classified fifteen of these forts and castles as world heritage sites.) Every year, hundreds of Black Americans make visits to these slave castles along the coast because these places symbolize the passage of their ancestors to the Americas. They are greeted by an evocative quiet, interrupted only by the sound of waves crashing.

Most of these visits are to the coast of Senegal, where there is a large slave castle on Goree Island, and to Ghana, where a large number of slave forts are located. In Ghana, four castles are the tourist favorites: El Mina Castle, Cape Coast Castle, Fort Coromantine, and Fort Metal Cross at Dix Cove. The Ghana forts are within 125 miles of Accra, the country's capital.

These visits are more like pilgrimages. Black Americans come face-to-face with the fact that their ancestors suffered in these oppressive dungeons. They stand where they stood. They grope in the dark as they once did. They feel the dankness of the tiny holding rooms. At El Mina, they walk across a large courtyard into which slaves were herded and priced. They imagine them being stripped and examined, as slave trader Canot wrote, "without regard to sex, head to foot . . . eyes, voice, lungs, fingers and toes were not forgotten." The psychological impact is intense.

Filmmaker Haile Gerima took this as the theme for his 1993 *Sankofa*—the story of Mona, a Black American fashion model who is possessed by spirits lingering in the Cape Coast Castle in

Ghana. "Sankofa" is the name of a mythical bird who looks back—whose motto is "Pick it up if it falls behind you"—and is a symbol of the ability to learn from the past. Ancestral spirits send Mona on a journey into slavery—and then into liberation.

Even for castle visitors who are not descendants of Africa, a hushed silence prevails; this is a moment of self-recognition for all. Sometimes in the silence you can hear a few crying.

These visits are reminders of the sheer tenacity of the captives. "We survived," as a Black American woman said simply after a recent tour.

## Of all the Africans brought across the Atlantic, how many arrived in North America?

Around five hundred thousand to six hundred thousand Africans were brought into North America beginning with the first shipment in 1619. Officially, the United States banned the slave trade in 1808, but after the ban, some slaves were brought in secretly—probably about a thousand a year for the whole country. The total includes these arrivals.

Thus, only about 5 to 6 percent of the ten million slaves scattered all over the Americas were brought to North America. This figure comes from scholars of the various slave trades in the Americas.

When people first hear this figure, they can hardly believe it. They think, this can't be all; there must have been more. It is hard to fathom how so many Black Americans—over 32.5 million by 1994—could come from this small original group. Many people assume that the slave ships must have brought in tens of thousands of Africans who were never tracked.

Sometimes it is thought that the stream of incoming Africans was similar to the European immigration to America in the late nineteenth century, when fifteen million people arrived between 1890 and 1914. But both of these impressions are wrong.

Besides the figure of five to six hundred thousand arrivals, we have other intriguing information about the trade in Africans to North America:

- The greatest number of Africans were brought into North American slavery between 1690 and 1800—over four hundred thousand. These arrived in the Chesapeake region—Virginia and Maryland. In 1680, there were slightly over 4,500. By 1700, 13,000. By 1770, about 100,000. When the American Revolution began in 1775, almost 200,000 Blacks were living and laboring in these two colonies.

- With less than 6 percent of the African slaves brought to North America, the bulk were taken to the Caribbean and South America; smaller numbers went to Central America.

- Fewer Africans were brought to North America than to the little island of Cuba.

- The Caribbean islands absorbed four million Africans, or 40 percent of the total. Brazil took in another 3.6 million, 36 percent of the total. To maintain and expand these colonies' economies, Africans had to be imported at a steady rate.

- Why were thirteen times more slaves sent to South America and the Caribbean than to North America? Brazil, Cuba, Jamaica, Barbados, and Saint Dominique had enormously high mortality rates for Africans. Tens of thousands of Africans died on the harsh plantations because of the taxing work required to produce sugar and because of cruel field overseers. It was also hard to replenish these slave populations naturally because more males than females were brought into the colonies.

### Did all the African populations in the Americas fare similarly?

A most intriguing fact has come to light about the general development of African populations in North America. The original population of six hundred thousand Africans grew at a high rate of natural increase. This was unique, especially when compared to the other slave communities.

For over 150 years, this high population growth continued. By 1820, the original population had grown to almost 1.8 million. By 1830, it was almost 2.3 million. In 1861, the year the Civil War started, there were 4.4 million Blacks.

This meant that Blacks in British North America were increasingly native born—born in places like Virginia, Georgia, and New York rather than Senegal, Central Africa, and the West African hinterland. As America developed, a greater percentage of Blacks were native born than were whites. Blacks born in Africa—then called "saltwater Negroes"—became a minority.

In Alex Haley's *Roots* (1976), when Kunta Kinte was brought from West Africa to the Maryland plantation in the 1760s or 1770s, he was treated by other Blacks as an oddity. Outwardly, the other plantation Blacks had changed. African cultural habits were being mixed with local traditions; for example, Blacks had been expressing themselves for some time in forms of English. Haley's portrayal was accurate in its depiction of the new majority of native-born Blacks and the minority of incoming Africans.

Hugh Jones, an English mathematician in the colonies, observed in 1724, "The Negroes are not encreased by fresh supplies from Africa and the West India Islands." By Jones's time, Africans in bondage were producing a considerable progeny.

Right off, most people explain this fact in two ways. First, they say the high natural increase is proof that slavery in North America was relatively benign—that Black slaves were given reasonable nutrition, a work routine that was not too harsh, and little interference from masters. According to this logic, Blacks were treated sufficiently well for them to produce large numbers of offspring. This idea of a gentle American slavery is not generally supported.

Second, people argue that slaveholders systematically forced breeding among Blacks. This is a most serious argument. Slaveholders had much control over their slaves and their relations—including, for example, the approval or disapproval of slave marriages. Slaveholders also had a special financial interest in high levels of fertility among slave women. Some increases in slave populations obviously came from individual slaveholders' manipulations. But evidence of a major scheme of forced breeding is hard to find.

Other explanations work better. During the first century of North American slavery, more African men than women were in the slave population. But the ratio between men and women began to even out, and with that came the possibility for couplings of various types. The balancing of men and women occurred around 1740. Obviously, this helped the population to grow.

Fertility among women was also high during the first centuries of slavery. Most women had children. Slave women started having children at an early age, and there were short intervals between births.

It was also true that slave men and women bonded with each other early on. These were not families in our sense; they were influenced by a wide range of African ideas of family. Important in these bondings was the African love of progeny. All of these factors helped produce a high, consistent population increase.

This increase had several consequences. North American slavery became more profitable. Planters knew that they would get a lot of slaves by purchasing a few. Black women became more valuable in the system of American slavery. Not only was their work in the fields valuable but their value went up further because they were the source of more slaves.

By 1825, Blacks in America comprised 36 percent of all the people of African descent in the hemisphere, though they had begun as 6 percent. America held the largest number of Blacks as slaves in the world. It had become the world's premier slaveholding power.

## Who were the "twenty Negars" of 1619 in Virginia?

On August 20, 1619, twenty Blacks were brought to the English colony of Jamestown, a small town on the waterways in the area we now call the Tidewater. They were the first Blacks to arrive in English North America.

When the twenty Blacks put their feet on Virginia soil, Jamestown was young. The colony was founded in May of 1607 and was England's first permanent foothold in the New World. It was run by the Virginia Company, which planned to bring men and women from England to raise crops for sale back home. Things had not gone well in the early years. Disease and death had stalked the colonists. Relations with the Powhattan Indians had been unpredictable.

But things were slowly picking up. Tobacco was becoming a highly profitable cash crop. In 1619 there were other changes as well: the first women were brought to the colony, along with hundreds of new men, and the first English legislative assembly was convened in a Jamestown church. (This would become the House of Burgesses.) To this situation were added the new Blacks.

John Rolfe, a rising tobacco farmer in Jamestown, made a laconic note about their arrival in his diary: "About the last of August came in a Dutch man-of-warre that sold us twenty Negars." It almost appears that he was expecting this to happen. "Negars" was one sixteenth-century name for "Negroes," and the term occasionally was used to refer to other nonwhites as well. The Dutch warship stopped at a place called Point Comfort along the coast. Captain Jope of the ship had run into hard luck and needed to sell the "Negars" to get provisions for his crew. The Blacks were put in a smaller boat and brought upstream to Jamestown, where they were handed over to their new owners.

These twenty Blacks are, in a metaphorical sense, the ancestors of the current Black population in this country.

Here is their story: They were brought from the Caribbean—not directly from Africa—where they had probably been field hands. They had been sold before. It became common for Blacks to be brought from the Caribbean. A later Jamestown official said that "Negroes . . . brought to Virginia [during this time] were imported generally from Barbados for it was very rare to have a Negro ship come to this country directly from Africa." It wasn't until the 1790s that slaves were routinely brought directly from western Africa.

Some of these early Blacks had Spanish first names. Most of them were adults, though it is unclear how many were men and how many were women. George Yeardley, the colony's governor, and Abraham Piersey, soon to be one of the colony's richest men, bought the bulk of them.

The new arrivals were not sold as slaves. They entered the

colony as servants, probably as "indentured servants." They could go on to own land and farm it. Slavery—stripping persons completely of freedom—had not yet begun in North America. Jamestown had been trying to use a mixture of English men and some Indians as labor. Now they would try Blacks.

Jamestown is an impressive site even today. Located at the water's edge, amid marshes and peninsulas, with flocks of fowl flying overhead, the recreated settlement sparks the imagination, making visitors reflect on the earliest days in our national history. In 1990, the Jamestown gallery gave a prominent place to the Black role in this first settlement's history.

## What was indentured servitude, and what did it have to do with early Black American slavery?

Many English colonists—both men and women—came to Jamestown for a new life overseas, inspired by recruiters' promises of an easy livelihood. Usually poor, they contracted with the Virginia Company to work for an English planter for a number of years—typically four to ten—in order to pay for their passage, food, and clothing. Their contracts were called "indentures."

Indentured servitude was active in Jamestown during the colony's eighteen-year existence (1606–24). After that, it expanded throughout early Virginia and the other English colonies. Hundreds of thousands of English men and women came into the new colonies through the system.

The life of English indentures was no utopia. During their contract, the planter was their provider and master, feeding and sheltering them. Planters felt justified in extracting heavy labor from them. "Barbarous usuage" of servants by "cruell masters" was common. A London report said planters in "Virginia do much neglect and abuse their servants with intolerable oppression and hard usage." Planters also treated indentures as commodities: their contracts were bought and sold; they were willed to heirs and even gambled away. Thomas Best wrote to his brother in London complaining, "My Master Atkins hath sold me for 150 sterling like a damnd slave as he is for using me so baselie."

Yet the English who survived and completed their contract were free people—probably poor and disgruntled, but free to farm or practice a trade. Often, English indentures were provided fifty acres of land—"head rights," as they were called. Some indentures moved up politically. In 1629, six members of the Virginia Assembly, the colony's legislature, had previously been indentures.

The status of Jamestown's first Blacks from 1619 appears to have been similar to that of English indentures. When Jamestown compiled its census in the 1620s, the first Blacks were listed. Although they worked for several families, they were not slaves. Slavery, by name, did not exist. Some them went on to be landowners. After a while, they were joined by a trickle of new Blacks, with whom they had contacts, marriages, and a social life. They stitched together a small subculture.

Unfortunately, indenture did not continue for the Blacks. It was not a secure status for them. Within just twenty years after 1619, Blacks were being considered differently and treated selectively.

In 1630, Hugh Davis, a white man, was "soundly whipped" for "defiling his body in lying with a Negro [woman]." For "getting a Negro with child," a white man was made "to do public penance for his offense at James City church in the time of divine service." The Black woman was whipped. White servants received slightly better food than the Blacks. Robert Beverley, an early historian of Virginia, wrote, "Some distinction is made between them [the English and Black servants] in their clothes and food." A Virginia law urged masters to furnish arms to their families and servants, "excepting Negroes." In 1640, John Punch, a Black man who had run away, was sentenced by a court to serve his master for life, while his white fellow escapees were punished with just four years of service.

The Blacks and English were heading toward different historical destinies. Planters preferred English workers, and if they could not get them, then the Irish would do. Soon, the supply of English servants could not keep pace with the need. Planters then purchased more Blacks from incoming ships, and Blacks became increasingly the bulk of farm labor. A local writer noted that Blacks "are called slaves, in respect of the time of their servitude, because it is for life." Blacks were edging toward slavery.

Some early Black slaves in Virginia, however, were still upwardly mobile in the 1640s and 1650s. They were remarkable people. Using a mixture of industry and cunning, Black men negotiated with their owners to work for their freedom—this was called "self-purchase." Anthony Johnson worked himself free, married an African woman, and became a landholder of 250 acres. Through thirteen years of work, Francis Payne purchased his freedom by buying two white indentures for his mistress. He also bought his wife and children's freedom.

From these days forward, there would always be a few free Blacks in Virginia. Yet most Blacks—those already present as well as new arrivals—were now singled out for slavery. By 1700, Black slaves outnumbered white servants four to one.

## Was slavery legal?

No legal category for slaves and slavery existed in English common law. Colonies' legislatures, therefore, devised new laws as slavery unfolded.

Legislators felt that they needed to regulate Black laborers. They also needed to deal with the special problems generated by the presence of Blacks, such as sexual relations between whites and Blacks. They needed to clarify the ambiguous status of Blacks. Also motivating these laws was the growing number of Blacks in the colonies. Law by law, Blacks were written into the statute books as slaves.

Maryland was the first colony to enact laws preparing Blacks for slavery. Beginning in 1639, the laws of Maryland—a colony of a few hundred people—were quick to close off the possibilities of freedom for Blacks. There had been a few Black indentured servants, and a few of them had become free. But the first legal decree required that Black servants serve for life. In 1644, the assembly voted that Black servants who became Christians could not be freed. (Obviously, prior to that, some Blacks had discovered that if they became Christians, they might become free.) The assembly also decreed that "divers freeborne English women forgetful of their free Condition and to the disgrace of our Nation do intermarry with Negroe Slaves" would be required to serve as slaves until their husbands died. Any children they had "shall be Slaves as their fathers were."

The Virginia assembly passed similar laws. In 1660, it defined punishments for "any English servant" who ran away "in company of any Negroes." This law responded to the white and Black servants who became comrades in escape. In 1662, a law was passed stating that if a white man and a Black woman had a child, the child "shall be held bond or free only according to the condition of the mother." The purpose of this decree was to stamp out the idea that biracial children might be considered free. In 1667, Virginia established that if a slave was baptized as Christian, "the conferring of baptisme doth not alter the condition of the person as to his bondage or ffreedome."

One of the severest laws was the Virginia statute of 1669 that exempted from punishment a master whose slave "should chance to die" while the owner was "correcting him." This law encouraged masters to discipline Blacks harshly. The law reasoned, "It cannot be presumed that the premeditated malice (which alone makes murder a felony) should induce any man to destroy his own estate." That is, no master would be so foolish as to destroy his own property deliberately; if a slave died while being punished, it was an accident.

It would be decades before the Virginia and Maryland laws on slavery were complete. But they were the basis of later laws that fixed Blacks in slavery. These later laws included the Alabama slave codes, which consisted of numerous statutes forbidding every conceivable activity by slaves; leaving a plantation without a pass, meetings of more than five males, nighttime hunting, and keeping a dog are just a few examples.

### What was a "slave" in the North American colonies? How were they acquired? How were they valued?

Imagine being a planter in South Carolina in the 1750s, on your way to Charleston—with some overseers—to buy two slaves for the farm.

You have read advertisements in the *South Carolina Gazette* for several slave auctions. Merchant houses old and new—J. and B. Savage, William Stone, Austin and Laurens—were vying for buyers in a prosperous economy. Africans, according to a board of trade official, had previously been bought with rice but "are now purchased for ready money." "Negroes are sold at higher prices here," he wrote. "This province is in a flourishing condition."

Like other planters, you are particular about the kind of slaves you purchase. South Carolina planters preferred "a talle able People." They liked Africans from the Gold Coast or West Africans from the Gambia region because the latter were useful in rice farming.

Next imagine how you feel as you travel home with your purchases, a woman and a man, both probably good field workers. The woman, small but sturdy, will be a good bearer of children, which means you will eventually have more slaves for work or sale. After dinner at home, you enter your purchases in the farm's account books. Maybe tomorrow you will give the slaves names. You now own them, plus all of their offspring, for life.

Such actions were routine for white landholders. For decades, slavery was entirely accepted in America. When we read the letters of slave merchants, it is striking that buying a person was considered little different from buying a mule or a wagon.

This was the way of the world in places like Virginia, North Carolina, and Georgia. Even New York, a Dutch colony until the English took it over in 1664, had a brisk slave market. In 1750, it had a slave population of ten thousand in a population of seventy-five thousand people, most of them living in New York City.

The African slave in the American colonies was a piece of property, like land, a house, a shop, or a horse. An inventory from 1784 for a Virginia farm listed the slaves "Old Charles," "Timmy a Boy," and "Black Hannah" alongside "Horses 10" and "Hogs 34." A slave could be bought, sold, and separated from his or her relatives. On an owner's death, a slave was usually passed on to an inheritor in the owner's will. Women slaves could produce offspring for the owner's heirs. Planter Sion Spencer wanted to acquire "young Negro wenches for the benefit of the increase of my Estate and Children." He was thinking ahead.

Because slaves could be passed from one person to another, they were often referred to as "chattel"—movable possessions. Yet no matter how many moves a slave experienced, he or she was considered property for life. In the first colonial slave laws, the Latin phrase *durante vita,* meaning "during life" or "for life," expressed this fact.

Slaves were considered valuable commercial property. It could cost a lot to buy a slave. Throughout the history of American slavery, wealthy planters had an advantage in their ability to acquire a number of slaves. However, there were cycles during which the price of slaves fell; an oversupply of slaves or a drop in a region's crop value could decrease the cost of slaves. Usually, however, the price of slaves would go up. During the 1830s and 1840s, at the height of American slavery, a healthy slave could bring hundreds of dollars, maybe even thousands.

Planters recognized the value of their slaves and sought to preserve their investment. In the days when masters owned both indentured servants and slaves, masters often worked indentures harder than they did slaves. An overworked indenture would soon leave a master and might be replaced. But an exhausted slave stayed and was ruined for life. This simple economic logic did not, however, prevent the violent abuse of slaves.

Slaveholders were the dominant force in the creation of American slavery. But the Africans who were made into slaves also managed to shape the institution in many different ways, as we will describe later. In South Carolina advertisements during the 1750s, slaves were nothing but numbers: from the slave ship *Emperor,* "Two Hundred and Fifty fine healthy Slaves"; from *Two Friends,* "sixty very likely new Negroes", and from the *Molly,* "twenty-four new Gold Coast Negroes." But Africans were so much more, especially as cultural beings, and they brought that to bear fully on American slavery.

## What was a plantation?

For a while in the seventeenth century, a plantation was any type of placement of English people overseas: "a settlement in a new or conquered country." By the eighteenth century, plantation had a specific meaning; according to the *Oxford English Dictionary*, it was "an estate or farm in . . . a subtropical country on which crops are cultivated by servile labor."

In American history, the word *plantation* is often synonymous with the farming system of the South; hence, we have the expression "the plantation South."

The master and his family were the hub of the plantation. Often, masters saw themselves as Old Testament patriarchs, presiding over a domain that included family, flocks, lands, and slaves, ruling all with firmness and generosity. But to become such a baron, they first had to be successful money-earning farmers.

The plantation was a hard-driving financial institution. It was created for a specific purpose—to raise crops to be sold, crops that could be made into products useful to people living hundreds, even thousands, of miles away. Whether this crop was tobacco, cotton, or something else, American plantations supplied a national and a worldwide market. This economic imperative molded the work and leisure routines of all people on the plantations—from the masters, their families, and the overseers to the Black slaves.

Plantation masters had to manage a complex enterprise. They had to know about the day-to-day running of the plantations and simultaneously keep their eye on the ever-changing marketplace. Overseers were used as the masters' intermediaries on larger plantations, but they required supervision. Slaves were at the bottom of the plantation's pyramid of power.

Yet the slaves' power position was a paradoxical one: although they were the lowest class, they held the key to the plantation's success. The slave population's labor and growth could elevate a farmer, while a mismanaged slave population could destroy a plantation's financial prospects. Masters could force slave compliance. But more frequently than they cared to, they had to be skillful diplomats in their dealings with slaves in order to secure high-quality, consistent labor.

A plantation was also a physical space, an assembly of houses, barns, yards, and fields. A map of Hermitage Plantation in Georgia showed fourteen different divisions, including a "mansion," "slave huts," hospital, and grounds. On most plantations, the master's family lived in the main house, or "big house." A separate area was set aside for slave dwellings, which ranged from small communal stone houses to log cabins to plank-covered frame houses to rows of separate slave buildings.

John Vlach, historian of plantation architecture, says that early in American slavery, masters and slaves were not so separate. But later, masters' houses and slave quarters were distinctly set apart. Letitia Burwell, a planter's daughter, described her sense of plantation divisions: "Confined exclusively to a Virginia plantation during my earliest childhood, I believed the world one vast plantation bounded by negro quarters." Former slave Harriet McFarlin recalled a Florida plantation: "Colonel Chaney had lots and lots of slaves, and all their houses were in a row, all one-room cabins, off to the side."

Five main crops were produced by the early American plantations: tobacco, wheat, rice, indigo, and sugar. By the 1650s, five decades after Jamestown's establishment, five million pounds of tobacco—once considered "Indian weed"—were being sold to England. Tobacco reigned as a plantation crop in Virginia and Maryland. Later, wheat was adopted as a Virginia crop.

In coastal South Carolina and Georgia, rice and indigo were the main plantation crops. In Louisiana, it was sugar. During the nineteenth century, cotton became the premier crop of Georgia and South Carolina plantations and many other places. After Eli Whitney's invention of the cotton gin in 1793, cotton gave new life to American slavery. Each of these crops required a different way of working.

Masters tried to create a self-sufficient, compact world on their plantations. "Supply all your needs yourself" was one planter's advice to another. At Gunston Hall, a large eighteenth-century Virginia plantation, there were many industries, all using slave workers: carpentry, blacksmithing, leather curing, shoemaking, metalworking, and weaving. Other plantations had brick making, cabinetmaking, and shoemaking. All "the coarse cloths and stockings used by the negroes and some of finer texture worn by the white family" were made on the plantation. A slave "distiller" made apple, peach, and persimmon brandy. Nomini

Hall, another Virginia plantation, contained a mill, a spinning building, and ironworks. Many plantations had smokehouses, dairies (for keeping milk and making butter), ice houses (for storing ice), and loom houses. The aim of all plantations was, as Virginia's William Byrd II, a major planter, wrote in 1726, to "live in a kind of independence."

Plantations varied in size. Until the American Revolution, many plantations had fewer than twenty slaves. Later, cotton plantations in the lower South, such as in Georgia, had an average of thirty-five slaves. In Louisiana, sugar plantations might have one hundred slaves or more. In 1860, five years before the end of slavery, the South had 46,274 plantations. Some 20,789 of them had between twenty and thirty slaves. But it is interesting that there were only 2,300 plantations across the whole South with more than one hundred slaves each. Most plantations were small or medium size.

The plantation reflected the order that the master desired for his world—his sense of hierarchy, values, and taste. His choices dominated. For centuries, though, the plantations were also the principal meeting grounds for Africans and Europeans. This was the place where they unconsciously influenced one another. Under these circumstances, Africans and Europeans began a history of cultural exchange that is special to American history.

## Were plantation life and slavery depicted accurately in *Gone with the Wind*?

The movie *Gone With the Wind,* based on the 1936 novel by Margaret Mitchell, came out in 1939. Using the new Technicolor process and replete with showy sets, sumptuous costumes, and a legion of polished performers (such as Olivia de Havilland, Clark Gable, Vivien Leigh, Leslie Howard, Hattie McDaniel, Butterfly McQueen), the film was an instant success, attracting millions of viewers. Even today, it has major audience appeal. For generations of Americans, it is "the key plantation movie."

In the novel and the film, Tara is a grand cotton plantation and house. At Tara live a happy extended family, embracing both whites and Blacks, although the Blacks are slaves. Tara's owners are gentle managers of their Blacks, and the slaves live vicariously through the charmed lives of their white overlords. Beyond the big house, hundreds of slaves till Tara's soil. It is a slave utopia—until the ugly Civil War inconsiderately intrudes. This was slavery at its height, the so-called "mature" slavery of the 1850s, not the smaller, less well-developed system of the seventeenth and eighteenth centuries. Once the war starts, Tara's slaves are nostalgic for their contented world. After the war, an aged Mammy even returns to support her mistress Scarlett.

*Gone With the Wind* is a great film. But as history it presented a series of seductive myths:

Plantation houses were not all grand. Grand plantation houses like Tara did exist in some places. From Virginia's Bacon's Castle (built in 1665) to South Carolina's Crowfield (built in 1730) to Kentucky's Grey's estate (built in 1843), they were their region's pride. Some states, like Louisiana, could boast plantation architecture of great beauty, such as that found at the Burnside, Belle Alliance, Chretien Point, and Oaklawn plantations. But most slaveholders' houses fell far short of pillared mansions. Austere and moderate in size were the rule. An old photo of South Carolina's Stirrup Branch plantation shows a "big house" of only two stories with a broad front porch. Others were simple farmhouses.

The big house did not hold one big happy family. The film's scenes of bonding between slaves and masters simplified and romanticized reality. Often their mutual loyalties were great. But there was also slave resistance. Slaves sometimes slowed work down, they "misunderstood" tasks, they pilfered, they picked up information and passed it on in the form of subversive rumors, they did as they pleased. Most slave resistance consisted of feigning sincerity and concern, while masking their true feelings.

The big-house slaves were not isolated. Particularly unlikely is the film's image of slaves in the big house who had ties only to the master's white family, rarely having a conversation or encounter with any slave beyond their domain. In fact, slaves created links, even on large plantations and across status lines. Back of the big house, another culture was constantly growing, with its own music, folklore, dance, and crafts. Big-house slaves would have been drawn into this sphere.

Mammy did not exist, but skillful big-house Black women did. Hattie McDaniel's Mammy garnered her an Academy Award, a first for a Black American. On the evening of her award, she told a reporter, "I love Mammy. . . . I understood her because my grandmother worked on a plantation not unlike Tara." Her magical acting made Mammy seem historically credible. In the film, Mammy was the anchor of the family—wise, vigilant, the speaker of cautions and keeper of protocol, chief mourner, and generous bearer of mirth. Selflessly, she served the white family. Having no children and no sexuality, she was capable of undivided devotion. The film was wrong in painting this picture.

As cultural historian Patricia Turner writes in *Ceramic Uncles and Celluloid Mammies* (1994), "implicit in each rendition (of mammy) was the notion that these thick-waisted black women were happy with their lot, honored to spend their days and nights caring for white benefactors." *Some* slave women played incredible roles in plantation big houses, negotiating disputes, imposing norms, moving families through crises. Being outsiders gave them a certain power. But such women seem to have been the exception.

As the credits come up for *Gone With the Wind,* Wilbur G. Kurtz is listed as the film's historian. He worked on the sets of Atlanta and the farms. Had he thought about the film's portrayal of slavery, he probably would have changed very little. The 1939 view of Blacks in slavery was just as the film presented, or probably worse. Today, a different view has emerged, which should allow us to see this great feat of cinema with new eyes.

### Was Malcolm X right when he divided plantation slaves into two groups—house slaves and field slaves?

Malcolm X's audiences always chuckled when he talked about the differences between the house slaves and the field slaves. He used a story of how the two groups reacted to the burning of the master's house: the house slaves wrung their hands, cried, and said, "Massa, we so sorry *our* house dun burned down." House slaves saw the master's house as their property, too. Malcolm X loved to denounce this as delusional thinking. In contrast, the field slaves would rejoice. To them, the house's burning was a revenge for the master's evil ways.

Malcolm X was right in distinguishing between these two basic groups within plantation slaves. But he exaggerated their differences, introducing too many class distinctions into slave society.

Nearly all slaves—men, women, and older children—worked in the field. Few escaped this type of work; even the elderly had tasks. The Museum of the Confederacy in Richmond, Virginia, has old photographs showing large groups of slave field hands. The photos show that there were as many women as men, and even a number of children. One ex-slave recalled, "Even us kids had to pick hunnert and fifty pounds of cotton a day, or we got a whoppin'." Records from an 1830s Georgia plantation noted, "The field hands are divided into seven gangs;—three of males, and four of females." Every plantation had skilled Blacks, such as carpenters, blacksmiths, and quilters, but they were not entirely excluded from field work. At harvesttime, they too were called to the field.

Besides the field hands, a smaller slave group acted as house servants. A good part of their world consisted of the confines of the master's house and buildings. The house slaves, according to former slave Austin Stewart, were "the stars of the party." Some ate better food in the master's kitchen. Their work was lighter. Often masters saw the house slaves as direct symbols of their wealth. Masters "always expect to have strangers visit them, and they want their servants to look well." A little special treatment made house servants think themselves superior to the other slaves and brag about their position. Even so, they held a position alongside other slaves. House servants could also lord their status over some whites as well, judging particularly low-ranking whites severely and often calling them "lowly types."

Materially, house slaves were slightly better off, but they lived no easy life. They endured intense inspections by whites, were subjected to routine psychological abuse, and had to bow to the whims of exacting masters and mistresses. One recalled, "Never any peace for us in that house." Another remembered, "It was hard to be massa's and miss's helper; they could be so evil." Valets, personal servants, or "waiting men" to slave masters had it doubly hard because they were alone with their masters a lot and were on call constantly.

Women chief "housekeepers" had backbreaking chores. "Up and down, up and down, I had to keep going all the time," one housekeeper remembered. Another slave recollected that her Aunt Nancy "had always slept on the floor in the entry, near Mrs. Flint's chamber room, that she might be within call."

This explains why so many house servants—mostly men with names like Christmas, Romeo, or Will—abandoned their demanding masters, left their dubious superiority behind, and joined the runaway slaves from the fields. When all the evidence is examined, house slaves were not so privileged or loyal after all.

Still, field hands did the most demanding plantation work. They rose at the first sign of light and toiled until sunset with brief breaks for food and rest. One Kentucky woman recalled, "I worked till I thought another lick would kill me." In the nineteenth century, plantation masters tried to make slave work into a rational, efficient process, like that of an industrial factory or a machine. "Management" was the new concept. This trend meant even tougher schedules for slaves.

Despite their burdened lives, field hands had each other's company. They suffered, but at least they were together. On large plantations, their numbers were considerable, sometimes enough for a small village. Equally important, they created a culture around their work, just as they and their ancestors had done in West Africa while working. They could still lift a song, play with riddles and rhymes, tell stories, banter, mimic.

Often, they made up double-edged satirical songs about overseers and masters. In 1774 an English traveler, Nicholas Creswell, noted, "In their songs, they generally relate the usage they have received from their Masters or Mistresses in a very satirical stile and manner." Folklorist Roger Abrahams calls this "singing the master." He presents an example of such parody:

*Oh, Mr. Reid iz er mighty fine man,. . .*
*He plants all de taters,*

*He plows all de corn,*
*He weighs all de cotton,*
*An' blows de dinner-horn.*
*Mr. Reid iz er mighty fine man.*

Field slaves' work culture created a protective circle around them, bringing them closer together.

**What was a typical day like in the life of an American slave?**

Plantation labor routines extracted work from Blacks for which there was little or no reward. Most slaves were put through a day of strenuous labor. Indeed, the aim of plantation bosses was to keep slaves busy. A slave reported, "I was not permitted to be a moment idle." Even so, a European traveling through the South noted, "The white men are all the time complaining that the Blacks will not work, and they themselves do nothing."

Historian Ira Berlin has explained that slave work routines differed from season to season and depended on the crop that was the plantation's specialty. March to October were usually the months of the most intense work because that's when planting and harvesting took place.

WORKING FROM "SUNUP TO SUNDOWN"

*Mornings*

Horns, bells, or someone blowing a conch shell awakened slaves before sunrise. Former Virginia slave West Turner recalled, "Yes, Sir, I can hear it now. Ol' overseer used blow us out at sunrise on the conker shell—Toot, toot!"

After waking, slaves prepared their meals and fed the livestock.

They had to be at work by the time it was light, when as one slave said, "the sun gets to showing itself about the trees."

If there was a Black "driver"—a person who stayed behind the slaves to monitor them—he might visit slaves' houses to make sure that all had gone to fields. After doing this, the driver followed the slaves to the fields to push them to work harder.

Field hands, house servants, and skilled workers went to their different types of work. At harvesting time, all had to work in fields.

*Midday Break for Food*

"Ten or fifteen minutes [was] . . . given to them at noon" for eating.

Usually, slave rations consisted of cooked cornmeal (one pound per adult), slivers of cold meat, some fatback or bacon (one-half pound per adult), supplemented by molasses or sorghum and sweet potatoes. When available, beans, various peas, and greens were put into a soup. Today, there are claims that the food for slaves improved in the nineteenth century and that slaves had a sufficient diet. Slave children were usually fed cornmeal mixed with milk, sometimes served in a trough.

The food was eaten using tin pans and iron spoons. Slaves could be kept from their break and meals as punishment.

It has been said that slaves ate at tables, but this has been disputed. A plantation visitor wrote, "I never saw slaves seated round a table to partake of any meal."

*Afternoons*

After eating, work continued, usually until sundown, except during the main harvesting times when the hours were longer. Preparing cotton after harvest, "ginning" it, could keep men working until nine at night. Boiling sugar cane required eighteen-hour days. Overseers would build bonfires in the field to keep the work going.

*Early Evenings*

After the main part of the work, slave Solomon Northup stated in his biography, "the labor of the day is not yet ended, by any means. Each one must then attend to his respective chores. One feeds the mules, another the swine—another cuts the wood, and so forth."

Then they returned to slave quarters.

Slaves cooked their last meal of the day, using plantation provisions plus vegetables from their own patches. ("They plant corn, potatoes, pumpkins, melons, etc. for themselves.") During certain seasons, slaves used trapped animals and fished at night as a way of adding food to their diets.

*Nighttime*

Nights allowed life in the slave quarters to begin; there was time for relaxing, talking, making crafts (such as baskets, in South Carolina), sewing, fiddling (on the banjo), dancing, and storytelling (such as tales of animal trickery). But often women were assigned spinning and weaving as night work.

After five or six days of solid work, Saturday afternoons were free for household work. Sundays were days off on plantations. Slaves used this time to cultivate their own patches. A Louisiana law in 1836 provided free Sundays for slaves. Free time promoted the development of a separate life in the quarters.

**Was slavery different for women?**

"Ah was born back due in slavery, so it wasn't for me to fulfill my dreams of whut a woman oughta be and to do." Nanny, Zora Neale Hurston's character in *Their Eyes Were Watching God* (1937), speaks these words to her granddaughter Janie. Although a fictional character, she could easily have been speaking for a long line of enslaved Black women. When white folklore portrayed the slave woman, she was either a "Mammy" (a slave woman who took care of white families) or a "Jezebel" (a woman of unrestrained sexual desires, a temptress). In real life, she was neither.

Slavery fell on Black women with special force. Historians Anne Firor Scott and Suzanne Lebsock have said that slave women had triple duties: they bore children, worked in their own home in the quarters, and worked in the fields and at other plantation tasks. The slave system did not exempt slave women from the dreariest and hardest work. They were often assigned tasks considered inappropriate for white women. One planter claimed, "Women can do plowing, very well & full well with the hoes and [are] equal to men at picking."

On the smaller farms, such labor distribution was a certainty because fewer male workers existed to do the strenuous work.

Matilda Perry, after long years of working in the field, claimed she "use to wuk fum sun to sun in dat ole terbacy field. Wuk till my back felt lak it ready to pop in two." Women, old and young, went to the fields with men. Often, they helped one another, as Hannah Fabro recalled: "I cud pick coton so fas' aftah awhile I pick my 200 pounds an' help out old Aunt Julie. I'd put great big han'fuls in Aunt Julie's bag so she don't get whipt." Some older women ran the few plantation nurseries, where the slave babies were kept during the day. But the overwhelming percentage of slave women, estimated to be as high as 89 percent, worked in the fields, as Nancy Boundry asserted, "to split wood jus' like a man."

Harriet Jacobs, in her autobiography *Incidents in the Life of a Slave Girl* (1861), stressed, "Slavery is terrible for men: but it is far more terrible for women. Superadded to the burden common to all, they have wrongs, and sufferings, and mortifications peculiarly their own." But Black women summoned great psychological and physical resources to handle the oppression. They might have inherited some skills and the capacity to withstand hard work from the women's culture of Africa, where they were pivotal in cultivation, livestock care, basketry, weaving, textile making, religious leadership, and healing.

Marriages and couplings among slaves were a refuge for both men and women, providing solace, support, love, fun, talk, and the comfort of kin. One of the great passions among slaves was building family. But the families they built were of a wide variety. As slavery developed, women held together many small clusters of kin consisting of them, their children, and perhaps a few other relatives. Baby Suggs of Toni Morrison's *Beloved* (1987)—the grandmother who anchored the family—is not just a fictional creation. When marriages did occur among slaves, they were informal agreements, sanctioned by the slave community and sometimes by slaveholders but not by law. Unions might be between people on the same plantation or on distant farms.

When women were pregnant, masters allowed them a short time off from work. Three or four weeks prior to a child's birth, women were allowed to stay away from work. But a month or so after the birth of a child, they had to return to their full duties. Then the slave mother cared for her child in between the regular work routine of the plantation. Often, young and elderly slaves cared for children while their parents went to work in the fields. But there are stories of mothers taking a child to work and placing "it under a tree or [by] the side of a fence." Slaveholders sometimes linked hard labor for women to childbearing, for one said, "Labor is conducive to health; a healthy woman will rear most children."

If the slave mother succeeded in rearing her child to a reasonable age, she then confronted the problem of keeping her child from being sold and taken away from her. Every child was a walking cash box. Harriet Jacobs recalled, "My children grew finely; and Dr. Flint [her owner] would often say to me, with an exulting smile, 'These brats will bring me a handsome sum of money one of these days.'" Such threats of selling children revealed to all slaves how vulnerable they were. Women who wanted to escape from the plantations had to think about the miserable prospects of leaving their children behind. A slave's reminiscence said: "I stood at the gate many a time, thinking about her and whether to run to the woods." Little Eva escaping bondage with her child, as in *Uncle Tom's Cabin*, was not a common occurrence.

Sexual harassment and rape of slave women were continual problems on the plantation. Recollections from the plantations tell of women slaves being looked at, admired as possible sexual partners, and being used. One slave master told of "that pretty Black girl" who "was seeing the bed" of a relative. Slaves remember similar stories. "My cousin Eliza was a pretty girl, really good-looking. When the girls in the big house had beaus coming to see 'em, they'd ask, 'Who is that pretty gal?'" Slave women had to be quick-witted and wily to avoid sexual abuse. Many resisted aggressively. A slave mother advised her daughter, "I should never let anyone abuse me. I'll kill you, gal, if you don't stand up for yourself. Fight and if you can't fight, kick."

At least one woman did more. Recently there has come to light the story of a slave named Cecilia from a small 1850s Missouri plantation who killed the master, Robert Newsom, who raped her. Her case went to trial and became a national issue. But neither this nor an escape from jail prevented her execution four days before Christmas in 1855.

## RESOURCES

Abrahams, Roger D. *Singing the Master: The Emergence of African-American Culture in the Plantation South.* New York: Penguin Books, 1992.

Blassingame, John W. *The Slave Community: Plantation Life in the Antebellum South.* New York: Oxford University Press, 1972.

Botkin, B. A. *Lay My Burden Down: A Folk History of Slavery.* New York: Delta Books, 1994.

Breen, T. H., and Stephen Innes. *"Myne Owne Ground": Race and Freedom on Virginia's Eastern Shore, 1640–1676.* New York: Oxford University Press, 1980.

Chaliad, Gerard, and Jean-Pierre Rageau. *The Penguin Atlas of Diasporas.* New York: Penguin Books, 1995.

Conniff, Michael L., and Thomas J. Davis. *Africans in the Americas: A History of the Black Diaspora.* New York: St. Martin's Press, 1994.

Forbes, Jack D. *Africans and Native Americans: The Language of Race and the Evolution of Red-Black Peoples.* 2nd ed. Urbana: University of Illinois Press, 1993.

Gates, Henry Louis, Jr. *Classic Slave Narratives.* New York: New American Library, 1987.

Genovese, Eugene D. *Roll, Jordan, Roll: The World the Slaves Made.* New York: Pantheon Books, 1974.

Jordan, Winthrop D. *White over Black: American Attitudes Toward the Negro, 1550–1812.* New York: W. W. Norton and Company, 1977.

Kolchin, Peter. *American Slavery, 1619–1877.* New York: Hill and Wang, 1993.

Kulikoff, Allan. *Tobacco and Slaves: The Development of Southern Cultures in the Chesapeake, 1680–1800.* Chapel Hill: University of North Carolina Press, 1986.

Levine, Lawrence W. *Black Culture and Black Consciousness: African-American Folk Thought from Slavery to Freedom.* New York: Oxford University Press, 1977.

McElroy, Guy C. *Facing History: The Black Image in American Art, 1710–1940.* Washington, D.C.: Bedford Arts Publishing, 1990. Text with artwork.

Meaders, Daniel. *Dead or Alive: Fugitive Slaves and White Indentured Servants Before 1830.* New York: Garland Publishing, 1993.

Meltzer, Milton. *Slavery: A World History.* Updated ed. New York: Da Capo Press, 1993.

Smedley, Audrey. *Race in North America: Origin and Evolution of a Worldview.* Boulder, Colo.: Westview Press, 1993.

Vlach, John Michael. *Back of the Big House: The Architecture of Plantation Slavery.* Chapel Hill: University of North Carolina Press, 1993. Text with photos.

White, Deborah Gray. *Ar'nt I a Woman? Female Slaves in the Plantation South.* New York: W. W. Norton, 1985.

Wright, Donald R. *African-Americans in the Colonial Era: From African Origins Through the American Revolution.* Arlington Heights, Ill.: Harlan Davidson, 1990.

TEXTS FOR YOUTH

Lemelle, Sid. *Pan-Africanism for Beginners.* New York: Writers and Readers Publishing, 1992.

Sharp, Saundra. *Black Women for Beginners.* New York: Writers and Readers Publishing, 1993.

# 3

# Pursuing Community and Freedom

## The 1730s–1830s

1730: Out of 654,950 people in British colonies, 91,000 are slaves—19 percent in the North, 81 percent in the South

1739: First major slave uprising occurs at Stono, South Carolina

1741: Earliest known "Negro election day" is held in Salem, Massachusetts

1746: Lucy Terry is the author of the first known poem by a Black

1760: Britton Hammon writes and publishes a fourteen-page narrative of his life

1764: Massachusetts legislator James Otis argues slaves have a right to freedom

1770: Crispus Attucks dies, along with four others, in Boston Massacre

1773: Phillis Wheatley gains fame for her poetry

1775: First antislavery society in America organized in Philadelphia

1775: The War for Independence begins

1776: The Declaration of Independence is approved; section on slave trade is deleted

1777: Vermont abolishes slavery; other Northern states later choose gradual abolition

1783: Revolutionary War ends; ten thousand Blacks have served in Continental Armies

1787: Prince Hall of Boston receives charter for first Black Masonic group in America

1787: Framing of the Constitution of the United States; Blacks count as "three-fifths of all other persons"

1780s: Free Black communities grow, especially in the North

1793: Congress passes the first Fugitive Slave Act, making it a crime to harbor a slave or interfere with a slave's arrest

1794: Richard Allen of Philadelphia founds the Bethel African Methodist Church

1800: Gabriel Prosser—a slave blacksmith—plans an elaborate plot to take Richmond, Virginia, and overturn slavery

1808: Congress bans the importation of slaves

1822: Denmark Vesey, a free Black carpenter, organizes plot to seize Charleston, South Carolina

1822: Liberia is established as a colony on West African coast

1827: *Freedom's Journal*, nation's pioneer Black newspaper, appears

1829: *Walker's Appeal*, written by militant David Walker, is published in Boston

1831: Nat Turner's rebellion shakes Virginia

Tom Feelings, a talented author-illustrator of children's books, has written a clever history book, *Tommy Traveler in the*

*World of Black History*. His main character is an appealing Black youth, Tommy, who devours history books. After exhausting the local library's holdings, he tells the librarian, "Miss Brooks, I can't find any more books on Black history." She responds, "I think I can solve your problem. Doctor Gray on Vine Street has a private collection. I believe he'll let you read some of his books."

Within minutes, Tommy is at Dr. Gray's door. The doctor leads him to a large book-lined study. Tommy is dazzled. After a while, he chooses *Negroes in Early American History*. As it turns out, this is to be no ordinary reading experience. Tommy—assisted by his active imagination—is transformed into a time traveler and transported into the past.

He finds himself in June 1776. He lands at George Washington's temporary New York headquarters, where he becomes an errand boy for Phoebe Fraunces, Washington's Black housekeeper. Next, Tommy finds himself aboard a Boston-bound ship along with free seaman Crispus Attucks, a fervent supporter of the movement for American independence. Tommy is present at the famous Boston Massacre where Attucks is killed by British gunfire on March 5, 1770, making Attucks the first martyr of the American Revolution. Tommy also visits the young Frederick Douglass while the latter is still a plantation slave in Maryland.

The eighteenth and nineteenth centuries, through which Feelings sends Tommy, were crucial to the founding of the new American nation. They were also decisive in the shaping of Black America. Enduring elements of Black America were established during these years.

Thousands of Blacks were locked in slavery at that time. All colonies had slaves until 1777, when Vermont abolished its slavery. In 1790, the first federal census counted 697,897 slaves—19.3 percent of America's total population and a whopping 92 percent of all Blacks in America. This number would grow rapidly. Within these shackled thousands, a new cultural community, a new consciousness of collective self, was emerging.

At the same time, free Black life was developing in small pockets, mostly in towns and cities. In 1790, 59,466 Blacks were free. They were a very active group, starting churches, establishing the African Masonic Lodge, and forming mutual aid clubs.

Portrait painter Joshua Johnson opened a shop in Baltimore. Jarena Lee became a famous woman itinerant preacher. Books were published. During this period, which historian Donald Wright has called "an age of slave unrest," Blacks increasingly acted directly on behalf of their freedom: petitioning, running away, plotting rebellions. All the while, slavery grew stronger and more oppressive.

Feelings's Tommy traveled in a time when Black community and freedom were dominant themes. What he might have seen is explored in this chapter.

## Was American slavery all that bad?

When tours of Colonial Williamsburg began to give more attention to slavery's place in the town's history, a Black assistant director reported that the questions most often asked were "Was slavery all that bad? Was it as bad as people have made out?" Repeatedly, he found himself explaining what made slavery awful.

It is possible for people today to be honestly confused about the true nature of American slavery. There has always been a neck-and-neck race between the idealized versions of slavery and its actual history. From the 1820s on, a standard defense of slavery argued that slaves were well taken care of; they were provided with clothing, food, and shelter by compassionate masters. A corollary of this argument was that slaves were better off than free wage laborers in America's North and in industrial societies like England. Supposedly, slaves lived in a cocoon, a preindustrial welfare state.

Arguments like these were reflected in the popular culture of the times. Nineteenth-century paintings of slavery scenes, for example, presented a congenial, cozy atmosphere. Many of the paintings were brilliant art, but their technical virtuosity hid slavery's ugliness. James Clonney's *Waking Up* showed a sleeping Black man being gently teased by a white boy. The boy was probably his master. Edward Troye's *Richard Singleton* depicted an immaculately dressed South Carolina Black horse trainer in top hat and tailcoat holding a sleek thoroughbred. The trainer was a slave. Eyre Crowe's *Slave Market in Richmond, Virginia,* presented a scene of neatly dressed slaves calmly waiting to be sold. While traders hover in the background, one woman even manages a smile. None of the horror that slaves felt—and often expressed—at auction is conveyed.

Slavery cannot be exonerated of its many oppressions. Some recent studies claim that aspects of the institution (food, masters' power over slaves, work schedules, urban slaves' lives) were not as terrible as previously thought. Yet every recent reputable authority agrees that slavery was an inhuman system.

What, then, was its most severe aspect? In a 1993 interview in the *Paris Review,* novelist Toni Morrison said that when researching *Beloved,* "I spent a long time trying to figure out what it was about slavery that made it so repugnant . . . so indifferent." She concluded that slavery's worst aspect was the physical violence against slaves. In particular, she singled out "something that was never properly described—*the bit.* This thing was put into the mouth of slaves to punish them and shut them up without preventing them from working."

Morrison rightly focuses on violence as fundamental to American slavery. During its first century, humiliating tortures were routine: bodily mutilations, scaldings, bizarre haircuts, and brandings. Slaves were regularly whipped, struck, and shackled. A 1757 New York slave who ran away was advertised as a man who had "buttocks pretty well marked with the lash." Large plantations often had stockades and private jails for detaining slaves. Women had no protection against sexual violence.

Finally, there were the gangs of patrollers ("paterollers")—legal bands of white men who could search slave dwellings, apprehend slaves for alleged crimes, judge them, and punish them instantly. Slave stories pictured patrollers as demons while praising those who outwitted them.

Slaveholders talked a lot about their concern for their slaves, especially in the nineteenth century. They saw slavery as a positive good. But they did not abandon severe discipline. Mean punishments sent shock waves through slave communities, generating fear that was essential to maintaining the system. Down to slavery's last days, the slightest assertions by slaves were punished outrageously. In some states, a slave striking a white was listed as an offense punishable by death.

Yet violence probably was not the most repugnant aspect of American slavery. The breakup of families and relatives was considered the most tragic occurrence by slaves. "My mother was sole and took from my father when I was jes' a few months old. I never seed him till I was six" was how Foster Weathersby remembered a separation.

Slave families were informal arrangements that had no standing in law. Even though sanctioned by the community of slaves and sometimes encouraged by slaveholders, slave marriages were not protected unions. When financial interests demanded, slaveholders sold slaves from families or kin clusters to itinerant professional slave traders. One could be sold at any time. Pleadings did no good. The sale was permanent, and usually there was no time for good-byes.

During the 1820s and 1830s, when some four hundred thousand Blacks were sold from the upper South to slavery in the lower South, virtually every Black family experienced the permanent loss of relatives. This was the great interregional slave trade. Historian Michael Tadman has discovered that one in three slave marriages was broken up and that nearly 50 percent of all children were separated from at least one parent. Most slaves were sold at least once in their lives.

Understandably, the auction and the auction block—representing the moment at which slaves were sold—became dreaded symbols among slaves. Both were vivid in former slaves' memories: "I saw slaves sold. I can see that old auction block now. My cousin Eliza was a pretty girl. The day they sold her will always be remembered." And as Josephine Smith recalled, "I remembers seeing a heap of slave sales. . . . About the worst thing that ever I seed, though, was a slave woman at Louisburg who had been sold off from her three-weeks-old baby, and was being marched to New Orleans."

An old slave song ran, "Mother, is massa gwine to sell us tomorrow . . . Gwine to sell us down in Georgia? . . . Farewell, Mother, I must lebe you. . . . Mother, I'll meet you in heaven." In the spiritual "Many Thousands Gone," liberation meant "No more auction block for me, No more. No more. No more auction block for me, Many thousands gone."

### How did Blacks survive the cruelties of slavery?

In 1933, Charles Sydnor wrote, in *Slavery in Mississippi,* that "being a slave was not for the average negro a dreadful lot." In 1957, the official Virginia public school history textbook said, "Life among the Negroes of Virginia in slavery times was generally happy." Today, statements like these are no longer accepted. In fact, two new realities have taken their place: American slavery was harsh, and Blacks were highly creative in finding ways to cope with its cruelties.

Recognition of the creative strategies of Blacks in slavery was a long time coming. During World War II, the perception of Blacks in slavery began to shift. In the *Myth of the Negro Past* (1941), Melville Herskovits asserted that African cultural elements were alive and well among American slaves. He called them "survivals." Herbert Aptheker followed, in 1943, with a book stressing slave revolts. In 1949, Lorenzo Turner demonstrated that the slaves' Gullah language on the Sea Islands off South Carolina and Georgia had its roots in West Africa. With the great 1956 work, *The Peculiar Institution: Slavery in the Antebellum South,* by Kenneth Stampp, perceptions of slaves changed further. Stampp equated the mental abilities of the dominant whites and Black slaves. Highlighted also were slaves' courage, their hatred of slavery, and their resistance.

In the late 1950s, historian Stanley Elkins disagreed with these new views. He dropped an intellectual bombshell, proposing that American slavery, like the Nazi concentration camps, had created a docile, infantile, dependent character, a figure he named "Sambo." Elkins's idea spurred historians to undertake even more far-reaching researches into slave life. During the 1970s, this research paid off in works on slave communities and culture, slave rebellions and resistance, the Black family, the relationship between slaves and masters, slave religion, Black folk thought, and slave consciousness and its origins. These works helped us to understand how Black slaves experienced the world.

America is indebted to the authors of these books. Using oral sources, slave narratives, 1930s Federal Writers Project interviews with ex-slaves, as well as new insights from psychology and anthropology, these authors brought to light an active Black presence in slavery. Passivity no longer characterized the popular view of slaves' response to oppression. Black slaves were neither happy-go-lucky, grinning and cringing, white-mimicking creatures, nor were they a solemn lot, revolutionaries to a person, entirely independent of whites.

The new image of Black slaves—men, women, and children—showed them to be somewhere in between: complex people, making choices where possible, pursuing freedoms where possible, and all the while fashioning a resilient, generous new culture. By excavating Black slaves from past mythic debris, these scholars elevated a new and impressive group onto the stage of American history.

This is what they discovered about the ways Blacks survived:

- Black slaves developed a sense of community. They found time and reclaimed cultural space within slavery in order to build a sense of solidarity. Often, this community grew "behind the big house" or outside the master's view. A myriad of activities reinforced their community feeling: marriage ceremonies, holiday revelries, leisure-day gatherings, harvest festivals, dancing, singing, storytelling, religious meetings, and funerals. Recently, Roger Abrahams has identified the plantation corn-shucking ceremonies—the singing, dancing, comedy, and theater—as central to the slave community's formation, particularly to leadership among slaves. Black slaves, true to their African backgrounds, relished the intense experience of gathering together. The slaves' goal, in the eloquent words of Eugene Genovese, was to create a territory of their own "within the narrowest living space and the harshest identity."

- Black slaves often wore a mask. Out of necessity, they disguised their true feelings, hid their resistance, feigned humility, and deferred to their masters and mistresses. As ex-slave Elijah Green remembered, "I lied on myself." Another said, "We had to lie to live." When in the field away from overseers, slaves' songs boasted, "If you treat me bad, I'll sure run away." Slaves who went by night to other plantations to visit friends and family had to hide and lie. Slaves stole chickens and pigs, broke work tools, and lied about being sick. Mothers stole and cheated to protect their children. Secrecy and dissembling were tactics in a prolonged survival struggle.

- Black slaves worked hard to establish families. Previously, creating families in slavery was thought nearly impossible. Slave families did face seemingly overwhelming odds. They had to confront the awful prospect of being torn apart at any moment. But Blacks valued family as a source of love, comfort, and support, and they sacrificed to keep it operating.

  In imagining the slave family, one must see it as different from today's ideal. Generally, slave families were created between men and women on different plantations. These women were called "abroad wives." Fathers traveled to see their mates and children during weekend free times. Often husband and wife were married for long periods of time. Sometimes slaves entered into a few marriages over a lifetime. Often slave women headed families. Marriage ceremonies were eclectic, blending different fragments of customs and belief.

- Black slaves created an inspiring culture. It consisted of songs, dances, stories, philosophies of life, sayings, rituals, natural lore, theories of personality, and ideas of humanity. Most significant in this culture was Black religion. In the nineteenth century, slaves were massively drawn into Christian ideas and worship. Black slaves reinvented Christian teachings and practice to suit their own purposes. Whereas slaveholders saw the Bible as justifying the rule of masters, slaves took the texts as outlining their liberation and the destruction of masters' evil earthly power. Whereas white Protestant meetings were staid, slave assemblies were joyous, featuring call-and-response singing, in which a leader sang a main line that was answered by his audience. In sum, slave culture transcended the day-to-day pains of Black life. Sterling Stuckey has called this "the Black ethos" in slavery, which "prevented . . . [slaves] from being imprisoned altogether by the definitions . . . the larger society sought to impose."

### How did the culture of slaves and their institutions come into being?

While conducting a 1970 interview on South Carolina's Sea Islands, linguist John Rickford was told an unusual story by a middle-aged schoolteacher. She recounted how her great-grandparents had flown back to Africa after being whipped mercilessly on a plantation. This story is found throughout the Caribbean and South America. Cuban runaway Esteban Montejo noted in his autobiography, "The Negroes did not do that [commit suicide], Negroes escaped by flying. They flew through the sky and returned to their own lands."

The flying-slaves story provides a great insight into the origins of slave culture. It is a story constructed out of several dissimilar cultural parts: Central African ideas (from the Bakongo people) about sorcerers and crossing the impenetrable Atlantic Ocean (called *kalunga*); runaway slave motifs from America; European Christian ideas about leaving earthly trouble by flying to heaven; West African, Indian, and American ideas about flights during dreams.

Slave culture and institutions were like this story. They were dynamic syntheses consisting of what historian John Boles called "borrowings and blendings." Themes of life and cultural conventions taken from many African cultures were the basic materials, but slaves married them to other cultural elements. By the 1740s, after great numbers of Blacks had been born in America, this was particularly true.

It was also true that the whole of any one cultural or institutional product was always greater than the sum of its parts. When the flying-slaves story was told, it became larger, more imaginative, more dramatic, and more profound than it had been in any of its sources. It rose up as a grand myth, speaking to slaves on many levels.

One of the most famous drawings from plantation life shows marrying slaves jumping over the broomstick held by friends as well-wishers cheer from the sidelines. "Jumping the broom" became a popular marriage ritual for increasing numbers of slaves. This custom also mixed different cultural elements. On one plantation, ex-slaves reported they "didn't have to ask Marsa or nothing. Just go to Ant Sue and tell her you wanted to get mated." Aunt Sue was continuing the tradition of African female spiritual figures. She advised the couple in both Christian and Black folk beliefs, how "marryin' was sacred in de eyes of Jesus" and how it was "bad luck to tech de broomstick." While "folks stepped high [over the broom] 'cause de didn't want no spell cast on 'em," they also prayed for Jesus and God to bless the union. Other couples report that a Bible story was told before jumping the broom.

Cultural blending also appeared in Black slave traditions far from the plantation, such as in the funerals of New England towns. Blacks had absorbed the Christian idea of entering heaven after death. In Africa, the dead were also considered travelers, who had to be provided comforts for the next world. Thus, slaves put liquor, little gifts, and keepsakes in their friends' graves. A woman, Jin Cole, who "fully expected at death . . . to be transported back to Guinea," collected "treasures to take back to her motherland . . . colored rags, bits of finery . . . peculiar shaped stones, shells, buttons, beads, anything she could string." As in Africa, music, dancing, and singing accompanied the corpse's procession to the grave. To New Englanders, Blacks did not seem to "express so much sorrow" as merriment, which Blacks felt the deceased would have liked—but this merriment was too much for New England towns, apparently, which passed laws regulating these processions.

The synthetic power of the Black slave imagination was most visible in the multitude of stories and legends that slaves spoke into being. One prevalent category of these stories featured Brer Rabbit, Brer Fox, Brer Lizard, Tappin the Turtle, and Brer Alligator as the principal animal characters. Another framed contests of wit between the master ("Old Master") and a slave ("John"). All these tales had a trickster theme, where the weaker character competes against a strong adversary and is victorious. Interestingly, these tales were recastings of old African trickster stories put to new purposes. African tricksters—such the spider Ananse and the god Legba—underwent a mutation in American slavery into, say, the irrepressible Brer Rabbit. Now they leveled animal characters remarkably like plantation masters.

Slaves who heard Brer Rabbit's exploits saw him as an instructor in how to negotiate slavery and as a sign of their capacity. But it is often forgotten that Brer Rabbit stories revealed a character of immense intellectual power—a major player in the fields of logic, puzzles, puns, jokes, deflating wit; a kind of existential reveler who raised great issues about the nature of reality and experience. Philosophers who have read Brer Rabbit stories comment on his sheer intellectual energy. It was like having an eighteenth-century Oxford don on the plantation.

Black slave culture and institutions came into being through the work of versatile culture makers. For them, there was meaning and joy in making cultural products, and even greater joy in sharing them with their brethren and sisters. No other slave group has left such an impressive and all-encompassing public culture. For this reason, Toni Cade Bambara in "Broken Field Running"(1974) and Toni Morrison in *Song of Solomon* (1977) reached all the way back to use the African flying myth.

### What were "Negro spirituals"? Why were they important?

In 1990, at the peak of their reputations, opera divas Kathleen Battle and Jessye Norman put on a special performance of spirituals at New York City's Carnegie Hall. They were singing the songs that were once called "cabin songs" or "plantation songs."

Resplendently gowned, the two sopranos—each different in vocal and dramatic style—alternated between singing together and separately, sometimes even competing with each other. It was a virtuoso performance, moving from the enthralling, utopian "Oh, What a Beautiful City," to the gently questioning "Lord, How Come Me Here?," to the tiny gem "He's Got the Whole World in His Hands." Jessye Norman's rendition of "Ride On, King Jesus" was particularly affecting, reminding the audience of the spirituals' profundity. Her powerful singing brought out the song's rebuke to earth- and time-bound kings, such as plantation masters. King Jesus was the timeless monarch of the slaves' world.

Battle and Norman were continuing the tradition of great Black American classical singers who have been devotees of the spirituals. Four are best known: Roland Hayes (1887–1977), Paul Robeson (1898–1976), Marian Anderson (1902–93), and Leontyne Price (born 1929). The loyalty of these artists to spirituals has kept these songs alive and appreciated as among the world's greatest music.

The concert hall is a long way from the quarters and fields of the Southern plantations where these spirituals originated. When sung among slaves, these songs were communal, a group property, led by skillful singers. "Call and response" was the technique often used, in which the leader would carry the song, invoking answers from the group. This was a form of "congregational singing." Harmonies—both simple and intricate—embellished the lyrics.

At their most basic, spirituals were narratives, a form of storytelling. Often, their story line was advanced by succinct, powerful images. In the spiritual "My Lord, What a Morning," the words promise a morning where "You'll hear de trumpet sound, To wake de nations underground." These few words communicated images of a new day, a blasting trumpet, an underground people waking up, and the newly aroused "nations." All of these images spoke to the slaves' condition.

Because of the many meanings of spirituals' images, they could be used as elaborate coded messages—telling slaves to be alert and hinting of plantation events, such as escapes, master's moods, whippings, or births in the quarters. Harriet Tubman sang a spiritual alluding to her escape the night she fled her Maryland plantation.

How spirituals originated has always been a matter of interest. Spirituals were probably being created in the eighteenth century, but the nineteenth century saw their greatest output. Their authors are for the most part unknown folk poets and poetesses. When an aged Black woman was questioned about their origins, she said, "When Massa Jesus He walk de earth, when He feel tired He sit a-resting on Jacob's well, and make up des spirituals for His people." This does not explain spirituals' origins, but it does reveal that these songs were vehicles for Christian idealism.

The spirituals took the characters, events, and themes of the

Bible and made them into active, immediate extensions of the slaves' world. This drawing near of the Bible's drama was a daring psychological enrichment of the slaves' lives. The message was that Blacks were not alone. Jesus, John the Baptist, Moses, Hagar, Hannah, Elijah, and Gabriel were there too. The spirituals reminded Blacks that they had a long history, springing from the days of ancient Israel. When life was over, the spiritual "Oh, When I Git t' Heaven" promised "a starry crown," a seat "side de Holy Lamb," and in a reference to slavery's hard work, "Me and my God gwine do as we please." Much of this Christian content came originally from the eighteenth century's Great Awakening, a period of intense religious fervor primarily among whites that was extended to include slaves.

Years after slavery, the spirituals were transcribed for preservation. Their fame was spread by the Fisk Jubilee Singers and the Hampton Student Singers, whose concerts in the North helped build their colleges. Composer R. Nathaniel Dett (1882–1943) dedicated himself to the preservation and musical refinement of spirituals. But the songs had their rough times. Robert Russa Moton, in *Finding a Way Out* (1922), mentioned how minstrel satires of Negro spirituals made him skeptical of their value. Years later, at Hampton Institute, Samuel Chapman Armstrong convinced him they were "a priceless legacy."

They certainly had been to slaves. As singer-scholar Bernice Reagon noted, slaves used them as "a leverage, as salve, as voice, as bridge over troubles one could not endure without the flight of song and singing."

## What should American parents tell their children about slavery?

When Sarah Debro was interviewed in 1937, she was ninety, but she still harbored strong feelings about her North Carolina slave days. In a straight-shooting remark, she said, "I's seen and heard much. My folks don't want me to talk about slavery, they's shame niggers ever was slaves."

Talking about slavery is still taboo for many Americans. Either Americans minimize its severity or ignore it altogether.

Black Americans do not have the luxury of living in denial about slavery. But in the past, Blacks also tried to avoid the issue.

Writer Roger W. Wilkins (1932) has said, "For people of my generation, slavery was pushed away." Rex M. Ellis of the Smithsonian Institution echoes this: "Question any African-American on their thoughts about Mike Tyson, their concerns about Clarence Thomas, Marion Barry's future in politics, or the recent riots in Los Angeles, and the answers would be as diverse as the population. Ask the same African-Americans about the institution of slavery," he continues, "and most will have the same response: It is a blot on the American psyche as well as our culture, and is a subject that should be avoided at all cost."

American parents today are groping for a way to educate their children about this period in American history. "What can we tell our children?" is a frequent query at Black history talks. Will talking about slavery depress the spirit of our children? How does one explain that slavery existed within freedom-loving America?

For Black parents, the slavery issue is fraught with special difficulties. It pertains directly to most Black families' ancestors—their forefathers and foremothers. Kenneth Noble wrote, "While many white parents dread the moment when they must speak frankly to their children about sex, many Blacks find the subject of slavery similarly discomfiting. How does one explain why Africans were taken from their homeland, sold like animals, and then forced to do society's most menial and often degrading tasks?" Noble frames Black parents' dilemma exactly.

Even so, things have been changing. In fact, for some time now, the general American avoidance of slavery has been receding. Archaeologists like William M. Kelso and Theresa Singleton have been digging up slave quarters at plantations such as Virginia's Carter's Grove and Monticello. These quarters can tell us more about slaves' living arrangements, especially women's lives. Historic sites such as Colonial Williamsburg have put on "living history programs" presenting scenes from Black slave life. Wynton Marsalis has written a concert piece on slavery, *Blood on the Fields.*

Unexpectedly, Black families have been staging family

reunions on the plantations where their ancestors lived and labored. As early as 1977, the descendants of an eighteenth-century slave named Betty gathered in her name on the site of North Carolina's Bennehan Cameron plantation. In 1988, a much-publicized reunion was organized by Dorothy Spruill Redford at Somerset plantation, again in North Carolina. In 1994, the descendants of George Washington's Mount Vernon slaves and others held a reunion.

More than ever, genealogists are shaking Black family trees and probing plantation records. In 1984, after searching records for five years, Roland Freeman assembled 110 family members at Snow Hill, Maryland, where their ancestors had been slaves. All of these activities are gentle approaches to slavery.

Impressively, Black writers are helping in the reconciliations. Poet Lucille Clifton's *Good Woman* meditated on the history of a woman named Mammy Ca'line, who as a little girl was brought with her mother, sister, and brother to New Orleans from West Africa's Dahomey in 1822. Once in America, she and her sister were sold, separated from their mother, and marched to Virginia to be slaves. Mammy Ca'line was Clifton's great-great-grand-mother. Despite all that happened to her, Mammy Ca'line was still above hatred.

Playwright August Wilson in *The Piano Lesson* (1990) made a 1930s Pittsburgh family face its slave past through a quarrel about an upright piano's fate. On the piano's facade, their father had carved family members' faces and events. Boy Willie wants to sell the piano to buy the land down South that their ancestors tilled as slaves. Berniece wants to keep it as a link to the past. Soon, the ghost of their family's master, Sutter, comes to haunt them. To halt the quarrel, Berniece plays the piano, thus exorcising Sutter and, in a sense, defeating slavery. Boy Willie realizes he does not need land to feel whole. They keep Papa's piano.

From all these efforts at facing slavery come pointers for discussing it with our children.

- Tell the truth about slavery, about its severity and its assault on the human value of Blacks. But be prepared for children's reactions in retelling tales of great horror.

- Say that Blacks did not "deserve" their bondage. They were not enslaved because they were inferior or because they came from Africa, supposedly a backward place. They were enslaved because their labor was valuable.

- Stress that Blacks did a superb job of resisting slavery's harshness. Part of their resistance lay in developing a life-sustaining culture—a culture that is now a luminous part of the world's heritage. New knowledge of Blacks in slavery is freeing America to appreciate their powerful achievement.

- Honor the Black slaves' historical optimism; they looked to the future, even when no end to slavery could be seen.

This is what Clifton's Mammy Ca'line and Wilson's characters are telling us. For them, slavery had a place in their past, but these people kept its ugliness from overwhelming them and undermining their ability to live life fully.

### Was there slavery in cities?

In the early eighteenth century, a New York City citizen worried about possible slave unrest, since "the Number of slaves in the Citty of New York . . . doth daily increase, and . . . they have been found oftentimes guilty of Confederating together." In April 1712, that worry became real: New York experienced a slave rebellion. Erupting at night, the revolt was led by Gold Coast and Dahomey Blacks, who burned an outhouse and killed some whites. After a few days, the rebellion was put down, and many Blacks were punished.

At the moment of the rebellion, New York City had a total population of over 6,300, with around 945 Blacks (about 15 percent), most of whom were slaves. This episode points out an important reality: one of the best-kept secrets of slavery is that not all slaves lived and worked on plantations.

Across the South and in the North, slaves also lived in cities. One in ten of the South's slaves lived in cities. In 1830, there were slightly over 6,349 slaves in Richmond (39.5 percent of the total population), 15,354 in Charleston (50.6 percent of the total), 14,469 in New Orleans (31.5 percent of the total), and 2,330 in Washington (12.4 percent of the total). Before the American Revolution, Northern cities also had slave populations, but they were much smaller.

Anyone traveling in early American cities would have seen a variety of Black slaves. The majority did menial, unskilled labor. But they played other roles as well. In marketplaces, slaves often dominated. Black women cleaned houses, cooked, worked in taverns, hawked food on the streets, did laundry, were seamstresses and weavers. Men were tailors, boot and shoe makers, carpenters, blacksmiths, bakers, and butchers. In Charleston, slaves worked in nearly all of the city's occupations; among the masons, bakers, and blacksmiths, there were as many slaves as there were whites. Blacks acted as personal servants to the rich and powerful—as ladies' maids, nannies, valets, stage drivers, barbers, wig makers, couriers, waiters, and singers.

Certain street corners in some cities became known for slave entertainers; their songs, dances, and syncopated playing of bones, jawbones, fiddle, and banjo drew audiences. Crowds of slaves gathered in New York City's streets on Sunday when their masters went to church. In some cities' public houses, slaves and poor whites mingled, bought drinks by the shot, and created public nuisances together.

Most of these slaves were owned by city dwellers. Often, they worked in their masters' households, especially if they were women. As urban slavery developed, a city master might own slaves but "hire out" a few to other whites needing laborers. In this case, the slaves returned the wages they earned to the master, supplementing his income. Sometimes, a rural master would send slaves to the city to "find work." Slaves often negotiated their room, board, food, and pay. Generally, they were hired for six months or a year, but sometimes for only a week or a day. Hiring of slaves proved convenient and profitable. It allowed whites to secure labor on a piecemeal basis without buying a slave.

Still, city slavery was different. Cities had codes for regulating and punishing slaves, and bosses for hired slaves could be tough.

But city slaves discovered little openings in slavery. Frederick Douglass, a plantation slave who was "hired out" in Baltimore, recalled, "I sought my own employment, made my own contracts, and collected my own earnings. . . . Some slaves have made enough, in this way, to purchase their own freedom." And "a city slave . . . is much better fed and clothed, and enjoys privileges altogether unknown to the slave on the plantation." Occasionally, slaves lived in houses completely on their own and seldom saw their masters.

Family life was a bit better for city slaves. Living separately from one's master allowed households to form that were slightly autonomous. But in Southern cities, women could outnumber men, due to the preference for domestic workers. This meant that households were often headed by a female. In cities where ratios of men and women were nearly equal, families still faced the prospect that members might be sold. When they were, it was back to the plantations or, as in the 1830s, to the lower South.

In 1993, an auction at Stack's in New York City offered rare artifacts from urban slavery: fifteen metal servants' badges from the 1810s to 1850s Charleston. Southern cities like Savannah, Mobile, and New Orleans regulated slaves for hire by requiring masters to buy badges that slaves wore. Made out of thin copper sheets, the badges listed a license number and the trade the slave was allowed to practice. "Servant" was the most common badge category, but badges were issued for "Porter," "Chimney sweep," "Mechanic," "Fisherman," "BC" (for bread cart), and "Fruiterer" (fruit seller). These badges were a unique remnant of slavery.

## Who were the runaway slaves?

In 1763, the *New York Gazette* ran an advertisement for a runaway slave. It was accompanied by a symbol showing a man in flight. In 1768, Connecticut's *New London Gazette* presented another symbol, showing a fleeing man, this time with a bundle on his head and a stick in his hand. Eventually, the symbol of a fleeing woman in a long dress was added for female runaways.

These were stick figures, showing only the minimal details

about this remarkable breed of humans. Determined to gain their freedom, they were heroes and heroines to the slaves they left behind and a menace to slaveholders' financial interests. When a slave ran away, his or her original price was lost, as was the upkeep cost and the value of the slave's labor. Unless caught, runaways would steal, burn property, and convince other slaves to join them. Lose two slaves and one's estate could be in financial ruin. Runaways also gave the lie to slavery. As historian John Hope Franklin has observed, "the very existence of such a group is a dramatic reminder that all was not well on the great plantation or even on the modest farm."

Fleeing slavery was often the beginning of a life fraught with peril and hardship. Runaways had no easy road to freedom. Yet slave songs urged runaways to keep going, to outwit their pursuers: "Run, nigger, run, De patteroll git you! Run, nigger, run, De patteroll come! Watch, nigger, watch, He got a big gun."

Too little is known about runaways. Freddie Parker, in *Running for Freedom* (1993), his labor-of-love work on North Carolina's fugitive slaves from the 1770s to the 1830s, has presented a deeper view of them.

Of the nearly 2,700 North Carolina runaways, 2,179 (81 percent) were men and 482 were women. More women probably wanted to flee but couldn't because they had children. Keeping their children alive and safe was too much of a challenge if they ran away. Some did take their children with them. In 1793, Poll, only seventeen, took her fourteen-month-old son, Hardy, to freedom. Men had advantages in escaping because they often had been issued passes and had traveled more than women. Furthermore, their skills allowed them to work outside of farms.

Most of the runaways were in their twenties and thirties. The older one got, the more the chances of becoming a runaway diminished. Yet older slaves tried to escape too. Three coastal towns with sizable Black populations—Wilmington, Bern, and Edenton—were known as towns that harbored runaways.

Slaves of all ranks ran away—skilled workers and field hands, house servants, and regular workers. They fled for a variety of reasons. The majority yearned to get to the North or Midwest to permanent freedom. Others sought to restore family ties with relatives who had been sold away. Some escaped to avoid being whipped. Others wanted to prevent their master from selling them. Still others left just to get away from the oppressive workload of slavery; they often fled at sowing or harvesting times. Many of the runaways had served under several masters; thus, they knew different areas, had ties with distant folk, and knew the layout of the countryside.

To apprehend runaways, slaveholders ran advertisements in local newspapers, offering rewards and giving minute descriptions. Ironically, it is these descriptions that put flesh on the symbols of the runaways. Certain features were standard in these notices: name, sex, estimated age, height, complexion, clothing, scars (brands, mutilations, African facial marks or filed teeth, smallpox scars), physical handicaps; whether they were "country born" in America, born in Africa, or had been West Indies slaves; their fluency in English (whether they spoke it well, spoke "broken English," spoke pidgin from the West Indies, or an African language); and lastly, their ability to disguise their mission.

To be successful, the runaway had to be something of an actor or an actress. Advertisements regularly warned that runaways were "artful," cunning, would "pretend to be free," might acquire a pass or forge free papers. Peter Epps described runaway Austin as "one of the shrewdest negroes." He would "invent many plausible tales to prevent his being apprehended." Runaways put on smiles, appeared humble, and suppressed their anger. Many runaways were light-skinned and used this to suggest that they were free or to pass as white. A few runaways were accompanied by white friends or lovers; they pretended to be the whites' slaves. A runaway lived a thespian's life.

Fugitive slave laws were enacted in 1793 and 1850, and severe punishments awaited captured runaways, but thousands of Blacks continued to "steal a little freedom."

### What was the Stono Rebellion, and why was it important?

At Stono, some twenty miles west of Charleston, South Carolina, in September of 1739, the first major slave rebellion occurred. The rebellion shook the colony. Although it was put down, it took months for the colony to return to a somewhat stable state.

South Carolina lagged behind the other English colonies; only in 1670 was Charleston established. The older colonies of Virginia and Maryland had prosperous economies based on tobacco. But it was some time before South Carolina found crops that could be cultivated for the overseas market. The lowlands of South Carolina's coastline were spawning grounds for malaria and yellow fever, which kept English settlers away. But once South Carolina discovered rice and indigo as cash crops and began to import massive numbers of Black slave laborers, it emerged as a prosperous colony. By 1775, almost ninety thousand Blacks had been brought into South Carolina.

When the Stono Rebellion took place, however, there were around eleven thousand whites and twenty-one thousand Blacks in South Carolina—the only colony with, as historian Peter Wood has termed it, a Black majority.

These numbers helped spur Black initiative and assertion. Black slaves held many specialized jobs: dairy workers, carpenters, gunsmiths, butchers, boatmen, fishermen, domestics with varied skills. (African skills added to their mastery in such areas as fishing and boating.) In Charleston, Blacks sold both goods and services to whites, which disturbed white merchants. Stealing by slaves grew steadily on farms, in the city, and at ports. Runaways increased. Some Blacks—each described as "a resisting or saucy slave"—boldly attacked whites. Whites sought to control Black maneuvers and resistance, often by cruel punishment and unusually harsh labor.

The rebellion took place as Blacks foresaw a bleak future of tighter control. On a Sunday morning, twenty slaves gathered near the Stono River with a slave named Jemmy as their leader. They began to march toward Spanish territory at Saint Augustine, where they envisaged gaining their freedom.

On their way, they made a bloody path. Their first stop was at Hutchenson's store, where they stole firearms and gunpowder, killed the storekeepers, and cut off their heads. Then they rushed to six houses, killing the families. The band grew—some slaves joining voluntarily and some by force—until their number rose to sixty. Reaching a field, they "set to dancing, Singing and beating Drums to draw more Negroes to them." Flags had been raised, and shouts of "liberty" went up. In time, they met some militia, whom they engaged and then dispersed. By day's end, twenty whites had been killed. After lying low, some of the group continued on to Saint Augustine. Days later, another group fought a determined battle thirty miles south of the previous fighting.

Recently, African historian John K. Thornton has reexamined this rebellion and discovered an interesting fact: the Stono's organization can be traced back to Central Africa. The rebels, he suggests, came from the Kongo kingdom, where there was a long, proud tradition of Christianity among Africans. This explains why the rebels were attracted by a future in Spanish-Catholic territory (today's Florida). Thornton argues too that the rebels had probably fought in the Kongo's wars, thus learning how to use firearms. Their dancing came from the traditional training of Kongo warriors. Their banners were also used by military units there. Finally, he says, their guerrilla tactics of fighting and then disappearing and reappearing to fight again resembled the tactics used in Central African wars.

The Stono Rebellion posed a major challenge to South Carolina's future as a slave state. For months and years afterward, it scared the colony's whites. As a report said, it "awakened the attention of the most Unthinking." Now it is known that African ideas provided the backbone for the uprising.

### Who was Lucy Terry?

For too long, Lucy Terry has been an obscure figure. Today, she is becoming known as the author of the first literary product by a Black American. A poem of twenty-six lines, written in 1746, was her pioneering work. Called "Bar's Fight," the poem memorialized an important local event in the history of Deerfield, Massachusetts—an attack by Indians on a group of white hay cutters. (*Bar* was a colonial English word for field or meadow.) The poem was not published until 1893, but for generations it was memorized, recited, and passed on by word of mouth.

Lucy Terry was brought to Deerfield from West Africa when she was only an infant. A slave of church deacon Ebenezer Wells, she was baptized in 1735 when she was about five years old. In 1756, she married Abijah Prince, an older free Black man, a landowner who had served in the French and Indian War (1744–48), and he purchased her freedom. Later she is referred to in local histories as "Luce Bijah," and their marriage was hailed as "a realistic romance going beyond the wildest flights of fiction." They had six children. After a while, they left Massachusetts for Vermont.

When Lucy Terry wrote her poem, she was sixteen. Deerfield was a frontier community where the contest between white settlers and Indian communities over land and expansion was being enacted. Terry's poem was an accurate narration of the Indian attack. Six men and one woman were killed in the attack, one person was kidnapped, one escaped. The poem evoked the raid's tragedies:

*August, 'twas the twenty-fifth,*
*Seventeen hundred forty-six,*
*The Indians did in ambush lay, . . .*
*Samuel Allen like hero fout,*
*And though he was so brave and bold,*
*His face no more shall we behold.*

*Eleazer Hawks was killed outright,*
*Before he had time to fight,—*
*Before he did the Indians see,*
*Was shot and killed immediately. . .*

*Eunice Allen see the Indians coming,*
*And hopes to save herself by running,*
*And had not her petticoats stopped her,*
*The awful creatures had not catched her,*
*Nor tommy hawked her on her head,*
*And left her on the ground for dead. . .*

For some, this poem raises a problem as an expression of Black American literature. It identifies with the white settlers against the Indians, even though these same white settlers were slaveholders. In addition, it has no Black subject. Yet this reveals the cultural choices in New England at the time and, not surprisingly, Blacks' desire to be included in the general community. Local historians noted that Deerfield was a tolerant society, as frontier societies often were, and its small Black population—"semi-independent property"—fared well. But evidence suggests a greater degree of separation. Terry used the poem to speak across the racial barrier to the major issues of her society. She wanted her voice to be heard in the center of things.

Lucy Terry Price lived until age ninety-one, dying in 1821. She was forever active and assertive. In 1785, after her property was damaged by whites, she persuaded the local government to provide protection. Her son attempted admission to Massachusetts's Williams College, and when he was refused, she took his cause to the board of trustees, making a three-hour appeal "quoting an abundance of law and Gospel." Once she argued a case before Vermont's Supreme Court. In her later years, she became a village raconteur, and her home attracted many visitors who were taken with her storytelling.

## How did the movement for American independence affect Blacks?

It is thrilling to read the petitions of early American Blacks seeking freedom. While other early Americans were speaking up for freedom, Blacks raised their voices too. Often signed by a single name ("Felix" or "Prince"), the petitions were well reasoned and their objective unmistakable.

As early as 1675, a Black Virginia servant claimed that his master kept him longer than his contract. He petitioned for freedom and compensation in "corne and clothes." In 1726, Peter Vantrump, a New York free Black, petitioned a court that he was fooled by "one Captaine Mackie" into going to North Carolina. There, he was made a slave. He asked "to be declared free."

As the War of Independence drew near, the number of such

petitions increased, and their language became more heartfelt. In 1773, Massachusetts Blacks asked the governor for "Relief" that would be "as Life from the Dead." In 1774, other Massachusetts Blacks protested they were "stolen from the bosoms of our Tender Parents," wanted "our freedoms" and their children's "lebety." Another petition called slavery "far worse than Nonexistence."

The period leading to the War of Independence witnessed an explosion of ideas and dramatic events. From the 1760s to the 1770s, the thirteen colonies went from being part of England's empire to a breakaway state preaching a new equality and freedom. It was a period of great change. Petitioning by Blacks shows that they had been waiting for such a period; they could identify with the new ideas. Finally, Blacks were hearing a public discourse that matched their own deep longings.

Crispus Attucks (1723–70)—for many only a stock Black history figure—was cvidence of Black interest in the period's issues. In the *Narrative of the Boston Massacre,* he was identified simply as "a mulatto, killed on the spot, two balls entering his breast." Actually, Attucks was more: he ran away from slavery, became a proud free sailor and whaler, learned to read and write, and most important, had been active in anti-British agitation. He was a petitioner too, having protested unfair taxes in a letter to Governor Thomas Hutchinson. Attucks's leadership in the Boston Massacre crowd expressed his seamless vision of the future, one that wove his and his country's together.

Blacks numbered between five hundred thousand and six hundred thousand at this time. Some who lived in cities and towns read of the new ideologies in newspapers and pamphlets. But most could not read. They discovered the ideas by overhearing their masters' dinner talk, listening to tavern debates, or picking up on street discussions. What they absorbed was the concept of natural and inalienable rights for individuals—a concept that sprang partly from the European Enlightenment. On plantations, too, these ideas circulated among Blacks.

But the big question for Blacks was this: Did the lofty talk about rights embrace them? Even if it did not, Blacks quickly adopted the language of the times. In 1774, a "Grate Number of

Blacks" stated in a petition, "We have in common with all other men a *natural right* to our freedom."

During this period, it is significant that slavery was increasingly questioned by whites. Many felt that all the talk about freedom and the institution of slavery were incompatible. After 1850, the Quakers strengthened their opposition to slavery. James Otis, one of the early advocates of colonists' rights, also assaulted slavery. In 1764, he wrote, "The Colonists are by the law of nature free born . . . white or Black." Arthur Lee, son of a slaveholding family, in 1767 tried to convince the Virginia House of Burgesses to stop bringing in Africans. Benjamin Rush, later a signer of the Declaration of Independence, in 1773 said that slavery was a "national crime."

Unfortunately, this antislavery sentiment did not have much immediate impact. Patrick Henry's famous "Give me liberty or give me death" speech in 1775 compared England's treatment of the colonies with "chains and slavery." In the speech's coda, he overlooked the real slaves. In 1776, when the Declaration of Independence was read from public rostrums, Blacks drank in its words and its rich imagery. Their oral traditions carried the news of "all men are created equal . . ." and "life, liberty, and the pursuit of happiness." But Blacks were aware that the document was silent on their condition and fate. Even Jefferson's proposed section against the slave trade had been eliminated.

When the War of Independence started, the situation offered unprecedented opportunities for Blacks. Slaves left their masters in droves. Lord Dunmore, the Virginia governor, offered freedom to male slaves who joined the British forces. From this came the famous "Ethiopian Regiment." In time, tens of thousands of Blacks fled to the British side. Yet Blacks observed that British supporters were often major slaveholders. At the war's end, the British evacuated thousands of Blacks—perhaps as many as fourteen thousand—to Jamaica, England, and Nova Scotia, Canada.

About five thousand free Blacks served in the American military and naval forces, mostly in racially mixed units. Some states allowed slaves to serve later in the war. Quite a number of these people were very successful as spies and undercover agents. Slave Pompey Lamb spied on the British garrison before the Battle of

Stony Point. James Armistead was hired by the British to be their spy but instead spied on them. Blacks also helped capture some of the major British spies. Slave Saul Mathews reported on British defenses at Portsmouth, Virginia. About three thousand Blacks were military laborers, building bridges, fortifications, and camps.

Beyond this, Black armed fighters were decisive at the battles of Trenton and Princeton in 1776 and Rhode Island in 1778, and the mostly Black Louisiana Regiment fought well in Alabama and Florida (1779). French General Lafayette praised Black soldiers, as did George Washington. Even a Hessian officer called them "able-bodied, strong, and brave fellows." Many soldiers were given their freedom.

The war prompted Northern states to curtail slavery. Vermont abolished it in 1777, New Hampshire in 1783, Rhode Island in 1784. Pennsylvania enacted gradual emancipation in 1780, as did Connecticut in 1784. New York and New Jersey followed slowly.

## How should we think today of Phillis Wheatley?

In 1772 in Boston, Phillis Wheatley and her master John Wheatley tried to publish a slender volume of her poetry. Doubt faced them at every turn. No one believed that young Phillis—a Black—was capable of being an author. Later that year, a group of eighteen distinguished Boston citizens was assembled to quiz her. Both the governor and lieutenant governor attended. If she passed their exam, they would vouch for her book.

A year later, in 1773, when her *Poems on Various Subjects, Religious and Moral* was published in London, it opened with a two-paragraph "To the Public," which read:

> We, whose names are underwritten, do assure the World, that the Poems specified . . . were (as we verily believe) written by Phillis, a young Negro Girl, who was but a few years since, brought an uncultivated Barbarian from Africa, and has ever since been, and now is, under the disadvantage of serving as a slave in a Family in this Town. She has been examined by some of the best judges.

Below this statement, her Boston judges signed their names.

Phillis Wheatley was born in West Africa's Senegambia area in 1754. Now it is known that she came from the Islamic Fula peoples and had been slightly exposed to Islamic learning and Arabic—stimuli for her later literary achievements. She was brought to America when she was six or seven and sold at a dock auction in Boston. Because of her frailty, she was one of the last to be bought. Out of sympathy, Susannah and John Wheatley bought her, seeing her half-naked on the auction block. "Phillis" was the name of her slave ship. She assumed her master's surname as her own.

Slavery in the Wheatley household was more akin to that of West Africa than the harsh American kind: Phillis was considered a part of the household. Once the Wheatleys discovered her intelligence, they exempted her from hard household work. She was tutored in English, allowed to study Latin, and provided with writing materials. At sixteen, she was baptized. At thirteen, in 1767, she published her first poem in the *Newport Mercury*, "On Messrs. Hussey and Coffin," reporting on their narrow escape from a Cape Cod storm. Many a visitor came to the Wheatleys' home to marvel at "their Phillis," as she came to be known.

She was all the more remarkable because her creativity was compressed into a very few years. She literally consumed language, learning to read English and devouring the Bible after sixteen months in America. John Wheatley wrote, "She attained the English language to such Degree, as to read . . . the most difficult parts of the Sacred Writings, to the great Astonishment of all who heard her." At fourteen, she was translating the Latin poet Ovid. Alexander Pope and John Milton were her favorite poets. She reached her writing peak at twenty and died in 1784, barely thirty.

Phillis Wheatley's literary powers transformed her into a celebrity—and after her trip to England in 1773, that celebrity became international. The French philosopher Voltaire knew of her. "The African genius" is how a later memoir hailed her.

Today's readers might not find her poetry all that captivating. In her time, the style in poetry was neoclassicist—that is, people admired formal, strictly rhyming poems. Such poetry was not very personal, nor did it soar with high-flown imagery. But Wheatley was good at taming language and inserting her passions into formal cadences. This ability can be seen in these passages:

*Some view the sable race with scornful eye—*
*Their color is a diabolic dye;*
*But know, ye Christians, Negroes Black as Cain*
*May be refined, and join the angelic train.*

and

*Take him, ye Africans, he longs for you.*
*Impartial Savior is his title due:*
*Washed in the fountain of redeeming blood.*
*Ye shall be sons, and kings and priests to God.*

Here, in two different poems, she quietly proposed radical ideas for her time: despite proslavery Christian claims, she announced that God did not discriminate. "Negroes Black as Cain" were free to come to God. Though slaves, they could be elevated, made sons, royalty, and emissaries of God. Several of Wheatley's poems were forceful, particularly "Goliath of Gath," where she portrayed David as the ultimate hero.

Increasingly, Wheatley's life and poetry are getting a necessary fresh look. Her husband John Peters was not a ne'er-do-well as previously thought; he was an able free Black man and owner of a quality Boston house. During her trip to England, she seems to have become more independent. She knew more Blacks than was previously figured, corresponded with several, and was alert to slavery issues, as indicated in a letter from 1774: "In every human Breast, God has implanted a Principle, which we call Love of Freedom."

Phillis Wheatley broke through the barrier of literary silence for Black Americans. As critic Henry Louis Gates Jr. has noted, she started *both* the tradition of Black American writing and of Black American women's literature.

### What was Thomas Jefferson's relation to Black Americans?

Monticello, a mansion in the Palladian style, was built on a commanding hill near Charlottesville, in Virginia's Albemarle County. Today it seems perfect: the symmetry of its porticos, white dome, terraced gardens, its tranquillity. It lends a noble aura to the life of Thomas Jefferson (1743–1826).

Tour guides at Monticello speak reverentially of "Mr. Jefferson." And they should: he was an Enlightenment thinker, a naturalist, inventor, architect, people's delegate, governor of Virginia, minister to France, secretary of state, vice president, and most important, author of the Declaration of Independence and third president of the United States.

However, Monticello had another side, making it and Jefferson part of the imperfections of early American history. Like many men of his station, Jefferson was a slaveholder—actually, a major one, near the top in Albemarle. In 1774, Jefferson owned two hundred slaves, dispersed at his several farms. Seventy or more of them lived at Monticello.

Beneath Monticello were passageways, rooms, and a kitchen where house slaves worked. Farm plots, where field hands labored, radiated out from the mansion. From Jefferson's dining room, Mulberry Row—a slaves' quarters of fourteen-by-twenty-foot log cabins—could be seen. Years ago tours did not mention Monticello as a slave plantation. Today this is more in the foreground.

For a long time, many believed that Jefferson's slaveholding robbed him of true greatness and made him into a hypocrite. How did he live with the contradiction of proclaiming "all men are created equal" in the Declaration of Independence, yet owning slaves? Was he a moral schizophrenic? Answers of all types—yes, no, and in between—have been made to these questions. Since 1993 marked Jefferson's 250th birthday, his legacy has been much discussed, and questions of race and liberty in his world have raged.

Defenders of Jefferson have been quick to point out that one-third of the men who signed the Declaration of Independence were slaveholders. They say that when Jefferson was elected to

Virginia's House of Burgesses in 1769, he proposed a measure for freeing slaves, but it was rejected. His defenders state that he disliked slavery and "trembled" at the ruin it would bring future generations of whites. Blacks, he thought, would strike back at whites for "the ten thousand . . . injuries they have sustained." Defenders argue that in the 1780s Jefferson resisted slavery in the new territories of the West. Others say that Jefferson was a kind slave master and softened slavery's blows by keeping Monticello's slave families together.

None of this has stopped Jefferson's critics. They say he was good at worrying over these matters, but his practice was often harsh. He ran his farms with an accountant's eye to financial gains. He had slaves flogged to get their obedience, hunted down runaways, and sold slaves (161 from 1784 to 1794) when the plantation needed revenue. On his close servant Jupiter's death, Jefferson worried mostly about the loss of his skills. When slaves' children reached the age for working, they were put up for sale. Even when old, slaves were kept working. While some slaveholders, such as Virginia's Edward Coles, anguished over slavery and freed their slaves, Jefferson did not budge. George Washington freed his slaves in his will; Jefferson did not.

Further complicating Jefferson's reputation is the issue of Sally Hemings (1773–1835). The Hemingses were the most important slave family at Monticello. Betty Hemings was the family's matriarch, who came to Monticello with Martha Wayles when she married Jefferson. Sally was one of Betty's twelve children. Beautiful, Sally served as a personal servant to Jefferson's daughters.

When Jefferson went to Paris in 1785, he took Sally Hemings to care for his daughter Maria. In Paris, the Jefferson-Hemings liaison was said to have started. Returning from Europe, she was a Monticello house servant and proceeded to give birth to five children, all conceived when Jefferson was at the plantation. During the 1804 presidential campaign, the story of Jefferson's slave mistress was circulated widely in the hope that it would scandalize voters. But Black Americans also kept this story alive for years, as evidenced in the novel *Clotel* (1853) by William Wells Brown. The character Clotel is Jefferson's daughter.

Since the 1950s, the Jefferson-Hemings liaison has overwhelmed Monticello's hero. Two books have been crucial to this: Fawn Brodie's *Thomas Jefferson: An Intimate History* (1974) and Barbara Chase-Riboud's novel *Sally Hemings* (1979). Merchant Ivory's film *Jefferson in Paris* (1995) added fuel to the fire. All unleashed a torrent of protest. Yet it is not clear why discussion of the liaison has infuriated Jefferson partisans; this relationship was not any different from master–female slave relations across the South.

More difficult to explain than Jefferson's slaveholder status or his love of Sally Hemings are his views of Black intellect. In *Notes on the State of Virginia* (1798), he wrote, "In memory, they are the equal to whites; in reason, much inferior. . . . In imagination, they are dull, tasteless, and anomalous." And "never yet could I find that a Black had uttered a thought above the level of plain narration." He attacked "Phyllis Whately," saying meanly, "The compositions *published under her name* [emphasis mine] are below the dignity of criticism." He concluded, "Blacks . . . are inferior to the whites in the endowments both of body and mind." Jefferson's own dedication to reason failed him.

Jefferson's story reveals that in the eighteenth century, American freedom and American slavery were interlocked. The former condition was often defined by the latter: for many whites, to be free was not to be a slave. Freedom was freedom over others—in this instance, over Blacks.

## Why were people so captivated by Benjamin Banneker?

When Williams Wells Brown published his *History of the Negro Race in America* in 1883, he included a chapter on "The Negro Intellect." Benjamin Banneker was its centerpiece. He deserved the renown, for Banneker was one of the great eighteenth-century intellectual adventurers.

Born a free Black in rural Maryland in 1731, he created an ingenious life: as a nature buff, inventor, mathematician, surveyor and planner of the nation's capital, astronomer, and almanac producer. A doting grandmother taught him to read. At a local school that permitted a few Black children, Banneker got more training, but he was largely self-taught. On his own, he mastered spherical trigonometry and calculus. Leisure for him consisted of studying tough mathematical problems.

Forever busy, Banneker learned to play both the violin and the flute. He was an avid beekeeper and wrote a treatise on bees' behavior. He kept a journal of his dreams, obviously interested in their meaning. With the time left, he read voraciously from the Bible, history, biography, travel books. He was one of our new nation's polymaths—a person of great learning in several fields.

Banneker's first taste of fame came at twenty-two. Having only a borrowed watch as a model, he constructed a striking clock—the first one made in America. All parts were made by him, including the mechanism for striking hourly, and he meshed the hour, minute, and second machinery. The clock became well known, attracting visitors to his home. In building this clock, Banneker joined his age's noblest minds, which were preoccupied with the phenomenon of time.

In the years that followed, he was a modest gentleman tobacco farmer, a bachelor who divided his labor between managing an efficient farm and working on mathematical calculations. Mathematics often won out. It was noted that "his door stood wide open, and so closely was his mind engaged that . . . [people] entered without being seen." The work paid off, for in 1789, he predicted an unexpected lunar eclipse. Widespread interest followed. People from around the country sent Banneker mathematical problems. Returning the solutions, he often enclosed new problems, stating them in rhymes.

In 1791, when he was sixty, Banneker was invited to join a land survey for the federal capital of the new American nation. Work on the project gave Banneker practical field survey experience and allowed him to promote his ideas on the new city's organization. After several months, Banneker returned to his farm to work hard on an almanac.

Banneker's almanac contained his astronomical and tide calculations, weather predictions, plus little essays, proverbs, and poems by him and others. He prepared all the technical tables. Published in late 1791, the almanac sold briskly, especially after it was known that Banneker was a free Black. Yet he found his work doubted, just as Phillis Wheatley had. Eleven white supporters had to issue a statement "in Order to clear up any doubt that may Arise as to Benjamin Banneker (a Black man whose Father Came from Africa) being the Author of the Prefixed Almanac." To them, "this Negro [was] fresh proof that the powers of the mind are disconnected with the color of the skin." His almanacs were published annually until 1796.

One of the most intriguing episodes in Banneker's life was his letter in 1791 to Thomas Jefferson. At the time, Banneker was the most accomplished Black in postindependence America. Nevertheless, a letter to such an exalted figure was a departure for the quiet Banneker. Banneker saw Jefferson as friendly to Blacks and was encouraged by his past condemnations of slavery. In the letter, Banneker wrote, "You are a man far less inflexible in sentiments . . . than many others; and . . . well disposed towards us." He also saw Jefferson as having similar scientific interests. Yet he knew of Jefferson's slaveholding and disparaging of Black intellect.

Banneker's letter tried to enlist Jefferson as a more public foe of slavery. Using Jefferson's ringing endorsement of freedom in the Declaration of Independence, he confronted Jefferson with his own words. He even said that Jefferson and his political cohorts "counteract [God's] mercies, in detaining by fraud and violence, so numerous a part of my brethren under groaning captivity and cruel oppression." He recommended that "you and all others . . . wean yourselves from those narrow prejudices which you have imbibed with respect to [Blacks]." Accompanying the letter was a handwritten copy of his almanac—proof of Black intellectual abilities and that Banneker did his own calculations. Jefferson's response was cordial, but it sidestepped the slavery issue with an ambiguous statement. He promised to send the almanac to Paris's Academy of Sciences.

Today, Banneker's life has special significance, as young Black Americans seek to enter the fields of science and mathematics. Bob Moses, central to the 1960s Mississippi civil rights struggle, has returned to that state to run the Algebra Project. Just as Moses once mobilized for voter rights, he is now organizing a movement for mathematical literacy. Today's slogan is "We shall overcome, this time with algebra." For this new movement, Banneker's achievements provide a beacon.

**Where did free Blacks gain their freedom? How many free Blacks were there?**

Free Blacks go back a long way in the American past to the colonies' early years. Technically, a free Black was as self-governing as any white person. Thus, from the beginning, free Blacks were an American oddity.

Through various routes, early Blacks joined the ranks of free people. In Virginia's fledgling settlements, the first free Blacks had been indentured servants. "Self-purchase" was another route to freedom in Virginia and Maryland. In this case, a slave traded work and earnings for freedom. In New England, masters occasionally freed slaves for their "good and faithful service," as William Randall did Peter Palmer in 1702. Blacks also became free through their masters' wills. Occasionally, assertive Blacks appealed to courts for their freedom and won, as did the Massachusetts slave named James who sued in 1735. By 1750, a few thousand free Blacks were scattered unevenly throughout the colonies.

After this time, their numbers began to grow, dramatically in some places. In fact, a true free Black American caste developed between 1775 and the 1810s.

Two forces converged to cause this change. First, American evangelical Christianity, which erupted into the 1730s–1760s Great Awakening, condemned slavery. Churches urged their followers to free their slaves. Second, the American independence movement forced the recognition that slavery directly contradicted the ideals of equality and liberty. The war further undermined slavery. Sometimes, slaves who served on the American side were rewarded with their freedom. Thousands of slaves ran away during the war, with some entering permanently the status of free people. In the North, the war's end led to states enacting laws abolishing slavery, albeit gradually. A "manumission fever" took over in some places, except for the lower South.

Some interesting realities accompanied this increase in free Blacks. By 1805, every state in the North had made provisions for the emancipation of slaves; by 1830, only 3,500 slaves remained in the region. The upper Southern states of Maryland, Delaware, and Virginia made it easier for slaveholders to free their slaves in their wills. In Virginia and Maryland, a child born of a free mother could be freed at age thirty-one. Slaves continued to buy their way free. Once free, they often bought their wives, husbands, or children who were still in slavery. In Cincinnati, in 1835, 18 percent of the free Blacks had purchased their freedom.

From 1790, when the first federal census was taken, to 1830, the total population of Blacks tripled from 757,363 to 2,328,642. At the same time, the percentage of free Blacks rose from 8 to almost 14 percent. This increase seems heartening until we consider that it also means that in 1830, more than 86 percent of Blacks were still enslaved.

Free Blacks lived a precarious existence, making their way in a nation where slavery was still strong in the South and prejudice virulent in the North. But they were up to the challenge.

## What was life like for free Blacks? What did they accomplish?

In 1826, Baltimore's free Blacks described their difficult position: "We reside among you and yet are strangers; natives, and yet not citizens; surrounded by the freest people and most republican institutions in the world, and yet enjoying none of the immunities of freedom." They concluded, "Though we are not slaves, we are not free."

America had no real place for free Blacks. They were an anomaly. No general ideology validated them. At the whim of shifting prejudices and anxieties, they were highly vulnerable. In the North, free Blacks were prohibited from voting. Illinois, Indiana, and Iowa excluded them from juries and forbade their testimony in cases involving whites. Schools, hospitals, and cemeteries were either segregated or closed to them. Men struggled to break into the trades. From the 1830s to 1850s, Philadelphia was wracked by five riots targeting Blacks.

In the South, free Blacks were regarded as an even greater irritant. Often, they were forced to register annually with local officials. Mobile, Alabama, made them pay an annual fee to retain their status. Occasionally, they sued in court, but their rights were not secure. Worse was the possibility of being kidnapped and sold into slavery. Stealing free Blacks was illegal, but this law was not enforced. In 1827, an ad in the *Raleigh Register* announced, "Stolen three Girl Children of Color, Free Born . . . Some dishonest Person has taken them off, for the purpose of selling them as Slaves." Nancy Valentine, their mother, signed the ad. Rich free Black families in Charleston, New Orleans, and Richmond avoided some of these aggressions, but not entirely.

Sheltering themselves against prejudice, free Blacks often huddled in small neighborhoods—like New York City's "Little Guinea" and "Five Points." The 1820s saw these neighborhoods at fever pitch over a question vital to the future of Blacks in America: Should Blacks remain in America, suffer harsh treatment, or seek a haven of their own outside America? Should they flee or stay?

The American Colonization Society, started in 1816, had an answer. It proposed establishing a colony in West Africa, using private and government money. Free Blacks and freed slaves would settle there. In 1822, the society—supported by an odd collection of humane whites and slaveholders—founded Liberia, having bought land from local African chiefs. Some Blacks, like Massachusetts merchant Paul Cuffee (1759–1817), already had been advocating a return to Africa. So when the idea of Liberia was proposed, thousands of Blacks migrated from the states of Maryland and Virginia and the cities of Providence and Boston. Later, they would be called "Americo-Liberians," and they would number approximately twelve thousand.

With this, a major dream of Pan-Africanism was brought to life—the dream of Mother Africa gathering up her scattered, traduced children. From this point on, back-to-Africa movements gained force. But the majority of free Blacks stayed put, arguing that going to Liberia strengthened slavery. They favored working out an American identity. A Boston Black spoke for many: "This being our country, we have made up our minds to remain in it, and to try to make it worth living in."

That they did. Free Blacks were enormously energetic as the builders of communities. Denied access to white institutions, they turned their energies inward, producing an early outpouring of Black nationalism. For instance, free Blacks developed their own churches because of white churches' "nigger pews" and balcony "nigger heavens," where Black worshipers were forced to sit. Many of the institutions they built are still alive today. The most famous are the African Grand Lodge—a Masonic fraternal organization—chartered in 1787 by Prince Hall (1748–97), a Barbadian who emigrated to Boston, and the African Methodist Episcopal church founded in Philadelphia in 1794 by Richard Allen (1760–1831).

Schools were also established, many in Philadelphia and New York. Mutual-aid societies, which pooled the resources of their members, opened libraries, sponsored lectures, provided insurance, cared for widows, helped the poor and the sick, paid members' burial costs, and kept up cemeteries. Bowdoin College graduate John Russwurm (1799–1851) helped found the first Black newspaper, *Freedom's Journal*, in 1827. By the time of the Civil War, seventeen Black newspapers were being published.

A most important advance were the National Negro Conventions, held annually from 1830 to 1835. William Hamilton, an organizer, stated their goal: "Under present circumstances, it is highly necessary that free people of color should combine and closely attend to their own interest." By bringing free Blacks together, the conventions helped consolidate them into a national community, and they expanded Black leadership beyond the local arena.

Today, women are seen as dynamic contributors to the building of free Black communities. In Petersburg, Virginia, historian Suzanne Lebsock found women on the economic cutting edge in the 1800s to 1840s: they pursued jobs (as seamstresses, domestics, laundresses, tobacco factory workers), headed households, and owned most of free Black property. In other places, free women started schools in their homes, literary societies, temperance clubs, and charity circles. Free Black America needed both manpower and womanpower to build its world.

### Why was David Walker's *Appeal*—written in 1829—so important for Black history?

David Walker's *Appeal, in Four Articles* is one of the most important books ever written by a Black American. For the early nineteenth century, it was surprisingly militant. Walker (1785–1830) is reminiscent of Frantz Fanon (1925–61), the Martinique-born psychiatrist whose radical ideas shaped the 1960s anticolonial and left movements. Both wrote in a white heat. Both excelled at jeremiads. Like Fanon, Walker urged Blacks to revolt against their masters and to contemplate violence as a remedy. Both died early and under curious circumstances.

Walker's seventy-six-page book, published in Boston, was addressed to "the colored citizens of the World, but in Particular and very Expressly to Those of the United States." He wrote as an American pamphleteer, using essays to arouse the public. Since 1760, free Blacks had been publishing up a storm—narratives, poems, sermons, addresses, letters, Masonic charges, petitions, hymns, a few newspapers. Walker joined this prolific tide. He knew the printed word's power because he was Boston agent for the *Freedom's Journal,* the newly established New York Black newspaper. But he was unprepared for the reaction to his pamphlet, which went through three editions in two years. In the South, his little book was feared, banned, and confiscated.

The first Black strongly worded protest against slavery was penned in 1788 by the anonymous "Othello." Others followed. But when Walker's *Appeal* appeared, he set a new standard. His message was plain, urgent, and uncompromising. Walker—born in North Carolina to a slave father and a free mother—was radicalized by seeing two merciless whippings of slaves. Later, he settled in Boston and opened a secondhand clothing store. In Boston, he saw other awful incidents, convincing him that the future was bleak for Blacks unless they faced their situation squarely.

In his book, Walker often sounded like an Old Testament prophet, bringing news of calamitous times. He called on "men of color . . . to cast your eyes upon the wretchedness of your brethren." He wanted them to "rescue them and yourselves from degradation." For him, slavery had transformed Blacks into "beasts," "brutes," and an "animal existence." He urged, "If you commence [to fight for freedom], make sure you work—do not trifle, for they will not trifle with you—they want us for their slaves."

He proclaimed to America: "I tell you Americans . . . that unless you speedily alter your course, you and your country are gone!!!" He delivered a judgment: "For God Almighty will tear up the very face of the earth!!!" He went further, "I speak to Americans for your own good. We must and shall be free . . . in spite of you."

Walker had proposals for change. Blacks had to unify themselves. Free Blacks must link themselves to their poor enslaved brothers and sisters; there was no safety for them in remaining isolated and living above other Blacks. Racism and greed were the twin pillars of the slave system; they had to be defeated. Also, Blacks must not slip into passivity because slavery seemed so monumental, so permanent. Every Black's heartbeat must be a life pulse resisting slavery.

In 1830, Walker died mysteriously. It was believed he was poisoned. But no mystery surrounds his impact on Black Americans at the time. He taught his readers that they could not accommodate oppression, that they must be outraged in confronting their miserable conditions. Once Walker unleashed these ideas, they entered into Black thought, surviving to the present day. June Jordan's recent essays of conscience, *Technical Difficulties* (1992), calling for "righteous rage" in the face of today's injustices are just one example of the continued vitality of Walker's stance.

### What was behind the famous trio of slave rebellions—Gabriel Prosser's in 1800, Denmark Vesey's in 1822, and Nat Turner's in 1831?

Messengers on horseback carried the word of strange happenings in Virginia's Southampton County in the last days of August 1831. A revolt by slaves had splattered blood across the landscape. When first planned, the revolt was to explode on July 4, commemorating America's exalting of liberty. Nat Turner's brigade had struck.

By the 1830s, Americans, especially Southerners—Black and white—had become quite familiar with major slave revolts. This awareness developed as early as 1791, when Haitian Blacks began their great revolution against their French masters. The island revolt shook the South. Newspapers carried many reports about the rebellion. Haitian slaveholders escaped to America with more news. The slaves they brought with them told Black communities what had happened. After Haiti, rumors of slave conspiracies sounded plausible.

And for good reason. In the summer of 1800, Gabriel (circa 1775–1800)—a slave blacksmith belonging to Thomas Prosser of Henrico County, Virginia—recruited fellow slaves to take Richmond. They would burn the city first, then seize a well-stocked weapons arsenal and take Governor James Monroe as a hostage in exchange for the abolition of slavery in the state. Gabriel Prosser and his allies took advantage of the remarkable freedom allowed the area's slaves to organize a most elaborate plot. As the governor stated, it embraced "most of the slaves in Richmond and the neighborhood." But slaves divulged the plan to whites. The military was summoned; many arrests were made; Gabriel and over thirty others were tried and hung.

In 1822, South Carolina's Denmark Vesey (1769–1822) organized city and plantation Blacks to achieve their freedom. Vesey was born in Saint Thomas in the Danish West Indies. He was owned by Joseph Vesey, a Bermuda planter, who settled in Charleston. Denmark Vesey—self-educated, widely traveled, and a speaker of several languages—bought his freedom in 1800 with six hundred dollars he won in a lottery. Through reading, he was aware of the widespread misery endured by fellow Blacks. He also was dissatisfied with his life as a freeman, for some people were freer than others. For several months, he and his comrades plotted to take the city's arsenals, guardhouses, and naval stores.

Unfortunately, a house servant betrayed him, and the plan was exposed. Over the next two months, 130 Blacks were arrested, and sixty-seven were convicted of trying to raise an insurrection. Thirty-five, including Vesey, were hung.

Nat Turner's 1831 rebellion had more impact than any other

slave rebellion of pre–Civil War America. He had a sizable group of supporters, between thirty and fifty, which grew as the revolt proceeded. They kept their plan secret. His supporters were loyal to Turner, calling him "General Nat." Within a small area, traveling to twenty-two homes, he and his band wreaked havoc. The rebels were stealthy, attacking late at night. They appeared fearless, invading masters' houses quickly. Their killings were swift and spared no one. Sixty whites were killed.

At the center of the revolt was the personality of Nat Turner. From his childhood, he was always thought special. Like an African child, his body at birth was searched for special marks on his head and chest, signs of his power and destiny. His mother claimed he knew of events before his birth—the sign of a prophet. He could read but had no recollection of learning how. As he approached the rebellion, he announced powerful visions: white and Black spirits darting across the night sky, blood falling from corn like dew. God was speaking to him. But his revolt was also helped by earthly forces: it was a time of economic hardship on plantations, meaning more work for slaves and less to eat.

The state militia was mobilized, and three thousand troops came out to subdue the rebels. Most of the slaves were killed, but Turner managed to escape. Virginia newspapers called him and his band every imaginable thing: "infatuated beings," "a parcel of blood-thirsty wolves," "deluded wretches." Six weeks later, Turner was captured, ragged but unbowed. A man who interviewed him later in jail said his "natural intelligence and quickness of apprehension, is surpassed by few men I have ever seen." Hanging was the punishment for Turner and his allies. With his death, he passed into history and folklore. Among free and slave Blacks, he was God's man, whose exploits were whispered across the upper South. Some even said the limb from which he was hung died.

## Why is Maria Steward mentioned more and more in Black history?

> O ye daughters of Africa, Awake, Awake, Arise! No longer sleep nor slumber. . . . Show forth to the world that ye are endowed with noble and exalted faculties.

> Methinks were the American free people of color to turn their attention more assiduously to moral worth and intellectual improvement, this would be the result: prejudice would gradually diminish.
>
> Men from England bought and sold me, Paid my price in paltry gold. But though a slave they have enroll'd Me, Minds are never to be sold.
>
> There are no people under heavens so unkind and so unfeeling towards their own, as are the descendants of fallen Africa.

These are some of the pungent remarks of Maria Steward (1803–79), a significant public figure in 1830s Boston. Born in Hartford, Connecticut, she later moved to Boston, where she married. She was a free Black. The sudden death of her husband and the stealing of her inheritance by businessmen forced her into a career of public speaking. She was one of the first American women to make a reputation as a public orator. Her appearances were mainly before Black audiences and to mixed antislavery groups. Two books of her ideas, one entitled *Meditations from the Pen of Mrs. Maria Steward* (1832), were published.

She was unique in her appeals to Boston's Black women. "How long," she asked, "shall the fair daughters of Africa be compelled to bury their minds and talents beneath a load of iron pots and kettles?" She wanted a Black women's high school so "that the higher branches of knowledge might be enjoyed by us." Women, she argued, should join literary societies in order to enhance their skills.

One of her messages, often unnoticed, was that Blacks should help themselves. To her, all was not well among free Northern Blacks. Listen to her urgings to young Black men: "I would implore our men, and especially our rising youth, to flee from the gambling board and the dance-hall; for we are poor, and have no money to throw away. . . . Our fine young men are so blind . . . to the future welfare of their children as to spend their hard earnings." She saw that freedom had its perils.

Steward was unsuccessful in cracking the wall of indifference to the ideas of a woman public speaker. Her problems were complicated by the sternness of her message. Eventually, she left Boston and went on to be a teacher in New York City and Washington and a matron in a government hospital treating former slaves. For decades, Steward did not appear in history books; today, she has a place. Now she deserves a biographer.

### How do recent discoveries at Andrew Jackson's Hermitage Plantation remind us of Africa's hold on slave culture?

For several years, a team of archaeologists, coordinated by Larry McKee, has been digging away behind Andrew Jackson's Tennessee mansion in the ground once occupied by rows of identical slave cabins (*Science*, March–April 1995).

In many of the cabins, they have found square pits carved out of the clay soil beneath the floor—"hidey holes," for food storage and for objects best concealed from master and overseer. McKee's team has also discovered "three beautifully wrought brass amulets shaped like human fists," used "to bring good luck to ward off evil." Across the South, McKee says, digs are turning up objects that slaves employed in spiritual practices: quartz crystals, medicine vials, lumps of sulfur, cut silver coins, a pierced coinlike medallion, glass beads for security against witchcraft, prehistoric projectile points, gaming pieces made out of European pottery shards. These hiding places and talismans are reminders that nineteenth-century slave culture retained elements of African cultures, even as Africa was receding from memory.

For years, "survivals" and "Africanisms"—as retentions from Africa are often called—have been assumed to be significant in American slave culture. Sometimes the surviving item is infinitesimal—like a single blue hexagonal bead found in a Cumberland Island, Georgia, slave cabin, brought from Africa and perhaps given special powers here. Or the Africanism might be a whole complex of practices, such as South Carolina's Sweetgrass basket making, which has been practiced for more than three hundred years.

The list of African contributions to American slave culture is long and diverse: dance patterns; call-and-response singing; polyrhythmic music; drumming skills, still demonstrated at New

Orleans's Congo Square; household design; ironworking and crafts, such as decorating clay pipes and sculpting walking canes; pictorial quilt designs; prophetic traditions; work patterns.

In recent years, the linking of African cultures to Black American culture during slavery and, more generally, to overall American culture has been a fast-breaking news phenomenon.

Ever since historian Peter Wood startled people in 1974 with the news that West Africans' skills at rice cropping, indigo raising, and herding contributed to the economic success of the South Carolina colony, the field of survivals has been hotter than a bull market. In 1987, Wood excited medical experts by telling how Cotton Mather learned of successful smallpox inoculation from "my Negro-Man Onesimus," a West African. South Carolina whites got similar information from their slaves.

One of the more recent ideas, advanced by historian Mechal Sobel, suggests that the very pace and pattern of work in eighteenth-century Virginia was changed by African notions of time. Work was slowed by Africans. They paid more attention to the social aspects of work. Over the years, there arose an Anglo-African sense of "Southern slow time." It's possible that such Virginia aristocrats as George Washington and Thomas Jefferson were influenced by a sense of time and pace shaped by African immigrants.

One place where African culture definitely crossed into American culture was in the area of language. Africans came from rich oral cultures; from the beginning of the slavery era, they developed a Black English. Linguist David Dalby contends, "The effects of Black English and West African languages on the development of American English have never been adequately studied." But "many well-known Americanisms are in fact Africanisms." He cites "OK" as a leading example, and he adds such words as jazz, jitterbug, hep or hip, banjo, boogie-woogie, rooty-toot, jam (as in jam session), to jive, to tote, to bug some one, to dig (to understand), uh-huh and uh-uh (for yes and no), and bad-mouth. Dalby sees much of this coming from the Mandingo language, the tongue of many early African immigrants to America.

Perhaps the latest news of Africanisms in America has come from Richard Westmacott's research, reported in 1993, in Black yard and gardening traditions in Alabama, Georgia, and South Carolina. There, he found a long tradition of swept yards and yard life that reached back into slavery and is still present today. He traced this back to West Africa. Westmacott claims, "Almost everybody had swept yards, including the plantations, which were swept by slaves or servants."

Stay tuned for news updates of African survivals on the American front!

## RESOURCES

Andrews, William L. *To Tell a Free Story: The First Century of Afro-American Autobiography, 1760–1865.* Urbana: University of Illinois Press, 1986.

Aptheker, Herbert. *A Documentary History of the Negro People in the United States.* Vol. 1. New York: Citadel Press Books, 1951.

Berlin, Ira. *Slaves Without Masters: The Free Negro in the Antebellum South.* New York: New Press, 1977.

Blassingame, John W., and Mary Frances Berry. *Long Memory: The Black Experience in America.* New York: Oxford University Press, 1982.

Brown, Letitia Woods. *Free Negroes in the District of Columbia, 1790–1846.* New York: Oxford University Press, 1973.

Campbell, Edward D. C., and Kym S. Rice. *Before Freedom Came: African-American Life in the Antebellum South.* Charlottesville: University of Virginia Press, 1992.

Egerton, Douglas R. *Gabriel's Rebellion: The Virginia Slave Conspiracies of 1800 and 1802.* Chapel Hill: University of North Carolina Press, 1993.

Greene, Lorenzo Johnston. *The Negro in Colonial New England.* New York: Antheum Books, 1968.

Gutman, Herbert. *The Black Family in Slavery and Freedom: 1750–1925.* New York: Pantheon Books, 1976.

Hogg, Peter. *Slavery: The Afro-American Experience.* London: British Library Board, 1979. Text with photos.

Honour, Hugh. *The Image of the Black in Western Art: From the American Revolution to World War I.* Vol. 4. Cambridge, Mass.: Harvard University Press, 1989.

Huggins, Nathan Irvin. *Black Odyssey: The African-American Ordeal in Slavery.* New York: Vintage Books, 1990.

Mellon, James. *Bullwhip Days: The Slaves Remember—An Oral History.* New York: Avon Books, 1988.

Nash, Gary B. *Race and Revolution.* Madison, Wisc.: Madison House, 1990.

Onuf, Peter S., ed. *Jeffersonian Legacies.* Charlottesville: University of Virginia Press, 1993.

Patterson, Ruth Polk. *The Seed of Sally Good'n: A Black Family of Arkansas, 1833–1953.* Lexington: University of Kentucky Press, 1985.

Parker, Freddie L. *Running for Freedom: Slave Runaways in North Carolina, 1775–1840.* New York: Garland Publishing, 1993.

Piersen, William D. *Black Yankees: The Development of an Afro-American Subculture in Eighteenth-Century New England.* Amherst: University of Massachusetts Press, 1988.

Porter, Dorothy. *Early Negro Writing, 1760–1837.* Boston: Beacon Press, 1971.

Rozelle, Robert V., Alvia Wardlaw, and Maureen A. McKenna. *Black Art, Ancestral Legacy: The African Impulse in African American Art.* Dallas: Dallas Museum of Art, 1991.

Sobel, Mechal. *The World They Made Together.* Princeton, N. J.: Princeton Press, 1987.

Stuckey, Sterling. *Slave Culture: Nationalist Theory and the Foundations of Black America.* Oxford, England: Oxford University Press, 1987.

Upton, Dell, ed. *America's Architectural Roots: Ethnic Groups That Built America.* Washington, D.C.: Preservation Press, 1986.

Westmacott, Richard. *African-American Gardens and Yards in the Rural South.* Knoxville: University of Tennessee Press, 1992.

Wood, Peter. *Black Majority: Negroes in Colonial South Carolina.* New York: Alfred A. Knopf, 1974.

Wright, Donald R. *African-Americans in the Early Republic: 1789–1831.* Arlington Heights, Ill.: Harlan Davidson, 1993.

TEXTS FOR YOUTH

Hamilton, Virginia. *Many Thousand Gone: African-Americans from Slavery to Freedom.* New York: Alfred Knopf, 1993. Illustrations by Leo and Diane Dillon.

Hamilton, Virginia. *The People Could Fly: American Black Folktales.* New York: Alfred Knopf, 1985. Illustrations by Leo and Diane Dillon.

Harris, Joel Chandler. *Jump!: The Adventures of Brer Rabbit.* San Diego: Harcourt, Brace, Jovanovich, 1986. Illustrations by Barry Moser.

# 4

# Black Americans Head to the War

## The 1840s to 1865

1841: Trial of the African rebels who took the slave ship *L'Amistad*

1830s–1840s: Abolitionists—Black and white, women and men—organize

1840s: Several great slave narratives are published

1841: Frederick Douglass gives first antislavery talk in Massachusetts

1843: Former New York slave Sojourner Truth begins traveling to lecture and preach against slavery

1846: Norbert Rillieux obtains patent for sugar-refining process

1846: Dred Scott files suit for freedom in Saint Louis, Missouri

1847: Abolitionist Frederick Douglass begins publishing a newspaper, the *North Star*

1849: Harriet Tubman escapes from slavery in Maryland, starts leading others to freedom

1850: The Fugitive Slave Law is enacted, toughening the 1793 act

1852: Harriet Beecher Stowe's *Uncle Tom's Cabin* is published

1859: Harriet Wilson's *Our Nig* is published, one of the first novels by a Black

1859: John Brown conducts bold raid on Virginia's Harpers Ferry arsenal

1860: Abraham Lincoln's election as president precipitates South Carolina's secession

1861: With attack on Fort Sumter, South Carolina begins Civil War

1863: The Emancipation Proclamation is issued, allowing Blacks to serve in Union Army and Navy

1863: The Fifty-fourth Massachusetts Regiment storms South Carolina's Fort Wagner

1865: The war ends, freeing a vast Black slave population

On May 24, 1854, Charlotte Forten, a sixteen-year-old free Black, began keeping a diary. Her goal for the diary was strictly personal; it was to be just "a pleasant and profitable employment of my leisure hours to record the passing events of my life."

For a decade, until 1864, she filled ornate little books with remarks. Increasingly, though, she commented on public matters. This was understandable. During these years, America was passing through a titanic clash over the future of Black men and women.

Charlotte Forten (1837–1914) was no ordinary young woman. The Fortens of Philadelphia were members of the free Black elite. They sent her to excellent integrated schools in New England, where she excelled at poetry and trained for teaching.

The Fortens were also strong proponents of abolition and equal rights for Blacks. Growing up, Charlotte met the leading figures of the antislavery movement. As a teenager, she followed race issues. One of her greatest joys was attending abolitionist speeches and rallies, to receive, as she put it, her "antislavery food." In 1862, in the midst of the Civil War, Charlotte went to the Sea Islands of South Carolina, where she taught some of the first freed slaves.

The diary Charlotte created is a unique historical record, taking us right into the heart of the times. Her diary reminds us that a series of conflicts, starting in the 1840s, tore the nation apart and led to the Civil War in 1860. It reveals how Blacks—even those living on plantations—were intensely aware of the national debate about slavery's future and the war. Her writing tells us of the mighty abolitionist campaigns, in which Blacks and whites, men and women, joined together to end slavery. She explains how average people, normally spectators, were swept up in the national conflict.

When Charlotte started her diary in 1854, four million Blacks lived in slavery, and another half million lived precariously as free people. When she finished writing in 1864, these Blacks were months from the Civil War's end and freedom. This transformation was one of the greatest of the nineteenth century—indeed, of world history.

## What was the "*Amistad* incident," and why did it provoke such a clamor?

On January 7, 1840, an unusual trial began in New Haven. The trial was about the "*Amistad* incident" or "*Amistad* mutiny." The trial became a public spectacle. Crowds descended on the courthouse. Yale students and faculty allied themselves with the defendants. Reporters flocked to the scene. Different nations took sides. A phrenologist studied the skull of the chief defendant.

The "*Amistad* incident" began with an African mutiny on a slave ship in the Caribbean, and most of the events occurred outside the United States. But the drama's crucial last episode, a legal battle, was enacted in three American courts. Eventually, the case would go all the way to the Supreme Court.

In 1838, a group of Mende were captured in West Africa, in today's Sierra Leone. They were sold to Spanish traders, who took them on a barbaric journey to Havana, Cuba. Only fifty-four of them survived. From Havana, they were shipped on the Spanish schooner *L'Amistad* (meaning "friendship") to Port Principe, Cuba, on June 28, 1839.

Four days into that voyage, the Africans revolted, killing the captain and three crew members. For nearly two months, the Africans tried to get the remaining crew to take them back to Africa.

In August, they were still sailing but were nowhere near their homeland, and they were running out of supplies. Soon, villagers along the New York coast reported seeing "a vessel of suspicious and piratical character," "a mysterious long black schooner." It was the *Amistad.* On August 26, 1839, the Africans came ashore at Montauk Point, Long Island, in search of food. At this point, the United States Navy seized the *Amistad* and took the Africans to a New Haven jail and a trial to decide their fate.

Leading this daring rebel band was a man known as Cinque, the "master spirit" of the group, whose whistle caused "the Blacks to spring around him." His tenacious leadership had held together the "Ethiop crew." Once on trial, he gave a remarkable performance, "evincing uncommon decision and coolness."

What should be done with the Black rebels? That was the big question.

Antislavery clubs quickly raised money to provide them with legal help. The case against the Africans was argued by a United States district attorney whose position was that they were still slaves, pirates, and murderers. But it was decided that since the *Amistad* had been taken on the high seas, the Africans could not be charged with murder.

Soon the Spanish government entered the dispute. In a letter to President Martin Van Buren, Spain argued that American courts did not have rights over the case. The Africans should be returned by the United States to Cuba. Van Buren was about to do so when a second court in New Haven ruled that Cinque and his comrades were freemen and could not be enslaved. This decision was appealed. Now the case went before the Supreme Court.

Former president John Quincy Adams eloquently argued the case for the Africans. The Court, persuaded by his defense, set the Africans free on March 8, 1841. A year later, the Africans were returned home.

Cinque emerged a hero from this contest. Several portraits of him were made and circulated among antislavery advocates. In 1839, the *New York Sun* published an image of him entitled *Joseph Cinquez, the brave Congolese chief, who prefers death to Slavery.* A year later, John Sartain painted a remarkable portrait of Cinque, with noble good looks, cloaked in a tunic, holding a lance, and standing in front of an African scene. Now he was a free man in art, just as in life. Art continues to grace Cinque and his Amistad comrades. Writer Thulani Davis has recently announced that she has constructed an opera about the incident.

**Why has Ann Plato, obscure for many years, been resurrected recently?**

Black history books hardly mention Ann Plato. In fact, little is known of her life: she was born in Hartford, Connecticut, was a free Black, was a devoted Congregational church member, and was active in the Black community.

To these few facts should be added that Plato was an enterprising woman. In 1841, she rose out of obscurity by publishing—probably with her own funds—a small book entitled *Essays: Including Biographies . . . Prose and Poetry.* Shrewdly, she persuaded the famous Black abolitionist Rev. James W. C. Pennington to write the book's introduction; his name helped ensure that the book would be taken seriously. With this book, she became a pioneer of the writing tradition by Black women in America.

*Essays* was a little gem of a book consisting of sixteen pithy essays and several thoughtful poems. Today, Plato's book would be classified as "self-help."

Her essays were the book's highlight. With titles such as "Education," "Benevolence," "Decision of Character," and "Employment of Time," they told the reader how to develop a life of discipline, determination, and handwork. These topics echoed Benjamin Franklin's practical advice on living productively that he published in *Poor Richard's Almanac.* For example, in

"Eminence from Obscurity," she gave her readers examples of people who had lifted themselves out of poverty.

Not much of the book dealt with Black Americans. Only occasionally did she address political issues. One poem, "To the First of August," praised the emancipation of slaves in the British West Indies on August 1, 1838. A verse ran, "And when on Britain's isles remote,/We're then in freedom bounds,/and while we stand on British ground,/You're free,—you're free,—resounds." Her poem "The Natives of America" told the American Indian story from when they were "a happy race" to when "beggars you will become."

Plato knew what she was doing by creating this kind of book. For a few years, it sold well, mainly to Blacks, but it also had crossover appeal for whites. *Essays* sold because it offered sincere, trustworthy advice, which was greatly needed by ordinary people living in the Northeast as the region changed from a mostly rural society to one centered around cities and factories. Plato's sermonettes helped these people—both Black and white—to meet the new challenges.

Critic Blyden Jackson criticized in a *History of Afro-American Literature* (1989) Plato's essays, saying they are written with "all the wit, verve, and originality of a hopelessly spinsterish schoolmarm." Jackson underestimated the quality of her writing. No credit goes to Plato for being moderately popular. Had she worked in Philadelphia or New York, she undoubtedly would have had greater success.

The recent republishing of *Essays* is resurrecting Ann Plato as an early Black American woman writer and as the first Black American self-help specialist.

### Who were the abolitionists, and what did the abolitionist movement want?

Abolitionists advocated a complete end to American slavery. They wanted "every chain [to] be broken and every bondman set free." From the 1830s to the Civil War, America saw a new energy directed toward freeing the slaves.

Abolitionists were a new crowd of intense, strong-willed, moral idealists. They saw slavery as an evil that had to be eradicated in order for America to fulfill its promise as "liberty's kingdom." In pamphlets and speeches, they depicted slavery as a serpent: a cunning, cold-blooded, slimy reptile, with its body coiled around the country, strangling it. Like the legendary saints who killed dragons, they pictured themselves beheading the snake of slavery.

When abolitionists arrived on the national stage, they overshadowed the old antislavery movement that had been founded in the eighteenth century, mostly by Quakers. These were good, fervent, often effective people, but timidity robbed the old movement of its power. Some Quakers even owned slaves.

Nor had the old movement dislodged slavery in the South. Although the old movement was centered in the South, it had greater results in the North, where after the Revolutionary War, states began to free slaves. But the Northern emancipation of slaves moved slowly. New Jersey, which banned slavery in 1804, still had eighteen slaves in 1860, the year the Civil War began.

The abolitionists were a rare species. No nineteenth-century American movement attracted a better breed of intelligent, tireless, courageous, high-minded political strategists. In the 1830s and 1840s, they realized that slavery was not disappearing; rather, it was entrenched and just waiting to expand.

Not too long ago, as historian Benjamin Quarles points out, history books focused on the great white leaders—mostly men—of the movement, such as Elijah Lovejoy, William Lloyd Garrison, Rev. Theodore Parker, Charles Sumner, and John Greenleaf Whittier, to name a few. Quarles writes, ". . . truly these were notable figures." In 1837, Lovejoy, an abolitionist preacher and journalist, was shot to death in Illinois by a proslavery mob. Garrison was a fiery orator and the editor of Boston's *Liberator.* In his first statement in the *Liberator,* he wrote, "I am in earnest! I will not equivocate! . . . I will not retreat a single inch!" Parker used his pulpit to preach against slavery. Sumner championed the cause as a Massachusetts senator. Whittier was, in a sense, the poet laureate of the movement.

But many talented Blacks were abolitionist leaders as well.

Quarles wants more recognition for them. Some of them, such as William Wells Brown and Frederick Douglass, had been slaves. Wells became a writer and a famous antislavery lecturer in Britain. Douglass was "the Negro lion" of the movement, a remarkable writer, orator, and politician.

Many others came from the North's free Black communities. Wealthy businessman Robert Purvis gave money. The Luca family, a singing group, performed at rallies. Philadelphia's William Still, who helped Blacks fleeing slavery, lectured on his work. In Boston's West End, home to almost two thousand Blacks, many strongly supported the movement. Lewis and Harriet Hayden's home became a Boston stopping place for seventy-five escapees.

Women were everywhere in the abolitionist cause. Prominent Black women were especially engaged, including Frances Ellen Watkins, a Baltimore schoolteacher who became the lecturer for the Maine Anti-Slavery Society in 1854. That year, she published poems that became very popular among abolitionists–"The Slave Mother," "The Fugitive's Wife," and "The Slave Auction." Bostonian Sarah Parker Remond lectured in America and England. Less well known but still important were Boston hairdresser Christiana Bannister and Philadelphia educator Sarah Douglass.

Of course, Sojourner Truth's speeches and songs had revolutionary impact. And many white women joined their Black sisters, often coming to the movement from other reform causes. Suffrage advocate Elizabeth Buffum and women's rights supporter Lucretia Mott were examples.

But the abolitionist movement was not one great big happy family. Tensions arose among so many talented people. Black abolitionists and their white comrades had serious conflicts. Blacks often worried about the control that whites exerted over antislavery meetings and lectures, and they quarreled with William Lloyd Garrison's commitment to nonviolence. They did not want to rule out violence as a means to destroy slavery. Disagreements like these resulted in abolitionist clubs being organized along racial lines. Douglass said abolition "is emphatically our battle; no one else can fight it for us."

Abolitionism was a significant movement, but by itself it

could not undo slavery. In fact, the abolitionists confronted much opposition—even in their own backyard, the North—from people who were proslavery, or anti-Black.

However, the movement accomplished three things. It kept the moral ugliness of slavery before the public. The movement intensified the sectional division of the country—the slave South versus the free North—that laid the groundwork for the Civil War. Lastly, the movement gave the nation its first model of interracial and male-female cooperation on a wide scale. It served as an example for the civil rights movement more than a century later.

## Why is Frederick Douglass considered a great man?

Douglass was one of the most impressive Americans of the nineteenth century. Since his life covered most of the century, from 1818 to 1895, and since he was very active as a public figure from his twenties on, he had an immense influence on his times.

During his nearly eighty years, he rose from slavery to become a renowned orator, the toast of two continents; the author of a best-selling autobiography; a prominent abolitionist leader; a newspaper publisher; an influence on President Lincoln; and, after the Civil War, a United States official and representative to foreign countries. As he once wrote, he had "lived several lives in one."

Throughout his life, he advocated rights for Blacks, but he also embraced women's rights, Irish freedom, world peace, and school and prison reform. Because of his advocacy of so many charged issues, Douglass was often the center of controversy. Threats came his way. He was beaten by gangs. His public appearances brought out hecklers and mobs.

Douglass was born Frederick Augustus Washington Bailey on a plantation in Tuckahoe, Talbot County, Maryland, in February 1818. He had an older brother Perry and two older sisters, Sarah and Eliza. His mother was Harriet Bailey, a slave owned by Aaron Anthony. He thought his father was Anthony, but this made no difference in the treatment he received.

His life was hard. Although young, he worked in the fields. He was "so pinched with hunger" that he had to scramble with dogs for scraps. Sent to another plantation, he saw his mother only five times during his life. At their last meeting, he was seven. She brought him a ginger cake shaped like a valentine—a gift of great love. Until he was eight, he never had a bed to sleep in.

In 1826, he was sent to live with new masters, Hugh and Sophia Auld in Baltimore. Now he was a city slave, with new options.

Sophia Auld helped him to learn to read. After overhearing Hugh Auld tell his wife that reading spoiled slaves, Douglass worked extra hard at mastering the skill. The day he was able to read, he said, "Light . . . penetrated the moral dungeon where I had lain. . . . I wished myself . . . anything but a slave." At twelve, he saved fifty cents to buy a copy of the *Columbian Orator,* a book of speeches, and began practicing oratory.

Unfortunately, he was sent back to the South in 1833 to a plantation owned by Edward Covey, notorious for "breaking" slaves. After many severe beatings, one day Douglass stood up to Covey and fought back. It was a transforming moment, a reclaiming of his manhood. Covey never lashed him again.

Returned to Baltimore, he planned an escape to New York, and in 1838, he did escape, by train and boat, with his future bride, Anna Murray, a free Black. Next he moved to Bedford, Massachusetts, where he made his first contacts with abolitionists. In 1841, he made his first great public speech at a rally. That same year, he chose a new name, Douglass.

Frederick Douglass was an uncommon sight on the antislavery lecture circuit: only twenty-four, he was broad-shouldered, with a head of bushy hair and a rugged face; he was articulate, possessing a deep voice like "a trumpet in the field." The fact that he was a fugitive slave added to his appeal. Douglass later surmised, "Many came from curiosity to hear what a Negro could say in his own cause." Many Black and white abolitionist leaders had not lived under slavery. Douglass could speak of slavery's evil from his own experience.

During the 1840s, Douglass's fame increased dramatically. His 1845 autobiography, *The Narrative of the Life of Frederick*

*Douglass: An American Slave,* sold forty-five thousand copies. The following year, he toured the British Isles, drawing great crowds. The *North Star,* his newspaper, was started in 1847, bearing the motto "Right is of no Sex—Truth is of no Color—God is the Father of us all, and we are all Brethren." In the late 1840s, he gave hundreds of speeches. All this made him a celebrity.

As the national debate over slavery heated up, Douglass became uncompromising in his views. He opposed new federal laws that returned fugitive slaves. He gave refuge to escaping slaves. His 1852 speech, "What to a Slave Is the Fourth of July?," was an outspoken attack on American hypocrisy.

He wanted Abraham Lincoln elected in 1860. But once Lincoln was president, Douglass attacked his compromises with the South. Douglass welcomed the coming of the Civil War. He was thrilled by the Emancipation Proclamation, which freed the slaves in 1863, and he worked hard recruiting Black soldiers for the Union Army.

When the war was over, Douglass's life entered a new phase. He fought for full voting rights for the newly freed Blacks and for racial segregation in schools and other public places. In 1877, he was appointed U.S. marshal for the District of Columbia, and he used his post to find federal jobs for Blacks. Later, he was a representative to Haiti.

In 1895, Douglass died in Washington of a heart attack, after addressing the National Council of Women. Thousands viewed his body at the Metropolitan African Methodist Episcopal Church.

### Why did narratives of slaves' lives become so popular?

Stories of slaves' lives were very much in demand in pre–Civil War America. In an age passionately engaged in the issue of slavery, slave narratives were a unique form of American autobiography, offering the voices of individuals who had escaped from slavery's evil. As the spirituals said, these people had "escaped from the lion's den" and from "a fiery furnace."

Many slave narratives were unwritten accounts told in the North's meeting places—in town halls and Black churches, at rallies, and in parlors. Antislavery groups sponsored these meetings. A man would tell how his wife had been sold; beatings were described; ruthless overseers were portrayed; a woman would tell of her child's death. Ending their stories, slaves told of their conversions to Christianity, saying triumphantly, "Lord, I don't feel no-ways tired."

Crowds were moved by this simple, affecting drama. Tears flowed at these meetings. For one speech, "the audience continually increased each successive evening" that it was given. The antislavery troops were mobilized. John A. Collins, an officer in the American Anti-Slavery Society, understood this impact: "the public have itching ears to hear a colored man speak, and particularly a slave."

Just as many people read these narratives in books. Even weekly newspapers ran slave stories. The decade of the 1840s was the great era of the American slave narrative. In 1845, Frederick Douglass's *Narrative of the Life of an American Slave* appeared. It was extremely well written, filled with damning episodes, and won over its readers. Other talented Blacks wrote their stories. William Wells Brown's sold eight thousand copies in eighteen months. The *Narrative of the Life and Adventures of Henry Bibb* was published with horrifying drawings of harsh plantation life. Milton Clark's famous narrative was accompanied by questions and answers on slave life. Women contributed about 12 percent of the slave narratives.

For the most part, slave narratives helped the antislavery cause. They told the world about the experience of being human property; they were living evidence against the slaveholders' claim to the right to continue slavery.

### Why was the Fugitive Slave Law of 1850 considered by Frederick Douglass to be "horrible and revolting"?

In a Rochester, New York, speech on July 5, 1852, an emotional Frederick Douglass denounced the Fugitive Slave Law as "horrible and revolting" and as the "most foul and fiendish of all human decrees."

Abolitionists of all stripes agreed with Douglass. The law represented a real setback for the antislavery cause and a big victory for the South. Blacks trying to escape slavery would now face even greater hurdles.

The law enacted by Congress applied to all states and to Washington, D.C. A new class of federal officials called "commissioners" was created by the law. Their function? They had the power to capture, arrest, and return runaway slaves to their masters. Commissioners would be paid ten dollars for each runaway they returned. Anyone assisting slaves to escape—as the Underground Railroad workers did—could be penalized heavily. Persons trying to stop an arrest were also subject to fines.

The Fugitive Slave Law was a reaction to the North's welcoming attitude toward escaped slaves. States in the North were passing laws protecting the liberties of runaway Blacks. Some state legislatures told their officials not to capture runaways. As the antislavery movement and the Underground Railroad spread, the North looked like the Promised Land to the South's Blacks. This worried the South. The North had to be made less inviting.

Slaves escaping from the South were not the only thing on Congress's mind in 1850. In fact, the Fugitive Slave Law was just one part of a bigger legislative deal, called the Compromise of 1850. The need for the compromise was created by the United States' victory over Mexico in 1848. New territories—California, Utah, and New Mexico—were brought into the Union. With their entrance, a volatile question arose: would they be free or slave states? Debating this question nearly undid the Union. The Senate's John C. Calhoun of South Carolina, Henry Clay of Kentucky, and Daniel Webster of Massachusetts came to the rescue with a new plan.

On the surface, the North did well under the compromise. California would be a free state. Settlers in Utah and New Mexico would vote on slavery. Any new states formed would do the same. Slave trading in the nation's capital would cease, but slaveholders could keep their slaves.

However, the antislavery fighters saw the compromise differently. Its inclusion of the Fugitive Slave Law was a capitulation to the South. The law signified greater acceptance of the South's immoral way of life, rooted in slavery.

## How did Blacks react to Harriet Beecher Stowe's *Uncle Tom's Cabin* in 1852?

Harriet Beecher Stowe's intention was to pull back the cotton veil that hid slavery from ordinary Americans' view. To do this, her book was first published in forty weekly installments in the popular *National Era.* She wanted to awaken the public's "sympathy and feeling for the African race, as they exist among us . . . under a system so . . . cruel and unjust."

The most important character in the novel was Tom, a slave who has suffered greatly. One scene shows Tom beaten within an inch of his life. But Tom is a "moral miracle" because he does not hate his oppressors. He is friendly, hardworking, loyal, and a Christian who is passive and has "the facility of forgiveness." For Stowe, Tom typified the entire Black race with "their gentleness, their lowly docility of heart . . . and their childlike simplicity of affection."

When *Uncle Tom's Cabin* appeared as a book, it sold three hundred thousand copies in the first year. This runaway bestseller became a major weapon in the campaign against slavery.

Both whites and Blacks read *Uncle Tom's Cabin.* Of course, most of its Black readers were living free in the North. Twentieth-century Black intellectuals—the most famous being James Baldwin in 1949—have criticized Stowe, especially for Tom's portrayal. But what did Blacks of her own time think?

Frederick Douglass, ex-slave and leading abolitionist, called the novel "the master book of the nineteenth century." Repeatedly, he defended her against the charge of inaccuracy. He declared his "reverence for her genius." Douglass went farther, saying how indebted Blacks were to her novel. "She who had walked, with lighted candle, through the darkest and most obscure corners of the slave's soul, and had unfolded the secrets of the slave's lacerated heart, could not be a stranger to us."

J. Sella Martin, another former slave, identified with Stowe's rendition of slaves being auctioned. He said, "Mrs. Stowe ... [has] thrown sufficient light upon that horrible agency of slavery."

William Wells Brown, famous escapee and a novelist himself, recognized the importance of Stowe's book to the antislavery cause: "*Uncle Tom's Cabin* has come down upon the dark abodes of slavery like morning's sunlight ... awakening sympathy in hearts that never before felt for the slave."

Not all Blacks sang Stowe's praises, however. Henry C. Wright, a Black abolitionist, did not like Tom's humble behavior. Tom's Christianity should have "beget self-respect," causing him to take his owner's "money ... horses ... clothes ... to aid him to free himself." Other Blacks were even more critical. Martin Delany, an early Black nationalist, also was disturbed by Tom's passive character. Some slaves, had they suffered like Tom, "would have buried the hoe deep in the master's skull."

Other Blacks took Stowe to task for suggesting that Black colonization outside the United States was the solution to slavery. One of the novel's characters, George Harris, was sent to Liberia (in West Africa) to settle. In 1853, Blacks at the American Anti-Slavery Society meeting in New York proposed a condemnation of the novel's colonization idea.

In sum, the Black community was divided over *Uncle Tom's Cabin.* Realizing that the novel helped the antislavery cause, leading Blacks supported it. But criticisms were voiced as well.

## Harriet Tubman was a famous "conductor" on the Underground Railroad. What does that mean?

Harriet Tubman was a leader—in fact, the most important woman leader—in the Underground Railroad.

The Underground Railroad was not an actual railway. This was simply a term borrowed from the era's swift and powerful new means of transportation. In reality, the Underground Railroad was a secret system that took slaves out of the South and into the North to freedom.

A whole railroad language was used to talk about the escapes.

People who went south for the slaves, who hid and fed them in the North, were called "conductors." The runaway slaves were "passengers." They traveled at night. During the day, runaways were hidden in city homes, churches, on farms, and sometimes in caves, forest nooks, or remote territory. The hiding places were the "stations" of the railway. There they were fed hearty food, their wounds were washed and bandaged, their hopes of freedom were sustained.

Harriet Tubman was the most celebrated Black "conductor." She was born to Harriet and Benjamin Ross around 1820 on a plantation in Dorchester County, Maryland. As a child, she was struck on the head by a two-pound iron bar hurled by an overseer. Seizures and spells of unconsciousness plagued her after the attack. Still, she was made to go to the fields, to lift and to plow.

On a pitch-black night in 1849, Harriet Tubman and her two brothers fled the plantation because they heard they were headed for the auction block. She left behind her husband, John Tubman, who refused to go with her. In a little while, her brothers gave up and returned to their slave cabins. But Harriet, who had been injured as a child, suffered seizures, and was thought a half-wit, continued and made it to the North.

Flight and freedom were so exhilarating for her that she repeatedly went into the South to bring out slaves. In twelve years, she made nineteen trips. Whole families were brought out by her. She toted a pistol to "encourage" the weary and disheartened who wanted to turn back and who might betray her whereabouts.

They said she would threaten that "dead niggers tell no tales; you go on or die." She knew how to muffle a frightened child's crying. She employed clever disguises, dressing as a man, an elderly granny, a beggar. In some states, she knew the terrain and its secret hiding places better than the locals. And she was swift: on a good night, she could cover miles and miles, sweeping her charges along with her.

Harriet Tubman rescued more than three hundred people from slavery. When the Fugitive Slave Law of 1850 was passed, committing the federal government to helping return slaves, she began to take her groups all the way to Canada. As she recalled later, "I wouldn't trust Uncle Sam with my people no longer, but I brought them off to Canada."

Southerners hated her. They proved it by putting up large sums for her capture—first, two thousand dollars, then five thousand, and, at the height of her success, forty thousand dollars. For many Blacks, she assumed the legendary role of "Moses," leading her people out of bondage. She had followed the spiritual's urging: "Go down, Moses, way down into Egypt land."

When the Civil War came, she was the perfect person to be a spy for the Union Army behind enemy lines. For three years, she spied in South Carolina. With the war's end, she settled in Auburn, New York, and assisted the needy newly freed Blacks. When the National Association of Colored Women was established in 1896, she was one of its founders. She lived until 1913.

The Underground Railroad harassed and embarrassed the South. It announced to the world that many slaves felt no loyalty to slavery. Many courageous individuals contributed to its success: Levi Coffin, a Quaker, who brought nearly three thousand slaves to freedom; John Fairfield, who went to prison for his work; and William Still, a Philadelphia Black who sheltered an army of escapees. But Harriet Tubman was one of a very few slave conductors.

An old photograph of Harriet Tubman shows her standing near six slaves she has led to freedom. Wearing a long high-collared dress and a little straw hat, her gestures—her satisfied look and folded hands—seem to be saying: "Here beside me is my life's work, my treasure—rescued humans." In 1994, Congress designated March 13 "Harriet Tubman Day."

**Sojourner Truth blazed an amazing path across nineteenth-century America. What was her story?**

Born into slavery in upstate New York in 1797, Sojourner Truth was owned by six masters before she was thirty. Along with other New York slaves, she was freed in 1827.

Sixteen years later, in 1843, she was an itinerant Christian evangelist, traveling continually to religious revivals throughout the Northeast. Thus she began a great career as a public personage, as great as any in America at the time. In 1850, she published the *Narrative of Sojourner Truth,* her autobiography dictated to abolitionist Olive Gilbert. Before long, she was sharing antislavery platforms with the likes of Frederick Douglass and William Lloyd Garrison.

Six feet tall, very black, and possessing a deep voice, Sojourner Truth was a unique performing presence. In all her meetings, she presented herself as a person of stature. Once she was described as "a tall gaunt Black woman in a gray dress and white turban" who "walked with the air of a queen up the aisle."

In 1852, she electrified the Second National Women's Suffrage Convention in Akron, Ohio, with her famous "Ain't I a Woman?" speech. Today, this speech ranks as one of the nineteenth century's most remarkable public addresses.

When the Civil War came, she fearlessly visited many Union Army camps. By 1864, she had sufficient stature to get a White House audience with Abraham Lincoln. At the war's end, she adopted a variety of causes: voting rights for women and Blacks; lands in the West for freed slaves; campaigns against alcohol and tobacco. She died at her Michigan home in 1883.

To understand Sojourner Truth, it is necessary to look beyond her life's landmarks to how she created herself anew out of the past.

First, her name was an example of this self-creation. Her original slave name was Isabella Baumfree; since she had been sold many times, her surname was always changing. Once free, to shed her slave history, she dropped Isabella and, after a divine revelation, assumed the powerful "Sojourner." Later she explained, "When I left the house of bondage, I left everything behind. I wa'n't going to keep nothing of Egypt [slavery] on me. . . . And the Lord gave me Sojourner, because I was to travel up an' down the land." Her last name also was given to her by God. She was "a person who has got the 'Truth' always present."

To make a point, she used basic, everyday experiences in arguments. She was a great communicator for the abolitionist cause, reaching the working classes, the unskilled, and the poor—groups not always embraced by the movement's leaders.

She was right in saying, "I can't read a book, but I can read de people."

Religion played a major role for her. As a child, her mother told her, "There is a God, who hears and sees you. . . . When you are beaten, or cruelly treated, or fall into trouble, you must ask help of him." Throughout her life, this idea of a personal God—a God to talk to, who would give directions and could be summoned—stayed with her. Repeatedly, she said God told her what to do.

Sojourner Truth kept her eyes on one vision of the future: "God still lives and means to see the Black people in full possession of all their rights."

**Why did the Anthony Burns case keep Boston in turmoil for ten days?**

When Boston printer R. H. Edwards issued a poster honoring Anthony Burns, he knew it would sell. To many Bostonians, Anthony Burns was a hero, and Edwards's poster presented him as an angel.

In a circle were scenes from his life as a slave, his escape on a ship, his work as a freeman in Boston, his humiliating arrest, and his departure from Boston in handcuffs. In the center of these scenes was Burns's handsome portrait, showing him clear-eyed, with charcoal skin, an easy smile, and dressed neatly in a velvet-collared coat, a vest, white shirt, and bow tie. The question suggested by the poster was clear: Why should this wonderful man be a slave?

Anthony Burns was born into Virginia slavery in 1834. His owner was Colonel Charles Suttle, who decided to hire out Burns to a mill owner in Richmond. In places like Richmond, masters customarily sent their slaves to work for others.

One day in February 1854, Burns, seeing an opportunity for escape, hid himself on a ship that took him to Boston, where he got a job. No one in Virginia knew of his whereabouts until Burns got in contact with his mother. Suttle knew then where Burns was and pushed Boston's federal officials to seize him. By

June, Burns had been arrested and held at the federal courthouse.

Under the provisions of the Fugitive Slave Law of 1850, Burns faced the prospect of being sent back to Virginia. The return of blameless Anthony Burns to slavery's clutches was too much for Boston to bear. Protesting crowds gathered daily at Fanueil Hall. At one point, Blacks and whites stormed the courthouse to rescue him, breaking down a door and killing a deputy marshal in the attempt.

To quell the mob, troops had to be called in. Rev. Leonard Grimes, a popular Black Baptist minister, raised over a thousand dollars to repay Colonel Suttle for his lost property. But nothing worked. Federal troops escorted Burns to a ship that returned him to Virginia.

Boston's turmoil reflected the explosiveness of the slavery issue in the 1850s. Many Northerners hated the South's intervention in its affairs. Abolitionists were willing to use violence to help runaways like Burns. Kidnapping is how they saw the enforcement of the Fugitive Slave Law. After the Burns case, federal officials had to think twice about seizing fugitives.

Rev. Grimes was finally able to buy Burns's freedom in 1855.

### Why did a Black man's legal suit, known as the Dred Scott case, become one of the most famous in American history?

Little is known of Dred and Harriet Scott, who filed petitions for their freedom from slavery in Saint Louis, Missouri, in 1846.

Dred Scott was about five feet tall, used a "mark" for his signature, and had no history of heroic action. Around 1795, Scott was born into slavery in Virginia. While in his thirties, he was brought by his master, Peter Blow, to Missouri. Even less is known about Harriet. She was also a slave when she married Dred Scott. Two of their four children survived. And little is known of just how the Scotts got started on their legal path. But once they started, they were tenacious.

What set up the case was a simple act by Scott's master, Dr. John Emerson, a U.S. Army surgeon. He took Scott from a slave state, Missouri, into a free state, Illinois. For two years, Scott worked as a personal servant. Then he was taken to Wisconsin Territory, which was also free. Finally, he was taken back to Missouri.

When Emerson died in 1846, Scott—with the help of white lawyers—sued Emerson's widow for his own freedom and that of his wife and two children.

According to Scott, this travel had freed him. His suit argued that "once free, always free." He had lived in a state where slaves were not held. Furthermore, he had resided in federal territory where slavery was forbidden by the Missouri Compromise of 1820.

The case was in the courts for eleven years. Scott won in a lower court in Saint Louis. That decision was reversed by the Missouri Supreme Court. Then a federal district court upheld the previous decision. Finally it reached the U.S. Supreme Court in 1857.

Chief Justice Roger B. Taney ruled that Dred Scott was to remain a slave. He declared unconstitutional the Missouri Compromise, saying that Congress had no right to prohibit slavery. Thus, he abolished the idea of free territories entirely. And if that were not enough, he declared that Blacks could not be citizens and therefore had no standing to sue in court. Dred Scott was a nonentity. Since the country's birth, Taney said, Blacks had "had no rights which the white man was bound to respect."

A firestorm of debate followed the issuance of this opinion. For Blacks, especially free Blacks, Taney's words were devastating. Inferiority and subjugation were to be their station in life. They were beyond the pale of civilization, a "degraded" and "unfortunate race." Nor could emancipated Blacks be elevated to citizenship.

Scott was freed two months later and continued to work until his death in September 1858. Harriet lived a few years after that. Several of his descendants attended a centennial ceremony in 1957.

The Dred Scott decision is regarded as a landmark in the nation's plunge into war. And it all started with a Black family's quiet insistence on using American law to assert their freedom.

**Why did the discovery of the novel *Our Nig* of 1859 become a major 1980s Black intellectual event?**

New York City's University Place Bookshop is a shrine to the Black studies field. As with all shrines, many people feel they must occasionally visit it to pay their respects to its massive book collection on Blacks and to its knowledgeable owner, Bill French.

Henry Louis Gates Jr., then a Yale professor of Black literature, might have been making such a pilgrimage on the day Bill French handed him a copy of H. E. Wilson's *Our Nig.* "Well, Mr. Gates, what do you make of this?" asked French. Gates inspected it and paid fifty dollars for the book.

Gates got the novel reissued in 1983. But that was the culmination of long and complicated literary detective work. Gates and a research team had to put some historical flesh on the author. Who was H. E. Wilson? Gates was worried because "nig" (short for "nigger") would hardly be a title chosen by a Black author. So, was this novel by a Black or a white?

Since the book originally was published in Boston, the researchers checked old Boston directories and censuses and found that a Harriet E. Wilson had lived in the city's Black district. Eureka! Later, they found that she was born in Fredericksburg, Virginia, in 1808, had moved to Philadelphia and then to Boston, dying there in 1870.

Gates made a lot of *Our Nig,* proving it to be the first novel by a Black person published in the United States. (A few by men had been published abroad.) It was also the first novel published by a Black woman. The find added a new dimension to Black writing. Wilson had pioneered a new generation of Black writers, who moved from poetry and autobiography to the novel.

*Our Nig* was not the story of Blacks on Southern plantations or the autobiography of a runaway. Instead, it boldly looked at the oppression experienced by a young Black woman in a white home in the heart of American abolitionism—New England, of all places.

"Lonely Mag Smith! See her as she walks with downcast eyes and heavy heart." Thus opens *Our Nig.* Mag, a lower-class white woman exploited by men, has fallen on hard times. But her luck changes when she meets Jim, a "kind-hearted African." By him, she has a pretty, vivacious girl, Frado. Then bad times return, and eventually she must leave Frado in the care of a white family, the Bellmonts.

Frado's life with the Bellmonts is the center of the novel. She is young and friendly but alone, and Mrs. Bellmont and her daughter Mary abuse her. Frado is turned into a servant, worked to the bone, fed the food of a prisoner, and treated like a toy by Mary. Her "room" is a cramped attic sleeping space. The Bellmonts even send Frado to work on their farm.

Frado manages to get free of the Bellmonts. By the novel's end, she has had a husband, but he is dead; and she has a child. She wants to write her story.

The great mystery of *Our Nig* is why it did not receive any attention in 1859. In all likelihood, it was because the book exposed the North's racial hypocrisy at a time when Northerners saw themselves as morally superior to Southerners. As the novel's title page said, Wilson showed "that slavery shadows fall even here." The North was not ready for this revelation.

Thousands have read *Our Nig* since its reissue, and it is now considered a masterpiece of nineteenth-century Black literature.

## Why is Harriet Jacobs's slave narrative transforming our view of slavery?

Since the vast majority of slave narratives were written or told by men, men have been thought to be the true representatives of slavery. It's as though the struggle against slavery was a fight for Black manhood and not Black womanhood.

Harriet Jacob's *Incidents in the Life of a Slave Girl,* published in 1861, helps set the record straight. The book was published more than a decade after the great period of slave narrative production, and it received only moderate public response. Then it fell into obscurity. Before long, her book was even considered a fake, written by an abolitionist, or perhaps it was simply fiction. Since Jacobs included some features familiar to novels, like created dialogue, her narrative lent itself to this charge. Silence enveloped her life and work.

In the 1980s, after six years of sleuthing, Jean Fagan Yellin, a professor of literature at Pace University, proved that Harriet Jacobs was a real person and that her narrative was based on her experiences as a slave. Jacobs, as it turned out, was born in 1818 and began her life in slavery near Edenton, North Carolina.

Jacobs's long narrative of forty-one chapters tells us what it meant to be a woman and a slave. The main difference in her story from those of the men lies in its emphasis on sexual exploitation. From the time she was a young woman, she was pursued by her master, who was determined to possess her as his sexual property.

Jacobs had two children by another man, hoping that this would deter her master. Her strategy failed. In desperation, she fled and hid in a secret place in her grandmother's shed. For seven years, she watched the world through a tiny peephole. Eventually, she escaped to Philadelphia, leaving her children and family behind.

There is another difference between Jacobs's story and those of the men. She describes her warm ties with kinfolk and friends—women and men. To be a good mother, to stay close to her grandmother, to find a true home for her family—these were all crucial to her. Men's stories presented more isolated figures. Frederick Douglass's narrative, for example, hardly mentions family or women, only occasionally referring to his wife Anna Murray Douglass.

Jacobs's title page carried a verse from Isaiah: "Rise up, ye women that are at ease! Hear my voice, ye careless daughters! Give ear unto my speech." Through Yellin's labor, it is now possible to hear Jacobs's voice reminding us that millions of Black women struggled in slavery to establish themselves as humans.

## Why is John Brown—a white man—considered to be one of the most important figures in Black American history?

In February 1989, *Ebony* magazine asked eighteen experts—mostly historians and mostly Black—to choose the fifty most prominent figures in Black American history.

All fifty final choices were Black. But John Brown got a number of votes. It was gratifying to have John Brown's role in shaping Black history acknowledged, because so often he has been branded a zealot, a traitor, and a cold-blooded murderer.

The historical truth is different. As much as any American, John Brown was a major crusader against slavery. His raid on the United States arsenal at Harpers Ferry, Virginia, in October 1859 was "a final thunderbolt" against the institution. Although the raid failed, it told the nation that men—white and Black men joined together—were prepared to die to end slavery.

Born in 1800, Brown grew up in northeast Ohio, where antislavery sentiment was strong. His home was a stop on the Underground Railroad. Meeting escaped slaves, Brown was deeply disturbed by their stories of persecution and injustice. In 1839, he came to a turning point and decided that he must fight slavery. It was a moral duty. His prayers implored God to bless his enterprise.

Some time passed before Brown's first major battle against slavery took place, in Kansas in 1858. Kansas was supposed to be a free state where slavery was not allowed. But in the 1850s, after many maneuvers and much violence, it looked like the state would become a haven for bondage. Brown and his sons went to Kansas to save it as a free state.

One dark night, they seized five ruffians who had attacked antislavery settlers, and killed them. Brown's attack ignited the territory. A local war followed. Men came to the fore to fight against the coming of slavery. In the end, Kansas remained free.

Brown had always cherished the dream of striking a blow against slavery in the South. By the spring of 1859, he had a plan for taking the federal munitions storehouse at Harpers Ferry (now in West Virginia). The scheme was first to take the arsenal. A band of twenty-two men—including six or seven Blacks and three of Brown's sons—was to spearhead the attack. Once word got out that Brown controlled the arsenal, it was expected that slaves in the surrounding area would rise up and join him. On October 16, 1859, Brown and his men made their move.

Only part of the plan worked. The arsenal was easy to capture, but the conductor of a passing train discovered what was going on. Brown mistakenly let the conductor go on to Baltimore, where he told of the attack. Militia, marines, and townspeople soon surrounded Brown. Ten of his men were killed and several captured. Brown was seriously wounded, captured, swiftly tried for treason, and hanged on December 2, 1849. Slaves did not join the attack.

In the *Ebony* poll, the experts voted for Brown because of his raid's legacy. What was it? Most important, Brown showed the country that a white man would sacrifice his life for the freedom of Blacks—not a common idea at the time. He understood that slavery would not disappear except through violence. In fact, John Brown's raid at Harpers Ferry might be considered the first battle of the Civil War.

## How did Blacks respond to the coming of the Civil War?

To protest the election of Abraham Lincoln as president, late in 1860 South Carolina and six other states seceded from the Union. In February of 1861, Jefferson Davis was inaugurated as president of the Confederacy in Montgomery, Alabama, the first Confederate capital. On April 12, the Confederate military attacked Fort Sumter, a federal fort in Charleston's harbor, and soon overwhelmed the federal forces there. Lincoln called out federal troops against the Confederacy, which soon numbered eleven states. The war was on.

Abraham Lincoln's initial war goal was to bring the South back into the Union, not to free the slaves. But Blacks knew that their future was a major question of the Civil War. The conflicts leading to the war had been clearly related to the issue of slavery. Four and a half million Blacks had one pressing question: will these events mean our freedom?

The turbulent events leading up to the war sent tremors through Black society. Blacks—free and slave, North and South—were profoundly affected by the crisis. The early mobilization of men and resources directly affected Blacks living in the South. Plantation slaves could see in their daily lives the war's immediate impact.

Northern free Blacks knew a great deal about the war's coming and the forces behind it. Frederick Douglass and other Black abolitionists welcomed it. Douglass wrote in May 1861, "For this consummation, we have watched and wished with fear and trembling. God be praised! that it has come at last." Northern Black communities wanted Black companies to join the fight. They sheltered numerous runaways who arrived as the war heated up. In Boston, places like Lewis Hayden's home and John J. Smith's barbershop received numerous fleeing Blacks.

Plantation Blacks did not have direct war news. To gather information, they had to keep their eyes and ears open. Blacks working on the South's roads and docks were the first to hear news in the war's opening days. They saw the troop movements and heard battle stories. Blacks collected news while working in prominent whites' homes, posh hotels and clubs, and at prestigious sporting events such as horse racing. These stories were passed along the slave grapevine to other Blacks.

Southern whites tried to keep Blacks ignorant or misinformed and painted the Union cause—especially the Yankee troops—as demonic. A slave woman recalled, "I wuz afraid of the Yankees, because Missus had told us the Yankees were going to kill every nigger in the South."

As the Southern war effort intensified, plantation slaves saw white society under stress for the first time. Basic goods such as flour, bacon, salt, beef, and tobacco were often scarce. Masters went off to war, leaving their wives to run the plantations. Frequently, they could not adequately feed themselves or their slaves. Amelia Montgomery wrote her husband at the front, "I have so little molasses, we can't give any more. . . . I do hate to have them [the slaves] begging me for meat." On some plantations, clothes and shoes were hard to come by. A slave child remembered that one winter, "we had no shoes and made tracks of blood in the snow."

Plantation Blacks often took advantage of this wartime situation. They satirized their masters in song:

*Has anybody seen my massa,*
*Wid de mustache on he face,*

*Go 'long de road some time dis mornin',*
*Like he gwine to leabe de place?*

With the master gone, slaves often stole. "It [the storeroom] has been robbed of nearly all groceries," wrote Kate Stone. Slaves also slowed down their work or refused outright to work at all.

As the weeks stretched into months, Blacks became more direct participants in the upheaval. Plantation slaves escaped to Union camps as the army penetrated the South. They were camp followers, sometimes traders, workers, friends of troops. Blacks were not yet allowed to fight in the Union ranks. Slowly, Lincoln and the Congress realized that Blacks could not be overlooked. As Frederick Douglass had predicted, the "war . . . would not be fought out entirely by white men." In 1863, when the Emancipation Proclamation prompted the admission of Blacks into the Union Army and Navy, his prediction came true.

## What was "the most daring and heroic adventure of the war" that Robert Smalls undertook in 1862?

When the news broke, the story seemed impossible.

Robert Smalls, a Black slave, had stolen a major Confederate gunboat and sailed it out of Charleston harbor and into the hands of the Union Navy. The boat, named the *Planter*, was a steamer, 140 feet in length, and had been converted from a boat originally used for cotton transport. Fitted with six large guns, it was perfect for patrolling the shallow waters around Charleston, providing defense against Union vessels. It was a valuable boat to steal.

Smalls had worked around ships for some time. When the war came, he was pressed into serving the Confederate naval installation at Charleston. While working on the *Planter*, he and other slaves hatched a scheme, as *Harper's Weekly* reported, "to place themselves under the Stars and Stripes rather than the Stars and Bars."

On a Monday evening, when the *Planter*'s white officers went ashore to spend the night, Smalls and eight other men, along with members of their families, gathered to execute the plan. In the early morning, they lit the fires under the boiler and steamed quietly out of the harbor. To confuse any onlookers, Smalls dressed in the captain's coat and hat and walked the ship's deck imitating his gait. When the *Planter* passed Fort Sumter, a Confederate bastion, Smalls gave the usual signal—two long pulls and a jerk of the whistle cord. The *Planter* was waved on.

When the ship was beyond the fort and headed toward Union naval lines, Smalls lowered the Confederate flag and quickly raised a white one. He did so just in time—the Union ships were preparing to fire. Instead, the boat was received cordially by the Union Navy.

Smalls continued to be the *Planter*'s pilot along the South Carolina coastline, and in 1864 he was made captain of the boat at a handsome salary. But Smalls brought more than a boat to the Union side. He defected with an unparalleled knowledge of the intricate coastline waterways and the hidden Confederate gun installations. He turned this information over to the navy.

After the war, he entered politics, serving first in the South Carolina legislature and, after 1875, spending five terms in the United States Congress. Probably his most famous remark, though, was made upon arriving on the Union side with the *Planter:* "I thought that the *Planter* might be of some use to Uncle Abe."

## What was the significance of the Emancipation Proclamation for Blacks?

President Abraham Lincoln had been pondering freeing the slaves for some time. But he never got farther than weighing the options. For many at the time, his pondering was similar to that of Shakespeare's *Hamlet*—serious but seemingly endless.

On July 22, 1862, he first floated among his cabinet the idea of freeing the slaves. The Union's military situation was weak, so the cabinet members advised him to wait. When he finally announced publicly his proposal on September 22, Union forces had just won the narrowest of victories at Antietam (in Maryland), one of the war's bloodiest battles. It was enough of a victory, though, for Lincoln to use it as the backdrop for his first version of the Emancipation Proclamation.

His proclamation said that slaves in the Confederate states would be "thenceforward, and forever free." But this did not take effect immediately. The slaves would be free if the rebel states continued in their rebellion. Lincoln gave the rebel states one hundred days—until New Year's Day, 1863—to decide whether to continue. In the meantime, Union officials would not arrest fleeing slaves.

One hundred days must have seemed like an eternity. Expectations were exceedingly high. Word of possible freedom reached the plantations quickly. The *New York Times* noted, "There is a far more rapid and secret diffusing of intelligence and news throughout the plantations than was ever dreamed of."

As New Year's Eve approached, the North's churches were packed with people waiting for Lincoln to sign his proclamation. Whites and Blacks talked, sang, prayed, and preached throughout the night. These assemblies were called "freedom watches." Would he sign? He had vacillated before.

Lincoln did sign. Telegraph wires hummed with the news. The reaction—especially among Blacks—was explosive. Marches wound through the cold streets of Philadelphia, Boston, and New York. In Washington, D.C., groups of former slaves who had escaped during the war gathered to pray and sing. In Norfolk, Virginia—a city already in Union hands—thousands of Blacks paraded behind a spirited band of drums and fifes. In her diary, Charlotte Forten wrote that Proclamation Day was "the most glorious day this nation has yet seen."

These days, historians argue that Lincoln was nudged toward issuing the proclamation by Black slaves who had left their masters, come to Union-held areas, and served in the Union camps as workers, spies, and guides. Ira Berlin leads in this opinion. He says these former slaves living and working in Union territory sent a message all the way up the line to Lincoln. The message was that freedom was a precious goal for the enslaved. Gradually, this influenced Lincoln, shifting his opinion to favor the slaves' idealism. Berlin's insight is on target.

No one was actually freed in the rebel states by Lincoln's proclamation. War had to do that. And slavery in the border states that stayed in the Union—Delaware, Kentucky, Maryland, Missouri—remained legal. This was a great disappointment.

So after the hoopla was over, what did the Emancipation Proclamation accomplish for Blacks? Even two years into the war, Lincoln had never said that slavery would not have a place in his future America. The proclamation, for the first time, committed the president and his government to the long-term eradication of slavery.

Moreover, it raised the future expectations among the South's slaves, making them a real problem for the Confederacy. Slaves began to show signs of "disorder and unfaithfulness" toward their masters. Whenever the Union Army came near the plantations, slaves joined their cause. Letters to newspapers complained, "The Negro has become a nuisance," "slaves sleep till noon," and "the Negroes care no more for me."

The proclamation did something else that was revolutionary. Lincoln allowed Blacks into the Union's military forces: "I further declare that such persons of suitable conditions, will be received into the armed service of the United States." Blacks had been yearning for this opportunity, a chance to fight slavery. Days after the proclamation, Blacks began to enlist. Four months later, they were in the field wearing the Union blue.

Reading the Emancipation Proclamation today, one is struck by how dry it is. It is brief and matter-of-fact, with no rhetorical flourishes. How could such an uplifting event—the freeing of four million souls—be described in such cold, legalistic language? Only in the last paragraph are the words "act of justice" used. Lincoln probably wanted to make the proclamation a part of the military campaign—"a military necessity," as it stated. With heavy rhetoric, Lincoln would have stirred up Northern whites who hated the freeing of the slaves. Wisely, he kept a low profile for his monumental act.

In years to come, slave folklore would put Lincoln on a plantation when he signed the document.

### Was Abraham Lincoln a "white supremacist"?

When Lincoln went to Richmond, Virginia, in April 1865, after his victorious army had occupied it, he was overwhelmed on the streets by newly freed Blacks. A Black army chaplain described the scene: "All the colored people in the world had collected." Lincoln was at the height of his popularity among Blacks. To them, he was "the Great Emancipator." Yells and whistles rang out. "Well, I suppose that God sent Abe Lincoln to 'liver us," a Black said. Lincoln was Father Abraham.

In the early 1960s, Black critics argued that Lincoln was unworthy of this show of Black homage. Lerone Bennett, a senior editor of *Ebony* and a historian, launched the first major salvo against Lincoln. "On every issue relating to the Black man . . . he was the very essence of the white supremacist . . . with good intentions." The great Lincoln was "a tragically flawed figure who shared the racial prejudices of most of his white contemporaries."

To Bennett, Lincoln had moved too slowly in abolishing slavery. Rather than being the Great Emancipator, he was the Great Procrastinator. When he finally issued the Emancipation Proclamation, Bennett said, it lacked "moral grandeur" and "freed few, if any, slaves." In sum, his historical standing was "pure myth." Bennett did make one important, but tiny, concession that Lincoln "grew during the war, but he did not grow much."

Two Lincolns can be discerned on the issue of Blacks and their place in American society. The first Lincoln was the Illinois lawyer and politician who ran for public office in the 1850s and became president in 1860. This Lincoln could not transcend common prejudices and declare Blacks equal to whites. In the 1850s, the notion of racial equality was an entirely unacceptable proposition to the majority of Americans. To survive as a politician, Lincoln could not be more racially democratic than his white voters.

His public words tell the story of the first Lincoln. In 1854, Lincoln said that if he had the power, he would free the slaves. Later, when he ran for the Illinois Senate against Senator Stephen Douglas, he had changed his tune. He said, "There is no reason in the world why the negro is not entitled to all the natural rights enumerated in the Declaration of Independence." But, he added, "he is not my equal in many respects." Still later,

Lincoln wowed the crowd by proclaiming, "I am not . . . in favor of bringing about in any way the social and political equality of the white and Black races." Driving the point home, he continued, "I . . . am in favor of having the superior position assigned to the white race." All the while, he predicted ruin for America as long as slavery kept its "house divided."

The second Lincoln was the president faced with the South's secession. His country looked doomed. Initially, Lincoln thought only of bringing the South back into the fold, even with slavery. As the pressures of war built, he had to confront slavery. At this point, Lincoln "grew," as Lerone Bennett claimed, beyond his early, narrow racial views. He became slightly more flexible, more teachable, and he began to see the war's grand moral dimensions.

Lincoln stepped away from his past. In 1862, Lincoln abolished slavery in the District of Columbia. This freed only three thousand slaves, but it cleansed the capital of slavery. With the Emancipation Proclamation in 1863, Lincoln finally embraced the idea that slavery had to go (even though he allowed it in pro-Union slave states). He dropped the offensive proposal that Blacks form a colony outside America. He opened the military to Blacks. This was a shrewd move. It put the South on the defensive and added needed manpower to the military.

Today it is believed that Blacks pushed Lincoln and the country to a higher moral vision of the war. Black leaders had urged Lincoln to let Blacks into the military. In the North, thousands answered the recruiters' call. In the South, they streamed into Union camps. Ragged, shoeless, carrying their few worldly possessions, they were a procession symbolizing determination more than anything else. In battle after battle, they fought heroically. These Black men lifted the war into a moral crusade against slavery.

None of this would have mattered, though, had the second Lincoln not been a great wartime president—firm against secession, a wily tactician, a master of timing, and a superb spokesperson for national ideals. Shortcomings he surely had, and they have to be faced. Yet it is also wrong to deny his immense skills. Perhaps it is time to emancipate Lincoln.

### Why is a monument soon to be built to honor the thousands of Black Union soldiers?

In a war filled with magnificent stories, it is hard to top the story of the Black troops in the Union Army. Today their history is being retold, changing the old assumption that the war was a contest fought by white men.

Black companies were formed first in Louisiana and South Carolina—in Union-held territory—in 1862. But these units were rare. Larger numbers of Black men began to serve in the Union forces only in 1863 after Lincoln's Emancipation Proclamation cleared the way. The proclamation stated that Black men would "be received into the armed service of the United States to garrison forts, positions, stations . . . and to man vessels of all sorts."

For Black men, entering the service was a chance to attack slavery and to prove themselves. They were fighting for their dignity. In the North, Black leaders went through Black neighborhoods exhorting young men to join. Posters urged, "Men of Color to Arms! to Arms! Fail Now and Our Race is Doomed. Our Last Opportunity has Come!"

Thousands answered the calls. Almost 180,000 Blacks enlisted in the army and more than ten thousand in the navy. From the North, 70 percent of the Black men between the ages of eighteen and forty-five volunteered for the army. Overwhelmingly, Black soldiers served in all-Black units under white officers. Only a few Blacks made it into the officers' ranks. All the troops were grouped as the United States Colored Troops.

Northern whites were opposed to enlisting Blacks. Even white Union troops were hostile. They asked whether Blacks had the intelligence and discipline to be soldiers. They wanted "a white war." People worried that Blacks with arms were potential bandits. Then there was the most damaging accusation of all: that once under fire, Black soldiers would be cowards and would run.

The South detested Blacks in the Union military. At first, captured soldiers were executed immediately. Jefferson Davis, the Confederate president, told states that they could decide whether to reenslave or kill them. White commanders of Black troops were also promised swift executions.

Black soldiers brushed aside threats and vilification. One soldier wrote, "Let us . . . raise ourselves from the mire. Let us be men." In the first year, they fought valiantly in some of the war's toughest battles. They were a commander's dream: brave, willing to charge, and stubborn in holding a position. Secretary of War Edwin Stanton wrote Lincoln praising them: "They have proved themselves among the bravest of the brave, . . . shedding blood with a heroism unsurpassed by soldiers of any other race."

Black troops saw action in nearly five hundred military engagements, forty of them major battles. Many Blacks were used as military laborers. Thirty-six thousand Blacks died while serving their country during the war.

Public occasions honoring the Black soldiers always drew large crowds. The New Orleans funeral for Captain André Callioux, a rare Black commander, drew the largest crowds ever. When the First Mississippi Negro Cavalry brought Confederate prisoners into Vicksburg, crowds of Blacks and whites showed up, but only the Blacks cheered. For the whites, the scene was the world turned upside down: Black soldiers on horseback, wearing the Union insignia, toting guns, and guarding captured white men.

Black battle zeal persuaded doubters. A Union general wrote, "On the return of the forces, those who had scoffed at the Colored Troops on the march were silent." Very occasionally, Confederates would even confess a little respect for their new Black adversaries. Now Americans will be able to salute these men at the capital's U Street monument.

### Black soldiers wrote many letters. To whom and why did they write?

For decades following the Civil War, the letters of Black soldiers sat unread in libraries and federal government collections. Then in the 1970s, inquisitive historians began to wonder what was in those dusty yellow bundles. What they discovered was a national treasure: an intimate look into the churning emotions, ideals, and disappointments of Black men who wanted to fight for freedom.

Like their white counterparts, Black soldiers were often avid and good letter writers. Many soldiers could read and write, especially free Blacks, among whom literacy was surprisingly high. Some even had college degrees and advanced training. Others learned to read and write in the military's regimental schools. But having few skills did not stop soldiers from writing letters. They wanted to express themselves.

Their letters were sent to friends and relatives, Black newspapers such as New York's *Weekly Anglo-African,* antislavery papers, government officials, and the secretary of war. They even wrote to President Lincoln.

Letters to government officials were sent because the soldiers were angry about their treatment. They complained of being forced into the service (through capture, intimidation, or torture). Letters talked about discrimination against Blacks, most of whom were kept in the lowest ranks until the war's last months. Letters told of Black soldiers forced to do backbreaking work for white regiments, reducing "them to the position of slaves again."

The letters' greatest complaint was the unequal pay given Black privates. At first, Blacks and whites both earned $13 per month, plus clothing or $3.50. All officers got more. In June 1863, this was changed. Blacks, regardless of rank, were paid $10 per month, and $3 could be subtracted for clothing.

This was an outrage, as a letter from New Orleans stated: "It is true the Country is in A hard struggle. But we All remember Mercy and Justice . . . we All Listed for so much Bounty, Clothing, Ration and 13 Dollars A month . . . the most has fallen short in all thes Things." Another letter asked, "We have done a Soldiers Duty. Why cant we have a Soldiers pay?"

In addition, soldiers who escaped slavery wanted to know about their families. Others, like Ruphus Wright, wrote to their wives: "I take the pleasant opportunity of writeing to you a fiew lines to inform you of the Late Battle we have had. we was a fight on Tuesday five hours. we whipp the rebels out." Another wrote his children: "now my Dear Children I want you to be contended with . . . your lots. be assured that I will have you if it cost me my life."

G. H. Freeman told a friend's mother of the young man's death: "I am sarry to have to inform You that thear is no dobt of his death. he Died a Brave Death in Trying to Save the Colors of

Rige[ment] in that Dreadful Battil . . . You and his Wife Sister and all Have My deepust Simppathy."

Others wrote to be released from service to care for families. At least one mother wrote the president about this: "Mr Abarham Lincon . . . I wont to know if you please wether I can my son relest from the arme . . . he is all the subport I have now . . . please answer as soon as you can . . . "

Above all, these wonderful letters reveal the steadiness of the Black soldier under the greatest pressures. As one letter said, "this and more too is what the colored soldiers has to endure . . . he receives them all, and is willing to fight."

### In what battles did Black soldiers especially prove their fighting ability and courage?

Just months after Blacks were allowed to enlist, the country had to admit that they were tough and high-spirited combatants, unwilling to retreat even under the fiercest fire, and that they sacrificed themselves repeatedly to advance the Union cause.

Out of hundreds of engagements, some battles were landmarks for Black soldiers:

- Port Hudson, Louisiana, May 1863. A report said that Black soldiers of the Louisiana Native Guards "confronted an entrenched enemy, charged repeatedly, braving artillery fire, and regrouping after repulse to mount further assaults." Severe losses were incurred and six Black flag bearers were killed.

- Milliken's Bend, Louisiana, June 1863. Louisiana and Mississippi regiments of untested soldiers defended a Union garrison with hand-to-hand combat and bayonet fighting. Blacks won a first victory against Confederates.

- Fort Wagner, Charleston, South Carolina, June 1863. The Massachusetts Fifty-fourth led the assault on impregnable Fort Wagner, guarding the sea approaches to Charleston. They were repulsed and suffered heavy casualties, but they established their reputation for bravery. (This was the climax of the film *Glory*.)

- Fort Pillow, Tennessee, April 1864. This was one of the darkest moments for Black soldiers. The surrender of four units did not spare their lives, not even the wounded; Confederate troops killed 195 in a wholesale slaughter. "Remember Fort Pillow" became the rallying cry among Black units.

- Petersburg and Richmond, Virginia, summer 1864. Black troops proved their bravery in charges against entrenched Confederates. More than seven hundred Black soldiers were killed on one occasion alone.

- New Market Heights, Virginia, September 1864: Three thousand Black soldiers attacked a major Confederate position near Richmond. One thousand are killed, wounded, or captured. They had faced "the very fire of hell." A special silver medal was given to two hundred soldiers for their bravery. Fourteen received the Medal of Honor. A bill before the 104th Congress will finally honor this battlefield and these brave soldiers.

- Saltville, Virginia, October 1864. New recruits from Kentucky, recent ex-slaves, proved "their mettle under fire." Over one hundred Blacks were killed.

- Battle of Olustee, Florida, February 1864. White Union forces were ambushed by Confederates. Black regiments were ordered to hold the line while Union forces regrouped and brought up artillery. Blacks achieved this while suffering severe losses.

## Was the 1989 film *Glory* accurate historically?

*Glory* was an almost-perfect synthesis of a noble story and first-rate acting. Morgan Freeman (Sergeant Major John Rawlins) and Denzel Washington (Private Trip) did some of the best acting of their careers, with Washington receiving an Academy Award for best supporting actor. Matthew Broderick created a believable portrait of the abolitionist commander, Colonel Robert Gould Shaw.

For the most part, *Glory* was true to the history of the Massachusetts Fifty-fourth—the first Black military unit raised in the North for the Civil War, largely the idea of Massachusetts Governor John A. Andrew, an early advocate for Black Union soldiers. Black leaders also made the company a reality by urging Blacks to enlist. Two of Frederick Douglass's sons were in the company.

The first part of *Glory* took the company from Massachusetts to the point at which they went to the battlefield, showing the hard times they endured. Also dramatized was the mixture of elation and worry soldiers felt as they went south to fight. Unfortunately, the film failed to indicate that most of the Fifty-fourth's soldiers were not slaves nor from the South but free men from the North. The film portrayed the soldiers' resentment about their unequal pay, though it could have explained this further. And it inaccurately presented the "contraband" soldiers—soldiers raised from Blacks who fled to the Union lines—as loose gangs of thieves, when in reality, they often performed well.

Once the film shifted to South Carolina, it focused on the central episode in the Fifty-fourth's history: the attack against Fort Wagner on a coastal island. In Civil War annals, this battle was one of the most important. Fort Wagner was a mighty fortification, a strategic point in Confederate defenses, bristling with arms, full of angry Confederates, terribly hard to reach and conquer.

The Fifty-fourth's part in the fight for Fort Wagner was accurately chosen as the film's central event. Traveling for two days without rest or food, the soldiers were weary when they arrived at the battlefield. Their commanders were offered the chance to lead the first assault on the awesome fort, a virtual suicide mission.

An eyewitness account said that when the Black soldiers and their officers assembled on the beach, Fort Wagner was silent. As soon as night fell and they began their approach, Fort Wagner "became a mound of fire, from which poured a stream of shot and shell." The Black troops did not relent but changed "step to the double quick" step, moving even faster toward the target. From this point on, the Fifty-fourth threw themselves into the fight, risking all. An officer later wrote, "There had been no stop, pause, or check, at any period of the advance."

Many of the Fifty-fourth died at Fort Wagner, with many more captured and wounded. Shaw, who had a premonition of his death, was killed as he crossed the fort's wall. Union troops did not prevail, but they put up an inspired fight. One section of the Fifty-fourth even briefly overran the fort's defenses. Years later, a Confederate officer at the battle confessed his admiration for the Fifty-fourth: "The negroes fought gallantly, and were headed by as brave a colonel as ever lived. . . . The Negroes were as fine-looking a set as I ever saw."

*Glory* missed showing a few poignant moments from the Fifty-fourth's history. It didn't show the presentation in Boston of the colors to the regiment—a great public event where the Colored Ladies' Relief Society presented them with specially made flags that they took with them into their later battles. The film didn't repeat the message the Confederates sent on the burial of Colonel Shaw: "We have buried him with his niggers." Lastly, the film didn't show William Carney, the Black soldier who saved the Fifty-fourth's American flag and returned to camp, saying, "The old flag never touched the ground, boys."

## RESOURCES

Adams, Virginia. *On the Altar of Freedom: A Black Soldier's Civil War Letters from the Front—Corporal James Henry Gooding.* New York: Warner Books, 1991.

Blight, David W. *Frederick Douglass's Civil War: Keeping Faith in Jubilee.* Baton Rouge: Louisiana State University Press, 1989.

Blockson, Charles L. *The Underground Railroad.* New York: Hippocrene Books, 1994.

Cornelius, Janet Duitsman. *When I Can Read My Title Clear: Literacy, Slavery, and Religion in the Antebellum South.* Columbus: University of South Carolina Press, 1991.

Douglass, Frederick. *Autobiographies.* Notes by Henry Louis Gates, Jr. New York: Library of America, 1994.

Horton, James Oliver. *Free People of Color: Inside the African-American Community.* Washington, D.C.: Smithsonian Institute Press, 1993.

Jacobs, Donald M. *Courage and Conscience: Black and White Abolitionists in Boston.* Bloomington: Indiana University Press, 1993. Text with illustrations.

Jordan, Winthrop D. *Tumult and Silence at Second Creek: An Inquiry into a Civil War Slave Conspiracy.* Baton Rouge: Louisiana State University Press, 1993.

Mabee, Carelton. *Sojourner Truth: Slave, Prophet, Legend.* New York: New York University Press, 1993.

Miller, Edward A., Jr. *Gullah Statesmen: Robert Smalls from Slavery to Congress, 1839–1915.* Columbia: University of South Carolina Press, 1995.

Nichols, Charles H. *Many Thousand Gone: The Ex-Slaves' Account of their Bondage and Freedom.* Leiden, the Netherlands: E. J. Brill, 1963.

Quarles, Benjamin. *The Negro in the Making of America.* New York: Macmillan Publishing, 1964.

Redkey, Edwin, ed. *Grand Army of Black Men.* New York: Cambridge University Press, 1992.

Shields, John, ed. *The Collected Works of Phillis Wheatley.* Oxford, England: Oxford University Press, 1988.

*Six Women's Slave Narratives.* Introduction by William L. Andrews. Oxford, England: Oxford University Press, 1988.

Yee, Shirley. *Black Women Abolitionists: A Study in Activism, 1828–1860.* Knoxville: University of Tennessee Press, 1992.

TEXTS FOR YOUTH

Porter, Connie. *Meet Addy.* Middleton, Wisc.: Pleasant Co., 1993. Illustrations by Melodye Rosales.

Porter, Connie. *Addy Learns a Lesson.* Middleton, Wisc.: Pleasant Co., 1993. Illustrations by Melodye Rosales.

Porter, Connie. *Addy's Surprise : A Christmas Story.* Middleton, Wisc.: Pleasant Co., 1993. Illustrations by Melodye Rosales.

# 5

# *Jumping into Freedom*
## *1865 to the 1880s*

1865– : Slavery's end brings social and economic transformations

1865– : Former Confederate states enact repressive "Black codes," mainly to control Black labor, but also forbidding Blacks to vote or sit on juries

1866: The nation's first Civil Rights Act is passed by Congress

1867: The Military Reconstruction Act passes

1867: First national convention of Ku Klux Klan held in Tennessee

1868: Adoption of the Fourteenth Amendment to the Constitution, guaranteeing citizenship to "all persons born or naturalized in the United States," thereby embracing Blacks

1868: Oscar J. Dunn becomes lieutenant governor of Louisiana

1868: Elizabeth Keckley, famous seamstress and dress designer, publishes her memoirs, telling of friendship with Mary Todd Lincoln

1869: George L. Ruffin is the first graduate of Harvard Law School

1870: Adoption of the Fifteenth Amendment, guaranteeing the right to vote to all male citizens

1872: Freedmen's Bureau goes out of existence

1875: The Civil Rights Bill passes Congress, prohibiting discrimination in public places

1877: Reconstruction is brought to an end

1879– : Exodus of Southern Blacks to Midwest and West

1880s–1890s: Literacy revolution takes place among Blacks

1800s: Bass Reeves is appointed U.S. marshal in Fort Smith Territory, Arkansas-Oklahoma border

1881: Atlanta's Spelman College, sponsored by Rockefeller family, opens

1883: Jan Ernst Matzeliger patents shoe-lasting machine

1889: Ninety-two lynchings are recorded

1880s: Harriet Powers—gifted narrative quilt maker—is at work

"A double-headed nigger"—just after the Civil War, the Blacks of Baldwin County, Georgia, gave Enoch Golden this nickname. It was an honorific, a sign of their deepest respect. Golden possessed the skills that post–Civil War Blacks valued the most. "He could read *and* write, *and* he knowed so much." One head could not contain all of what he knew.

Golden did not hoard his ability to read and write. During slavery, he had taught reading to anyone yearning to learn, a dangerous act in a South hell-bent on keeping Blacks illiterate. Later, he recalled that he had "been de death o' many a nigger 'cause I taught so many to read and write." Here, Golden was referring to the punishments inflicted on Blacks who managed to learn how to read and write. After slavery was over, he worked even harder, teaching deep into the night and rising before dawn to start again.

As freed Blacks walked through the South's ruins and into an unmapped future, securing a little education became a religion. South Carolina's Elizabeth Doggett, herself barely literate, opened one of the first schools and fought to keep it open despite violent "bushwhackers." Another South Carolinian, field hand "Father Haynes," raised money from desperately poor Blacks to build schools and hire teachers. Some South Carolina Blacks bought old schoolhouses previously owned by whites and moved them log by log over many miles. Their acts made freedom a reality.

As these brave people discovered, being free was going to be hard work. The old South had collapsed, and with it had gone a known way of life for Blacks and whites. Blacks hungered for land and a way to earn a living. Resentments against Blacks and their freedom simmered everywhere. Even with all these uncertainties, Blacks jumped into freedom. After all, they had been willing to die for it in the war.

### How did Blacks react to the end of the Civil War and of slavery?

Peter Randolph, a free Black, arrived in Richmond, Virginia—previously the Confederate capital—in April 1865, twenty-five days after the city's fall. He described brilliantly what he saw that day: "The scene that opened before my eyes as I entered Richmond can not be accurately described by word or pen. The city was in smoke and ashes. . . . The colored people from all parts of the state were crowding into the capital, running, leaping, and praising God that freedom had come at last. Many of the old [Black] people had prayed and looked forward to this day, but like Moses they were permitted to see it afar off, and not enter it."

New songs told of Blacks' high spirits. One said the Union armies had "driven the host of hell away." Another repeated, "My soul wants something that's new, My soul wants something that's new," almost as a chant.

Like Richmond, the South was chaotic at the end of the War. Vast areas in the Carolinas, in Georgia, and in central Virginia had been devastated. Roads were filled with troops. Defeated

"Rebs" and triumphant "Yanks" were finding their way back home. Wagonloads of the maimed were headed toward hospitals. People were living in the twilight—indeed, the last hours, the last minutes—of slavery.

Southern whites were watching their whole world turn upside down. Some feebly claimed slaves as "my own neggers." But their "pet institution" had been destroyed. Gone too was their government. Jefferson Davis had fled. The flag had been lowered in Richmond. The Confederacy's anthem could only be sung in secret. The Union would gradually consolidate its victory and establish rule across the South.

In the moments between the old system of slavery and the new order, Blacks made their first hesitant moves into freedom. Many huddled in camps; one Virginia camp was described as "a great mass of colored people, so many gathered on the suburbs and taken care of in the best way possible under the circumstances."

As time passed, Blacks grew bolder. Many left their plantations, creating a kind of floating population across the South. Blacks traveled ceaselessly along the roads, looking for food, medical help, relatives, long-separated wives and lovers, groups to join, Union troops, and Black military units.

Some were simply curious and traveled just to see the world that had suddenly opened up. "Right off, colored folks started on the move," recalled one older slave. "They seemed to want to get closer to freedom, so they'd know what it was—like it was a place or a city." Some searched for a place to put down new roots.

Some remained on their plantations and tried new work arrangements with their former masters. Samuel Agnew, a Mississippi minister, recorded in his diary that "hiring will now become the order of the day." But his father's offers were "not palatable to the negroes and . . . he will have to make them some concessions." It was harder to hire Blacks for work.

Others wanted their own land beneath their feet, land that they could farm themselves. In the 1950s, a former slave recalled, "We wanted to smell our own ground." They felt that if they did not leave the plantation, they were not living their freedom.

The fall of the Confederacy and the final defeat of slavery

changed Black life as nothing had before. As one slave put it, "the Big Freedom came." "It must be now dat de Kingdom am a-coming, and de year ob jubalo."

### Why did Virginia Blacks revel in the tale of Jefferson Davis fleeing Richmond dressed as a woman?

After the fall of the Confederacy, Virginian Black oral traditions judged the South's effort. Raconteurs told a special story about Jefferson Davis's final days as Confederate president. According to the story, he tried to escape capture by Union troops by pretending to be a woman. Storytellers said Davis got himself up in a great gown, a bonnet, and wig (maybe he wore a bonnet with a veil to cover his face), rouged his cheeks, and got into a carriage to be driven out of the city. Real women accompanied him. It was to be a great deception.

In the story, Richmond was in flames. Outposts were established on the city's edge to catch fleeing Confederate officials. Union troops were looking especially for Jefferson Davis. He had to be caught.

When the carriage carrying Davis arrived at an outpost, soldiers opened its door and asked the passengers' names. Davis answered in a voice closely imitating a woman's. Just as the soldiers were about to wave the carriage on, one noticed Davis's boots sticking out from under the dress. He was arrested.

Jefferson Davis did flee, but not as told by the folk story. On April 2, 1865, Davis was in church when a messenger came with a note from General Robert E. Lee that he should leave Richmond immediately because Confederate soldiers were losing at nearby Petersburg. If they lost, the Union troops would descend quickly on Richmond. Davis left the church, gathered his cabinet members, and went by special train to Danville, Virginia, a town southwest of Richmond. Davis, however, was not captured until May in Georgia.

The folk story, however, gave Blacks the last word on the Confederacy. With it, they were saying that Davis was a coward, that he would not stay and face his conquerors. The story also made him unmanly, a person who hid among women—indeed, *as* a woman. His high-pitched voice was a further indictment. This was a simple story with a wicked thrust.

Historians wrote their judgments on the Confederacy's death. Black oral culture wanted to have its say, too.

**What major changes occurred in American society because slavery ended?**

Between 1800 and 1850, the Western world witnessed several great transformations. Democracy was one of these changes, affirming basic rights for ordinary people. Overturning American slavery was an extension of this broad-scale change.

Slavery had already been abolished in the British Caribbean: in 1833 in Antigua; in 1834 in Jamaica; in 1837 in British Guiana. Political revolutions swept across Europe in the late 1840s. From Paris to Berlin to Bucharest, governments that were more democratic came to power on the backs of passionate uprisings. Even in faraway Russia, the serfs were liberated in 1861 after centuries of terrible oppression.

Liberating Blacks from American slavery was one of the monumental processes of world history. No wonder it brought about several important changes:

- It destroyed the power of the largest, wealthiest, most politically sophisticated slaveholding class in the world.

- It established that slavery, the ownership of humans, would not be tolerated in America. The United States Constitution's Thirteenth Amendment, adopted in 1865, stated that "neither slavery nor involuntary servitude . . . shall exist within the United States."

- It took billions of dollars in property from the Southern slaveholding class without compensation.

- It made citizens of Blacks. This was stated in the Fourteenth Amendment, adopted in 1868: "All persons born . . . in the

United States . . . are citizens of the United States and of the State wherein they reside."

- Blacks, in theory, were now free laborers. No master dictated their work to them. They could sell their labor, contract for work, and withhold their labor.

- The language of freedom that surrounded the war and liberation of the slaves imparted new vigor to American notions of freedom. As Abraham Lincoln said at Gettysburg in 1863, "it is rather for us to be here dedicated . . . that this nation, under God, shall have a new birth of freedom."

- For a brief moment, the ideal of Black equality had been given a boost. It was put forth as a possibility for the country's future.

With slavery defeated, America emerged on the other side of the war as a different nation.

**More and more Black Americans are celebrating "Juneteenth" as a holiday. What are they celebrating?**

June 19, 1865, is the date referred to as Juneteenth. It is said that this was the date that Blacks living in Texas learned of their emancipation, two months after the collapse of the Confederacy.

Generally it is believed that Texas slave masters conspired to keep slaves from knowing they were free. Some think that the slaveholders did this in order to get their crops in before the slaves left the plantations. One ex-slave argued, "The Negroes was freed . . . but the whites decided that he would not free him in the state of Texas until the crop was gathered." In fact, a lot of slaveholders from other parts of the South had earlier brought their slaves down to Texas to avoid the Yankees and to escape emancipation, because Texas was distant from the war and thought to be safe for harboring slaves.

So the arrival of freedom in Texas was a special moment to savor.

Emancipation holidays spread among Blacks in the post–Civil War period. This was a grass roots movement. No national commission set these holidays. Each community had its own formula for the celebrations and used different days.

Few places celebrated the emancipation on January 1, the date when President Lincoln issued the official Emancipation Proclamation. This occurred in the dead of winter, and celebrants wanted warmer dates. May 28 was used in some places, commemorating the day in 1865 when federal agents in the South announced the end of slavery. September 22 was also used, because on this date in 1862 Lincoln announced his intention to issue the proclamation. And August 8 was used, but not for any historical reason.

An example of these celebrations can be found in Richmond, Virginia, in the 1880s. Black merchants decorated their storefronts. Dances, such as the stylish "Freedom Ball," were held. Parades wound through the city's streets, with a marching band acting as a Pied Piper to children. Country folk came to town with goods to sell. Tables piled high with food drew hungry buyers. Leading Black citizens—both men and women—recited famous orations and poetry and gave speeches on what the proclamation meant.

Today, Juneteenth celebrations—taking place on or around June 19—are more popular than ever, particularly in the western states, where many Texas and southwestern Blacks have migrated. Rodeos, drill teams, rhythmic marching "step shows," baseball, pageants, Ferris wheels, and barbecues have converted them into all-day family affairs. But the core reason for the event remains: the celebration of that first sweet taste of freedom's elixir.

### What was the Freedmen's Bureau, and did it really help blacks?

In March 1865, a month before the Civil War ended, Congress created the Freedmen's Bureau. In the original legislation, it was called the Bureau of Refugees, Freedmen, and Abandoned Lands. The government did not have a plan for dealing with the Black situation or the condition of whites at the war's end. The Union government hoped that the bureau would be able to solve whatever human problems arose out of the war.

The bureau's first chief commissioner was Major General Oliver G. Howard of the Union Army, after whom Howard University was named. His staff, mostly civilians, were assigned to various Southern states. For the whole South, nine hundred agents were to administer the many programs, but they were not enough. From the start, the bureau also did not have sufficient funds to do a thorough job.

It was supposed to provide food, clothing, fuel, medicine, hospital care, and shelter to dislocated whites and Blacks. As the war's last campaigns ripped through the South, the need for basic support was immense. The South bore all the marks of a scorched war zone. To address the region's needs, the bureau passed out millions of food packets to both Blacks and whites. It built hospitals. Homes were set up for the aged. Orphans were gathered up from the roads and cities' alleys.

Part of the bureau's mandate was to establish schooling for Blacks. In this, the bureau was most successful, and it contributed to schools of all varieties. By 1870, 250,000 Blacks were in bureau schools. Some of the premier universities of today's Black America—Hampton, Howard, Atlanta, and Fisk—were assisted in their first years by the bureau.

One of the most difficult tasks for the bureau was trying to work out labor arrangements for ex-slaves. At first, Howard wanted to give them land from the almost one million acres that had been taken in the military campaigns or abandoned by whites. But this plan did not get crucial support from President Andrew Johnson, Lincoln's successor. As time passed, the bureau encouraged ex-slaves to accept work as wage laborers on former plantations. Soon these workers slipped into being tenants on white farms, working extremely hard for very little money.

The bureau was hated by Southern whites who did not like the idea of a government agency helping to raise the level of Blacks. As a white writer said, "the negro is a sacred animal." Yankees were in love with the ex-slaves, "like the Egyptians were about cats. Negrophilism is the passion with them." On the other side, some bureau agents collaborated with white farmers in trying to shackle Blacks to oppressive work.

Congress ended most of the bureau's work in 1869, though its help for Black veterans continued until 1872.

## What was Reconstruction?

On the day that Robert E. Lee surrendered to Ulysses Grant at the Appomattox Courthouse in Virginia, the Union government had no real plan for redeveloping the South. How was the defeated, rebel white South to be treated? As an archenemy? As an old crony with whom the North had had a momentary spat? Just as important, what should be done for the millions of newly freed Blacks? Should they be rewarded for their loyalty during the war, helped into a new life, or left to fend for themselves? Basically, Reconstruction was the federal government's attempt to handle these issues. It proved to be a large and controversial undertaking.

Reconstruction was both a period and a program. The period ran from 1865 to around 1877. As a program, the overriding objective was to remake the South after the Civil War.

Lincoln's successor, President Andrew Johnson, issued the first proclamations about Reconstruction in 1865. These proclamations were lenient. In a grand gesture, he pardoned the great majority of the Southern white rebels. The only people not pardoned were Confederacy officials and wealthy plantation owners. All the Southern states had to do was abandon slavery, never advocate leaving the Union, and pay their war debts. The federal government would not monitor the defeated states.

Blacks were left out in the cold by Johnson, unrewarded for fighting so valiantly in the war. Johnson did not want Blacks to vote either, saying it would "commence a war of the races." Coming so soon after the war, his plans drew a barrage of fire in the North.

Naturally, most Southern whites welcomed President Johnson's proposals; they did not want to change. Southerners were still bragging about their wartime dash and courage and were not the least bit repentant about the war. They were ecstatic when Johnson appointed old anti-Negro politicians to be state governors. Worse still, Southern states had enacted laws—the infamous Black codes—that still placed Blacks in labor bondage, forbade them to serve on juries, and prohibited them from voting. These codes reenslaved Blacks.

Congress took a dim view of both Johnson's ideas and the white South's stubbornness. Many people wanted to know whether the cataclysm of war had been worth it. Republicans, the dominant party in Congress, pushed for stricter control of the South and a real consideration of Black rights. Among them was a group called the Radical Republicans—men like Thaddeus Stevens and Charles Sumner.

Congressmen like these started the so-called Radical Reconstruction. First, they extended the life of the Freedman's Bureau. Then Congress passed the Civil Rights Bill of 1866, ensuring that all persons born in the United States would be considered citizens. Equally important, Blacks could enter into contracts, bring lawsuits, and were to receive "equal benefit of all laws."

In 1867, the second phase of Reconstruction began, the Radical or Congressional Reconstruction. It was much stricter toward the South. In 1868, Congress forged ahead to pass the Fourteenth Amendment to the Constitution, which prohibits states from taking rights away from citizens. Five military zones were created in the South. Each zone had a general at its head and was controlled by upward of twenty thousand troops. State and local governments were to be elected by the male population, including Blacks. This was called "manhood suffrage," and many Black men voted in these elections. In great numbers, they were Republicans.

Their votes, with those of white allies, propelled a lot of Blacks into office for the first time. Sixteen went to Congress. More than six hundred served in state legislatures. In courthouses across the South, hundreds of Blacks held office. Ranging from United States senators to justices of the peace, Black officials totaled about two thousand. This was the most visible and revolutionary gain of Radical Reconstruction. Owing to this new

Black political class, this period is often called "the Black Reconstruction."

Important problems were handled by the new biracial political class in the South. Some places got their first public schools. Taxation was made more equal. Laboring people were given more bargaining power. The new class of officials brought new technology (for example, railways) and new infrastructure (roads) to the South.

Radical Reconstruction was hated by many Southern whites. The idea of former slaves governing them was just too much for whites in the defeated region to take. White allies of the Blacks were denounced and given the derisive names of "carpetbaggers" (Northerners who had recently come to the South) and "scalawags" (whites born in the South but loyal to the Union during the war). To the white South, a new tyrannical class had seized power.

For these whites, all of the changes had to be reversed. In time, the Ku Klux Klan and other white supremacy groups began to terrorize Blacks. The "Redeemers"—a group that encompassed a broad range of white political interests—rose to redeem the South from Reconstruction, undoing Blacks and oppressing poor whites, as well. When the last federal troops left the South, Blacks were left on their own.

## Did the freed slaves really get "forty acres and a mule"?

No. If they had—all four million former slaves—they would have possessed a state almost the size of Texas.

Recently, more than twenty thousand Blacks, most of them from California, filed claims with the Internal Revenue Service based on post–Civil War legislation to provide former slaves with forty acres. Each claim was for $43,309. Two theories exist about the origin of this figure: one says that it is today's value of forty acres and a mule, and a second that it is the difference between the median income of white and Black households. The IRS denied the claims.

When Congress voted in March 1865 to establish the Freedman's Bureau—a law signed by President Lincoln—it was directed to divide the conquered and abandoned land in the South into plots of forty acres. These plots were to be rented or sold to the newly freed Blacks and the whites helpful to the Union cause. The army was to supply mules.

A few months earlier, General William Tecumseh Sherman had issued a field order that set aside a long strip of coastal land from Charleston, South Carolina, to Jacksonville, Florida, for exclusive use by freed slaves. Blacks quickly settled the area. Sherman promised that final title to the land would come from the president. This never happened. President Johnson gave the land back to white planters. To stay, Blacks had to sign labor contracts. Resisters were evicted.

Bayley Wyatt, a Yorktown, Virginia, freedman, responded to an eviction in 1866 by telling federal officials, "Our wives, our children, our husbands has been sold over and over again to purchase the lands we now locate upon. For that reason we have a divine right to the land." When threatened with eviction, some defended their holdings, built barricades, and had armed parties patrol their lands. An official later added, "The freedmen had the idea that they had a certain right to the property of their former masters, that they have earned it."

Ultimately underlying "forty acres and a mule" was the ideal of economic justice. Just because Blacks did not get their portion of earth and their beast of burden, that did not put an end to their desire for such justice. Black lore revived the notion of a gift of land at several times after the war, but nothing happened. According to W. E. B. DuBois's *Black Reconstruction* (1935), the whole episode "great discouraged them [the Blacks] and broke their faith in the United States Government." Still, Blacks maneuvered around obstacles, bought land, and by the 1890s–1900s, had accumulated millions in rural property.

## Who were the Black elected officials of the Reconstruction?

They were called "monkeys," their legislatures "zoos." In American history textbooks as late as the 1960s and 1970s, they were indicted as "uneducated," "corrupt," "self-indulgent" with public funds, spending taxpayers' money on French gold-framed mirrors, expensive cigars, and fancy spittoons. One historian wrote, "These Negroes . . . had been slaves all their lives, and were so ignorant they did not even know the letters of the alphabet." D. W. Griffith's landmark film *Birth of a Nation* (1915) pictured them as an evil perpetuated on the innocent South.

Contrary to this view, Reconstruction's Black elected officials were actually pivotal in restoring democracy to the South. For the most part, they were sober, responsible lawmakers who wanted to make the South into a new society, where both races could progress.

Recently, Eric Foner, a Columbia University historian, compiled profiles of all Black officials from the Reconstruction era—some two thousand men. Through this massive work, he overturned the old stereotypes about these Black lawmakers and discovered their remarkable traits:

- Blacks began to hold office in 1867 and continued beyond 1877, past the end of Reconstruction.

- Contrary to propaganda from the past, the great majority of these officials were literate. Often they learned to read and write while in slavery. Many had unique stories of how they had acquired these skills. One, for example, learned through listening to the lessons given to his master's children.

- The states with the largest number of Black elected officials were South Carolina (314), Mississippi (226), Louisiana (210), North Carolina (180), Alabama (167), and Georgia (108).

- The number of Black elected officials depended on the size of a state's Black population. In the 1870 census, the population of South Carolina was over 60 percent Black; Mississippi and Louisiana, over 50 percent; Alabama, Florida, Georgia, and Virginia, between 40 and 50 percent.

- Sixteen Blacks rose to the heights of American public life as members of Congress. South Carolina had six. Mississippi had two senators, Blanche K. Bruce and Hiram Revels, and one congressman. Alabama had three Black congressmen. Florida, Georgia, Louisiana, and North Carolina had one Black congressman each.

- P. B. S. Pinchback was briefly governor of Louisiana. In addition, Louisiana had three Black lieutenant governors, Mississippi had one, and South Carolina one. Many Blacks served as state secretaries of education.

- Most of these officials had been slaves. Some had been sold on the auction block, one eighteen times. Some had watched their closest family members being sold. They were survivors. Free Blacks were also numerous as officials.

- Many of these former slaves had been humble farmers, laborers, and storekeepers; others were carpenters, barbers, blacksmiths, masons, and shoemakers. Officials who were free Blacks had been ministers, teachers, editors, lawyers, farmers, laborers, and storekeepers.

- No group of public officials in American history has experienced as many violent attacks as they did. Of the two thousand, thirty-six were killed. Others received death threats. Forty-five were forced from their homes. Over forty were attacked, and some were shot at.

These people did what history demanded. As Alabama legislator James K. Green said, "the tocsin [bell] sounded . . . and we walked out like free men and met the exigencies as they grew up, and shouldered the responsibilities."

**Is it true that after the Civil War, Blacks underwent a "literacy revolution"?**

Yes. So many Blacks learned to read and write in this period that it was revolutionary. A Virginia white wrote, "The sight of these crowded schoolrooms . . . is one of the most patent evidences that we are in the midst of a revolution." In fact, this was the greatest expansion of literacy among any population in the world in the nineteenth century. Slavery had held back the Black demand for reading and writing. With slavery abolished, the Black demand grew to a rapidly moving stream, then a river, and finally crested in a flood.

In 1865, it was estimated that one in twenty Blacks could read—only 5 percent. Five years later, this had jumped to almost 19 percent. In 1880, 30 percent could read and write. In 1890, over 40 percent were literate. By 1900, half of all Blacks could read.

Booker T. Washington, the famous Black leader, recalled the passion for literacy: "It was a whole race trying to go to school. Few were too young, and none too old, to make the attempt to learn."

Prior to emancipation, enslaved Blacks learned to read and write through a variety of clandestine strategies. Blacks were aware that it was dangerous to be known to read. A Georgia slave was told that if his reading was discovered by whites, his fingers would be cut off. Leonard Black's master destroyed his first book, bought secretly, and threatened, "If I ever know you to have a book again, I will whip you half to death." Masters saw reading and writing as "spoiling the good niggers."

But nothing stopped slaves from becoming readers—not even hearing that their fellows had been hung. They learned from white boys and girls. They stole books. They attended secret schools. They studied a few pages after midnight while lying on their floor beds. They learned in plantation religious meetings. In cities like Baltimore, Charleston, and Richmond, slaves were taught by free Blacks in special schools.

After the war, more opportunities for learning to read and write appeared. There were also practical reasons for Blacks to learn to read: they had to understand labor contracts, deeds, and property titles. Almost every freed person faced trickery in contracts: your tiny house could be taken, your labor rewarded with next to nothing in pay, your harvesting of a man's crop could be repaid in spoiled meal and rancid meat rather than money.

To answer the demand for learning, many groups worked to form classes. Black women were some of the first to respond. Charlotte Forten Grimké, a highly educated Black woman, was a teacher in Charleston's Robert Gould Shaw Memorial School. Lily Ganderson, who had run a secret school during slavery days, was still tirelessly working through the 1870s. Black men were equally dedicated to teaching. In Goldsboro, North Carolina, "two colored young men, who but a little time before commenced to learn themselves, had gathered 150 pupils, all quite orderly and hard at work." Many white groups, including missionaries like the American Bible Society and American Tract Society, and many white teachers, especially talented women, were also pioneers in this movement.

Literacy had a magnetism.By learning to read and write, these Blacks penetrated the mystery of writing, unlocked it, and were able to tell others what messages the peculiar alphabet conveyed. Literacy became a part of status culture among Blacks. Soon, Black preachers had to be like teachers. Churches insisted, "Don't bring me no preacher who can't read."

These changes were set in motion by the marvelous moment when a newly freed person made the leap from just knowing the alphabet to decoding a sentence. As one said, "I was so happy when I saw that I could really read."

## When were the Black colleges started?

The origins of these institutions were modest. The first Black colleges were the result of small donations from missionary groups and the Freedman's Bureau. Often, local Black communities helped these fledgling academies by bringing farm produce to feed students, working on buildings, and raising money for them. Hams, chickens, and eggs were sold to benefit the colleges.

Most of the Black colleges in the South were started in the years after the Civil War. However, Lincoln University in Chester, Pennsylvania, was opened in 1854 to educate free and fugitive males. It is the oldest major Black college in America. Atlanta University, Richmond's Virginia Union University, and Raleigh's

Shaw University were established in 1865. Other important colleges followed: Fisk University (in Tennessee) in 1866; Talladega (Alabama), Howard University (Washington, D.C.), and Morehouse College (Georgia) in 1867; Hampton Institute (Virginia) in 1868; Tougaloo (Mississippi) in 1869; Benedict (North Carolina) in 1870; and Spelman (Georgia) in 1881.

Most students were extremely poor. More often than not, they came to college shoeless, with no possessions other than the clothes on their backs. Frequently they had walked for miles. Many Northern whites came to teach the eager pupils. At first, the colleges were just primary schools or secondary schools and taught only basic skills. College classes came later. At Hampton Institute, the very first students were not even taught in classrooms but under an oak tree, later named "Emancipation Oak."

For decades, whites ran these colleges, and the colleges' educational philosophies often reflected their founders' ideas of Black needs in the postwar years. However, Tuskegee, founded by Booker T. Washington, was from the beginning a school staffed entirely by Blacks.

In 1966 the Museum of Modern Art in New York exhibited forty-four rare black and white photographs of men and women in Hampton Institute's vocational classes in the 1880s. As at Hampton, most of these colleges stressed vocational education, training students in the practical aspects of life—how to establish a household budget, set tables, can foods, organize farm accounts—and in vocations like printing, carpentry, mechanics, and brick masonry. Fisk was different. Its administrators shunned vocational education and promoted a liberal arts education—literature, history, and the arts—for its students.

These colleges were—and still are—crucial in producing the new Black college-educated middle classes.

### Were Blacks welcome at white colleges?

Yes and no. A few Black men were accepted by white colleges in the North as far back as the 1840s. But most had an exceedingly hard time getting admitted.

Jonathan C. Gibbs (1828–74), from an elite family in Philadelphia, attended Dartmouth College in New Hampshire and graduated in 1852. Later he recalled being "refused admittance into eighteen colleges in the country because of my color." He went on to be Florida's secretary of state and superintendent of education in the 1870s, during Reconstruction.

Alexander Thomas Augusta (1825–90), born a free person in Virginia, wanted more than anything to be a doctor. He went from Philadelphia to California trying to gain admittance to a medical school and was refused everywhere. Eventually he went to Canada to the University of Toronto, where in 1856 he was awarded a degree. After serving as the first physician-officer in the Union Army's Black regiments, he was hired at Howard University's medical school—the first Black on an American medical school faculty.

Black women entered white colleges in the 1860s. Oberlin College in Ohio, the first college to admit white women, pioneered in the education of Black women. Mary Jane Patterson (1840–94) was the first Black woman in America to receive a bachelor of arts degree, in 1862. Later she was appointed the first Black principal of Preparatory High School for Colored Youth in Washington, D.C.; this became the famous Dunbar High School, and eventually became one of the greatest Black high schools in America.

Fanny Johnson Coppin (1837–1913) worked her way through Oberlin College, earning her degree in 1865. Coppin was an outstanding public educator in Philadelphia, running schools, bringing poor Blacks into the schools, building local clubs, writing newspaper columns, and spreading self-help ideas. She often was compared with Booker T. Washington for her efforts to promote self-reliance.

The most valuable contribution made by white colleges to Black life was in training Black women to be doctors—115 between 1865 and 1900, a fantastic achievement in a time of discrimination.

Rebecca Lee (1833–?), who "early conceived a liking for and sought every opportunity to be in a position to relieve the suffering of others," graduated from Boston's New England Female

Medical College in 1864. She is the first Black woman to earn a medical degree in the United States. Rebecca Cole (1846–1922) received her medical degree in 1867 from the Woman's Medical College of Pennsylvania. In 1870, Susan McKinney Steward (1847–1918) graduated at the head of her class from New York Medical College for Women. "Dr. Susan" was what her patients called her. All these women went on to work in community clinics and labored on behalf of women and children's medicine.

Not all white colleges and universities were open to Blacks, but many rose to the challenge of making Blacks freer through education.

### Were Blacks prominent in the development of the American frontier and the West?

The West, with its vast landscapes, unlimited skies, great and fascinating native peoples, and its sense of opportunity, captured the imagination of many Blacks, just as it did for whites. From the eighteenth century to well after the Civil War, thousands of Blacks took their chances and rode west. Black cowboys, estimated at over five thousand, were a familiar sight in the region's towns and plains. Out of this history has come larger-than-life men and women.

In the 1760s in the West, Jean Baptiste Point Du Sable was one of the earliest great trailblazers in our country's history. Born in Haiti and educated in France, he came into America through New Orleans when it was a Spanish-controlled city. For ten years, Du Sable traveled by horse and small caravan through the forests and immense prairies as far as the Great Lakes. He was able to go so far and endure these difficult journeys because his Native American friends helped him. The spot that really impressed him was the place Native Americans called "Eschikagou." In 1774, he built a trading post there. By 1800, Du Sable had sold his trading post and was off again. Eschikagou became Chicago.

Edward Rose followed, in the 1800s, as one of the first leaders of fur-trading expeditions across the Mississippi. He was among the "Mountain Men," a group including many Blacks. These Blacks got along better with the Native Americans than most whites did: "They could manage . . . better than the white men," said one frontiersman, "with less friction."

Jim P. Beckwourth (1798–1867) was a trader-traveler across the West. He fled there because he had a fight with a mean boss in Saint Louis. He lived among the Crow Indians, who called him "Morning Star." He discovered a path across the Sierra Nevada to California—called Beckwourth Pass—in the 1850s and explored the West until he died. In 1993, the United States Postal Service issued a stamp in his honor in its series on the "Legends of the West," placing him alongside the likes of Chief Joseph, Wild Bill Hickok, and Annie Oakley.

Men were not the only pioneers. Aunt Clara Brown purchased her freedom from slavery, made her way to Saint Louis, and there signed on as a cook for gold prospectors. In 1859, she was taken on the back of a wagon to Central City, Colorado, where she became a leading citizen and businesswoman, opening the first laundry. Known for her generosity to the poor, the Colorado Pioneers Association gave her official honors when she died in 1877.

California especially was a magnet for Blacks. William Leidesdorff (1810–48), from the Virgin Islands, was a major figure in the 1840s in California, when it was still part of Mexico. He came to California a wealthy man and there increased his holdings. Leidesdorff died at an early age, but before his death he was a major San Francisco property owner. He also owned a ranch and the famous City Hotel.

When gold was discovered in California in 1848, more free Blacks left behind their lives in the East. In two years, more than two thousand Blacks had come looking for gold-lined streams and gold-filled mountains. Diaries mention them as working hard at the gold sites. White miners resented the Blacks, claiming that they had special gold-detecting powers.

After the Civil War, the Black role in the West picked up. Congress created six Black army units in 1866. Nicknamed the "Buffalo Soldiers" by Native Americans, they patrolled from Canada to Mexico and from the Mississippi to the Rockies. It was a rough life for them; they had to eat poor rations and ride broken-down horses. Yet their desertion rates were a fraction of those of frontier white units. In 1993, a monument that had been proposed by General Colin Powell Jr. was dedicated to them at Fort Leavenworth, Kansas.

One of the truly remarkable Black figures to emerge from the West was Bass Reeves (1838–1910), called by historian Art Burton "a legendary lawman of the Western frontier." Bass was born in Lamar County, Texas; grew to be six feet two inches tall; and weighed 180 pounds. On his beat, he was recognized by his large black straight-brimmed hat and his two Colt revolvers. He was appointed a deputy U.S. marshal by Judge Isaac Parker in the Fort Smith territory, a badlands encompassing seventy-five thousand square miles. Reeves was fearless and, during his service, captured over three thousand outlaws. A police chief said of him, "The veteran Negro deputy never quailed in facing any man."

Black history in the West can only grow in importance, for it has barely been tapped and yet is so essential to broadening the Black past to all of America. Today, Oklahoma and Arizona are moving quickly to promote this recognition.

## How common were Black cowboys?

When the 1993 film *Posse* hit the theaters and a year later the United Postal service issued stamps honoring Wild West star Bill Pickett (1870?–1932), America began to get the message: Blacks had played a significant role in the West's history. Denver's Black American West Museum, created by the dedicated Paul Stewart, became a must-see for tourists.

In the museum, one learns an astonishing fact: of the thirty-five thousand men on the cattle drives in the late nineteenth century, more than five thousand were Black!

Blacks became cowboys because there was a certain harsh equality on the frontier. They did the same fighting, riding, roping, and drinking as the white cowboys. The job of being a cowboy was so hard and work so easy to find that there was not much energy or reason for discrimination. If discriminated against, a person could move on, find another spot to settle, and seek out a new, fairer employer. A Black cowboy's durability, toughness, and skill were almost more important than his complexion.

*The Life and Adventures of Nat Love, Better Known in Cattle Country as Deadwood Dick,* the 1907 best-seller, gave a rare view of the Black cowboy's life. Nat Love (1854–1921) was completely loyal to the rough life of the West. "I gloried in the danger," he wrote. Born in Tennessee, he traveled west to join the cowboys' life in Kansas, Texas, and Arizona. He became a first-rate range rider and a legendary marksman. In his book, he wrote, "I carry the marks of fourteen bullet wounds on different parts of my body . . . but I am not even crippled." He even knew Bat Masterson, Frank and Jesse James, and Billy the Kid. He had the life all cowboys aspired to. When he got married in 1890, though, he had to put the roaming life behind him.

The West produced larger-than-life Black characters, people who accumulated legend. "Cherokee Bill" was a scout for Cherokee Indians while a teenager and became an outlaw and a king of shoot-outs. At his hanging, just days before his twentieth birthday, when asked for his last words, he answered tersely, "I came here to die, not to make a speech."

Isom Dart had a reputation as one of the West's greatest horsemen. A man who knew Dart said, "No man understood horses better. I have seen all the great riders. . . . He could outride them any of them."

Bill Pickett (1870?–1932) originated "bulldogging"—a cowboy competition sport in which a steer is chased by a man on horseback and then wrestled to the ground. In 1971, he was voted into Oklahoma City's Cowboy Hall of Fame. The back of the stamp recently issued in his memory reads, "Bill Pickett: 1870–1932. Fearless Black cowboy, rodeo showman and rancher, said to have invented bulldogging. Both Will Rogers and Tom Mix served as his assistants."

Not the least of these personages was "Stagecoach Mary"—six feet tall, stocky, and often dressed in rough clothing—who made a life in tough Montana. Born in Tennessee, Mary Fields went to the West alone and became a gun-toting, sharp-tongued, no-nonsense, tale-telling, drinking, cigar-smoking, useful citizen, especially during her mail-carrying days. No matter what the weather, she delivered the mail.

### Who were "Pap" Singleton, Henry Adams, and the Exodusters?

After Reconstruction, Blacks lived in extreme conditions. They were mostly sharecroppers who were cheated out of their share of the profits by ruthless white farmers. Blacks who owned their land lived under mountains of debt. Political rights, such as voting, were threatened at every turn. Night riders terrorized Black homes. Many Blacks, therefore, sought an escape from the South.

When "Pap" Singleton looked at the condition of Southern Blacks in the 1870s, he thought, "Blacks had nothin' but their freedom an' it looked like they might soon lose that." Henry Adams was outraged: "The whole South—every state in the South—had got into the hands of the very men that held us slaves." He concluded, "We had better go."

Henry Adams and "Pap" Singleton were leaders of a Black migration movement to Kansas in the late 1870s. Just as Blacks had done in the 1820s when some went to Liberia in West Africa, they looked to movement as an answer. Singleton and Adams were different from many Black leaders of the time. They were not high-born, nor had they been elected officials. They were ex-slaves with limited education. But they were very capable grass roots mobilizers of ordinary people.

Kansas offered the promise of a place free of racial tyranny. John Brown had fought to rid it of slavery. It was on the frontier. Its spaces were wide and sparsely settled. Owning land was possible. Singleton ran ads in Nashville's papers, while Adams spread the idea in Louisiana. The news moved quickly to other states. In 1877, a Kentucky group set out for the plains and started a colony at a place called Nicodemus, after a legendary slave Nicodemus, named after the biblical figure, who came to America on the second ship and later bought his freedom.

Posters were distributed throughout the South about Nicodemus. "To the Colored Citizens of the United States," they read, "we are proud to say it is the finest country we ever saw. The soil is of rich, black, sandy loam. . . . The country . . . looks most pleasing to the human eye." Another poster said, "See what colored citizens are doing for their Elevation!" Thousands of people answered the call, moving also to Nebraska and other states. More than ten thousand people came to Kansas in two years.

"Exodusters" was the name given these migrants—from the Bible story about the Hebrews who fled the bondage of Egypt for freedom in Israel. They came by wagon, by train, and on Mississippi steamboats. Of course, Kansas was not paradise. For nights on end, migrants to Topeka slept in the overcrowded Floral Hall. Earth dugouts were their first homes. Prejudice existed there too. In the South, whites worried as their labor moved away. "Pap" Singleton and Henry Adams were the movers and shakers behind this dramatic outpouring of Black citizens. And this would not be last time that Blacks would leave the South en masse.

### Were there Black inventors? What did they contribute to America?

Black inventors created machines and devices that greatly helped American industry. Their breakthrough inventions made manufacturing more efficient or resolved technical problems that industries faced. As a result, American companies made substantial money from Black inventors' work.

Black inventors reflected a larger American ambition. As technology writer James S. Wamsley has noted, "inventive spirit and innovative genius have always been part of the American spirit." Certainly, this was the attitude in the nineteenth century, when the phrase "American ingenuity" became synonymous with American culture. Americans became a nation of tinkerers, and Blacks joined this great national movement. After the Civil War, Blacks were freer to use their talents for their own purposes. The years 1865 to 1900 were a great era for Black invention. But even before 1865, Blacks were invention pioneers.

In 1843, the first major industrial invention by a Black was patented by Norbert Rillieux (1809–94), a New Orleans native. Rillieux was one of the most privileged Blacks of his time, born a free person to Constance Vivant, a woman of mixed parentage, and Vincent Rillieux, a white engineer, who sent him to Paris's L'École Centrale for training to be an engineer.

His invention was the vacuum evaporation system. It replaced the old extremely hot and dangerous method of refining sugar—the open-kettle method used throughout the sugar-producing South. Rillieux's invention was more efficient, was not dangerous, and greatly increased production, making sugar cheaper. The granulated sugar we use today derived from his invention. It changed the American table, and it made him wealthy.

In 1848, a second great invention was developed by Lewis Temple (1800–54) for the New England whaling industry. What Temple constructed in his little dockside workshop was a toggle harpoon, which allowed the fishermen's harpoons to catch inside the whale. With the straight harpoon blade previously used, the whale had been able to break free. Whalers quickly recognized the importance of Temple's invention, naming it "Temple's toggle." In 1853, the fishing newspaper *Whalemen's Shipping List* pointed out the large quantities of whale oil obtained from catches made because of Temple's invention. Unfortunately, Temple never patented his harpoon, so he didn't get rich from it.

Three more major inventions came one after the other in the 1880s, the pinnacle of Black invention in the nineteenth century.

In 1882, Elijah McCoy (1843–1929) patented a machine called the McCoy Graphite Lubricator. Until McCoy's machine, trains had to stop periodically because their engines would overheat due to friction. The engines had to be lubricated by hand. McCoy's lubricator, when attached to engines, did this automatically, and the trains no longer had to stop.The term "the real McCoy" arose from the success of Elijah McCoy's invention. Managers of the railroads settled only for his lubricator—the real thing—rather than shoddy substitutes.

In 1883, Jan Ernst Matzeliger (1852–89) patented a shoe-lasting machine. Previously the upper leather part of a shoe was sewn to the sole by hand, by men who were called "lasters." This was costly, time-consuming work, but no one thought a machine could do it. Matzeliger studied lasters at their jobs and observed their hand movements closely, until he was able to get his machine to duplicate what they did. Matzeliger's invention was an instant hit with shoemaking companies—it could do a hundred pairs of shoes in the time a laster did fifty. Fortunes were made from his work. However, Matzeliger's tireless effort, spending night and day in cold shops, had undermined his health. He died of tuberculosis at age thirty-seven.

Granville T. Woods (1856–1910), known as the "Black Edison," was truly a world-class American inventor. In 1884, he invented a telephone transmitter that carried voices over long distances. In 1887, he invented a telegraph system for railways that allowed messages to be sent between moving trains and between moving trains and stations. More than fifteen inventions for electric railways were patented by Woods. General Electric and Westinghouse bought his patents.

The Black inventors of the nineteenth century were brilliant technicians, driven and hardworking, and they were insightful about what American industry needed. Their inventions succeeded because they were indispensable.

Washington, D.C.'s Anacostia Museum, part of the Smithsonian Institution, mounted in 1989 an exhibit on Black inventors. One of its most compelling features concerned "Enslaved Inventors: Hidden Contributors." It highlighted the fact that many inventions by slaves have gone unrecorded and were not given patent protection. New research shows a number of important inventions during slavery times. Elizabeth Keckley (c. 1818–1907), whose autobiography is now published, is credited with new methods of cutting and fitting dresses, which she taught to Washington, D.C.'s dressmakers. Benjamin Montgomery (1818–1877) invented a steamboat propeller.

And, a slave mechanic, "Ned," of Pike County, Mississippi, prompted a dispute within the federal patent office after his owner, Oscar J. E. Stuart, wrote in 1857 about registering Ned's labor-saving cotton scraper. Stuart wanted the patent awarded to himself, since Ned was owned by him. In 1858, a judgment was rendered: a patent could not be issued to Ned nor to Stuart. A slave could not hold a patent nor could his master receive a patent for a slave's invention. This held until the end of the Civil War. Because of this, we will never know the full extent of Black invention within slavery. But after slavery, Black inventors made up for this loss.

### What did nineteenth-century quilting contribute to Black history?

Kadella, Hannah, Aunt Ella, Aunt Margaret, Jane Batson, Jane Bond, Jennie Nevile Stroud, Mary Williams, William Dean—these are just a few of the names of important Black quilters from the days of slavery. Most of these people created their designs while living on plantations. They made quilts for the master's family and some for their own families, sewing on them at night after a hard day's work.

Black quilt makers were artists. Their work included cultural ideas and symbols from their own immediate backgrounds, as well as design patterns from Africa. Their appliqué quilts used small pieces of cloth that were sometimes cut into symbols (such as flowers, feathers, suns, crosses, or coffins) and then sewn onto larger pieces of cloth. This appliqué technique came from West Africa, where it was very popular, and had been passed down among Black women in America.

These quilts also featured irregular patterns. Often, the pattern would be a little lopsided, or colors would not harmonize exactly, or there would be a mixture of different designs or an unusual blend of bright colors. These irregularities gave the quilts a certain feeling of spontaneity; the designs appeared to dance.

Some quilts contained information in their symbols and designs. Often they created pictures that told Bible tales or showed people in their daily lives. Other quilts were intended to be protective charms, offering the person using them protection against sickness and evil.

By the 1880s, a most notable quilt maker was becoming recognized. Her name was Harriet Powers (1837–1911). Born a slave in Georgia, she married Armstead Powers, had three children, and together they lived as a farm family. She did not read or write; thus, her quilts were a means of visual communication for her. Only two of her quilts have survived. In 1886, Jennie Smith, an upper-class white woman, bought a Powers quilt. Powers sold the quilt for a few dollars because her family had fallen on hard times. It broke her heart, and she visited the quilt several times. She called it "the darling offspring of my brain."

It was understandable that this quilt was so dear to Harriet, for it is a version of the appliqué quilt, consisting of eleven panels, each containing many figures—animals, humans, suns, stars. Together, the panels told stories from both the Old and New Testaments, starting with the Garden of Eden and ending with the Nativity. (Today, it is preserved at the Smithsonian Institution.) Harriet Powers made her second quilt for Atlanta University. This was another spectacular artwork, consisting of fifteen panels of dense images—biblical symbols, like the whale swallowing Jonah; solar eclipses and meteor showers; and her favorite animals.

Harriet Powers and her forerunners in quilt making have made a tremendous contribution to Black American history. They transformed quilt making from a plantation chore into a form of artistic expression. In making these quilts, they also enriched Black women's culture.

## RESOURCES

Anyike, James C. *African-American Holidays: A Historical Research and Resource Guide to Cultural Celebrations.* Chicago: Popular Truth, 1991.

Aptheker, Herbert. *A Documentary History of the Negro People in the United States.* Vol. 2. New York: Citadel Press, 1951.

Berlin, Ira, Barbara J. Fields, Thavolia Glymph, Joseph P. Reidy, Leslie S. Rowland, Steve Miller, and Julie Saville. *Freedom: A Documentary History of Emancipation, 1861–1867.* New York: Cambridge University Press, 1993.

Berlin, Ira, Barbara J. Fields, Steven F. Miller, Joseph P. Reidy, and Leslie S. Rowland. *Slaves No More: Three Essays on Emancipation and the Civil War.* New York: Cambridge University Press, 1993.

Berlin, Ira, Thavolia Glymph, Steven F. Miller, Joseph P. Reidy, Leslie S. Rowland, and Julie Saville. *The Wartime Genesis of Free Labor: The Upper South.* New York: Cambridge University Press, 1993.

Blassingame, John W. *Black New Orleans, 1860–1880.* Chicago: University of Chicago Press, 1973.

Brotz, Howard. *African-American Social and Political Thought, 1850–1920.* New Brunswick, N.J.: Transaction Publishing, 1992.

Cooper, Ann J. *A Voice from the South.* New York: Oxford University Press, 1988.

Foner, Eric. *Freedom's Lawmakers.* New York: Oxford University Press, 1993.

Foner, Eric. *Reconstruction: America's Unfinished Business.* New York: Harper & Row, 1988.

Foster, Frances Smith. *Written by Herself: Literary Production of African-American Women, 1746–1892.* Bloomington: Indiana University Press, 1993.

Fredrickson, George M. *The Black Image in the White Mind: The Debate on Afro-American Character and Destiny, 1817–1914.* Hanover, N.H.: Wesleyan University Press, 1971.

Johnson, Cecil. *Guts: Legendary Black Rodeo Cowboy.* Fort Worth, Tex.: The Summit Group, 1994.

Katz, William Loren. *The Black West.* 3rd ed. Seattle, Wash.: Open Hand Publishing, 1987.

Litwack, Leon. *Been in the Storm So Long.* New York: Random House, 1979.

Moses, Wilson Jeremiah. *The Golden Age of Black Nationalism, 1850–1925.* New York: Oxford University Press, 1978.

Potter, Joan, and Constance Clayton. *African-American Firsts: Famous, Little-known, and Unsung Triumphs of Blacks in America.* Elizabethtown, N.Y.: Pinto Press, 1994.

Ritter, Joan, ed. *Collected Black Women's Poetry.* New York: Oxford University Press, 1988.

Skinner, Elliot P. *African-Americans and U.S. Policy Toward Africa, 1850–1924.* Washington, D.C.: Howard University Press, 1992.

Story, Rosalyn M. *And So I Sing: African-American Divas of Opera and Concert.* New York: Warner Books, 1990.

Wahlman, Maude Southwell. *Signs and Symbols: African Images in African-American Quilts.* New York: Studio Books, 1993. Text with illustrations.

Wells-Barnett, Ida B. *Selected Works of Ida B. Wells-Barnett.* New York: Oxford University Press, 1991.

Wiggins, William H. *O Freedom: Afro-American Emancipation Celebrations.* Knoxville: University of Tennessee Press, 1987.

Woodward, C. Vann. *The Strange Career of Jim Crow.* New York: Oxford University Press, 1957.

# 6

# Becoming Whole and New Again

## The 1890s–1920s

1892: Journalist Ida B. Wells begins her lifelong crusade against racism

1895: W. E. B. Du Bois becomes the first Black American to receive a Harvard Ph.D.

1895: Booker T. Washington, principal of Tuskegee Institute, delivers Atlanta Exposition speech

1896: U.S. Supreme Court upholds "separate but equal" in *Plessy v. Ferguson*

1896: The National Association of Colored Women, led by Mary Church Terrell, is organized

1898: Black regiments participate in Spanish-American War

1900: First Pan-African Conference takes place in London

1901: Booker T. Washington dines with President Theodore Roosevelt at the White House

1904: Boley, Oklahoma—a Black town—is established

1905: Du Bois calls conference of Black leaders—leads to Niagara Movement

1906: In "Brownsville incident," Black soldiers in Texas are accused of raiding town

1907: Cowboy Nat Love publishes bestselling autobiography

1908: Jack Johnson wins the heavyweight title in the 14th round against Tommy Burns in Australia and launches the most spectacular heavyweight career of all time.

1908: Los Angeles's Woman's Day Nursery, for children of working Black mothers, opens

1913: Woodrow Wilson presses segregation in the nation's capital

1914: Outbreak of World War I; the war ends in 1918

1914– : The Great Migration begins, bringing Blacks north

1915: Madam C. J. Walker establishes a hairdressing products empire

1916: Marcus Garvey starts Harlem branch of United Negro Improvement Association

1920s: The era of the Harlem Renaissance

When Toni Morrison reads from her novel *Jazz,* her smooth voice draws the listener into a story of Black hope and love gone awry in the New York of the early 1900s.

The novel is actually a history of Blacks as they entered the new world of the city. Her Blacks lived in apartments. While mostly middle class, they danced to bands like the Ebony Keys, had occasional leisure for card games, used cosmetics, and had their hair styled into silky marcelled waves. Rural life, fieldwork, and maybe even callused hands were memories.

*Jazz*'s characters were looking for a different wholeness and newness. According to Morrison, "the wave of Black people running from want and violence crested in the 1870s. . . . They were country people, but how soon country people forget. When they fall into love with a city, it is forever." She adds, "The minute they arrive at the train station or get off the ferry and glimpse the wide street and the wasteful lamps lighting them, they know they are born for it. There, in a city, they are . . . themselves: their stronger, riskier selves."

What they wanted was desired by many Blacks, who in the 1890s numbered nearly 7.5 million citizens, 11.9 percent of the nation. During the 1890s to 1920s, Blacks looked for a totally new, whole sense of self. For some, that new self was to be found in the North. For others, it lay across the plains to the west. Still others looked to art, music, writing, and philosophy. All the while, though, life was mean. Thousands of Black voters were removed from the rolls. Mob violence stalked Blacks. Jim Crow laws confined Black life. Overwhelmingly, Blacks were poor; nevertheless, businesses grew and an aristocracy rose.

As the twentieth century arrived and began to unfold, the pace of Black life quickened. An optimism about the future dared to rise.

## Who was "Jim Crow," and where did the term originate?

In 1828, a white actor and singer named Thomas Rice claimed that he saw and heard a Black man—someone living on the streets—sing:

*Step first upon yo' heel,*
*An' den upon yo' toe,*
*An' ebry time you turns around,*
*You jump Jim Crow.*
*Next fall upon yo' knees,*
*Then jump up and bow low,*
*An' ebry time you turn around,*
*You jump Jim Crow.*

Supposedly, Rice realized that if he imitated this street Black's strut and song, he would become a great success. In another of Rice's stories, he said the term came from an entertaining Kentucky slave named Jim, who was owned by a Mr. Crow.

Whatever the truth, Rice used "Jim Crow" as a part of his minstrel act. He would blacken his face, dress in beggar's rags, mimic what he thought was a Black shuffle, and sing "An' ebry time you turn around, You jump Jim Crow." In 1832, "The Ballad of Jim Crow" was published and became a tremendous hit worldwide. With this song about "Jim Crow," Rice popularized the term.

Pretty soon, "Jim Crow" became something of a stereotype of Blacks. Then the term began to designate the segregation of Blacks. It was used first in the North during the late 1830s and early 1840s. The *New York Mirror* used the phrase in 1837. In the 1840s, early segregated railway cars, dividing Blacks from whites, were called "Jim Crow cars."

It's not surprising that when special laws segregated Blacks in public places, these laws would be called "Jim Crow laws." After 1875, Black passengers were increasingly segregated on the railways, streetcars, steamboats, and ferries. Blacks traveled on the decks of boats while whites sat in enclosed rooms. Hotels, restaurants, and show halls catering to whites were all off limits to Blacks. Schools were segregated in Southern states after 1875. All this was the new "Jim Crowism."

The Supreme Court in 1896 agreed with this Jim Crow South when it upheld Louisiana's power to segregate railway cars, in *Plessy v. Ferguson.* Homer Plessy, a New Orleans shoemaker, and his Black backers had brought the suit. The Court approved "separate but equal" facilities for Blacks, making segregation constitutional. Only one judge, John Marshall Harlan, dissented. After this, Southern states enacted an orgy of segregation laws.

Today, the *Plessy v. Ferguson* decision is considered to be the Supreme Court's lowest point. The Court retreated from upholding the Constitution, specifically the Fourteenth Amendment. "Jim Crow" now stood for more than a minstrel's ditty; it had become a crime against Black freedom.

If there is a light moment in the history of "Jim Crow," it occurred in February 1961 during U.S. Senate nomination hearings of Robert C. Weaver for secretary of housing. Since Weaver was the first Black nominated for a cabinet post, his hearings brought out determined opposition. Chief among his tormentors was Texas Senator Robert Blakley who tried to link Weaver to communism. Blakely asked Weaver about a review of his *The Negro Ghetto* (1948); the review appeared in a left publication and was signed "J. Crow," as a joke. Blakley, not catching the joke, asked Weaver if he knew "J. Crow," to which Weaver replied, "I did not know he wrote book reviews." No one laughed, but Senator Blakley had taken a heavy hit.

### Did Blacks participate in the minstrel traditions, and did they change it?

Minstrel shows were the most popular form of American public entertainment from the 1850s to the beginning of the twentieth century. At first, minstrel shows were white affairs, consisting of songs, dances, jokes, and speeches. Then whites began to use burned cork to blacken their faces, and they gave themselves "Black features."

Eric Lott in his subtle study of blackface minstrelsy defined it as "an established nineteenth-century theatrical practice, principally of the urban North, in which white men caricatured blacks for sport and profit." These shows assumed that Blacks were premier songsters, musicians, dancers, and comics. White minstrel companies became famous singing and prancing "in the Negro style." So popular were the shows that Commodore Perry's sailors, when they went to Japan in 1853, performed one for the Japanese emperor. In America, the audience for ministrelsy was white, male, and working class.

James Weldon Johnson, the famous 1920s Black writer, described white minstrelsy "as a caricature of Negro life. . . . It fixed the tradition of the Negro as only an irresponsible, happy-go-lucky, wide-grinning, loud-laughing, shuffling, banjo-playing, singing, dancing sort of being." In fact, two stereotypes of Black males were offered in these shows: one the Southern plantation darkie and the second the Northern urban dandy.

Starting in the 1850s, Black performers were allowed to take part in the shows. After the Civil War, white audiences demanded

Black performers in order for the shows to be more "realistic," and it was not long before all-Black minstrel groups were formed. However, the basic stereotyped and often derogatory content of the shows changed little. Audiences rejected any major deviations. Still, the Black performers brought vitality to the shows, especially through new dances (such as the jig, buck-and-wing, and stop-time).

Between 1890 and 1900, the popularity of all-Black shows grew rapidly. Groups with flashy names thrived: Al Fields's Darkest America and the Harrison Brothers' New and Old South. A Harrison Brothers ad, in the *Indianapolis Freeman,* ran:

> Wanted: Colored performers, men and women. Men who can double in . . . band and stage. Real Black men and women. Good dressers on and off stage.

Minstrel shows gave training to Black performers like Bert Williams and to songwriters like Gussie Lord Davis (who published over two hundred songs) and James A. Bland (who wrote "Carry Me Back to Old Virginny," adopted as the Virginia state song). Bland appeared thousands of times and wrote more than seven hundred show songs.

As minstrel shows became especially popular in the 1890s, white and Black companies competed fiercely for audiences. Better financed and willing to harass their competition, white minstrel companies forced many Black-owned minstrel companies out of the most lucrative markets.

Black performers must have sensed that the minstrel scene was coming to an end for them. They began to depart from the form. Billy McClain began organizing gigantic shows, sometimes involving hundreds of singers and dancers. Sam T. Jack, an important burlesque theater owner and manager, decided that what was needed was a spectacular show glorifying the Black woman. His solution was *The Creole Show.* This smart, up-to-date show—more of a musical comedy than a plantation show—played in the major cities and was a big hit. The *Octoroons* in 1895 moved even farther away from minstrel shows, rejecting some stereotypes and presenting a greater variety of songs. A new Black entertainment image was being established.

### Why were the 1890s to 1910s called the "Black Women's Era"?

In a mere twenty years, so many talented Black women emerged in a spectacular variety of pursuits. Their goal? As one educator said, "we feel it is our duty to work for the ennobling of our race."

Combating school segregation, Black women worked hard to educate Blacks. Annie Holland (1871–1934) developed North Carolina schools and parent-teacher associations. Lucy Laney (1854–1933) made Georgia's Haines Institute one of the South's best high schools. Virginia Estelle Randolph (1874–1958) opened many Virginia schools and trained many teachers. Elizabeth Evelyn Wright (1872–1906) established the industrial and agricultural education school later known as Vorhees College in South Carolina.

Numerous clubs were formed by Black women to lobby for social change and civil rights. Through these clubs, Black women demonstrated their leadership. After a long period of development, the National Association of Colored Women (NACW) was formed in 1896. Mary Church Terrell (1863–1954), an Oberlin graduate, club movement leader, and the daughter of the first Black millionaire, served as its first president. A major step forward, NACW offered a national structure for Black women's organizations, gave Black women greater clout, and expanded their services.

Others founding clubs were Josephine St. Pierre Ruffin (1842–1924), a defender of Black women; Mary Morris Talbert (1866–1923), an organizer of Buffalo (New York) Sunday school teachers; and Fannie Barrier Williams (1855–1944), a committed Chicago social reformer. Williams's address entitled "The Intellectual Progress and Present Status of the Colored Women . . . Since Emancipation" precipitated much talk because of her position that Black women were "the only women in the country for whom real ability, virtue, and special talents count for nothing when they become applicants for respectable employment."

Black women writers were just as active. As poets, journalists, novelists, pamphleteers, playwrights, editors, and historians, they made a strong impression on an increasingly well-read public.

Delilah Beasley (1872–1934) was a California journalist who published *The Negro Trail-blazers of California* (1919), the first such history. Journalist Ida B. Wells (1862–1931), with her hard-hitting articles, led the fight against lynching, chain gangs, and sharecropping injustices. In 1891, she co-owned and edited Memphis's *Free Speech,* but because she revealed the whites responsible for lynching three Blacks, she could not stay in Memphis. She went to New York and resumed her moral advocacy there. Later, she moved to Chicago, where she campaigned for Black inclusion in the Chicago World's Fair in 1893. In 1895, she married Ferdinand Barnett, owner of the *Chicago Conservator,* who supported her political activities.

Pauline Hopkins (1856–1930) was a premier writer and a founder of the *Colored American Magazine,* printing short stories and serialized novels. Hopkins's novel *Contending Forces* (1900) was in tune with the times. It covered a family from the 1790s to after the Civil War and offered insightful social commentary on Black life. Hopkins's dream was that "we must ourselves develop the men and women who will faithfully portray the inmost thoughts and feelings of the Negro." She succeeded; her novel ranks as one of the best of the nineteenth century.

Many women found careers in the theater. Black women were able to break with minstrel acts before the men did because most minstrel characters were based on men. Anna and Emma Hyers were the first Black women to gain true renown on the American stage. Famous for their great voices and elaborate costumes, they toured widely in the West. Aida Overton Walker (1880–1914), a striking beauty, was already on her way to stardom in her teens. By 1900, she was a celebrity; three years later, she appeared before the British royal family at Buckingham Palace.

## How did rousing Black American gospel music originate and develop?

Of all the Black American musical forms, gospel music is the one that was most distinctly nurtured in the bosom of the community. It is also the music whose history closely parallels the evolution of Black religious culture.

Gospel forms came from slave communities. Spirituals contributed their inspiration and theology to gospel. The "shout" that is so typical of gospel—the powerful vocal proclamation—came also from slavery days. The concept of a singing, praying, preaching gospel congregation that becomes one—one vessel of the spirit—comes from slavery times. Lastly, the democracy of the gospel congregation, where everyone can participate and where every person's religious energy is valued, derives from slave religious meetings.

Yet, post-slavery Black religious traditions contributed as well to gospel. Jubilees—fervent mass singings—accelerated gospel's appearance. In the 1870s, traveling Black college choruses spread the tradition of Black choral music essential to gospel. The Fisk Jubilee Singers of Fisk University and Hampton Singers of Hampton Institute toured widely. Music entrepreneurs started other groups, like Orpheus McAdoo, whose Virginia Singers toured America, Europe, and South Africa.

Still, it was not until the nineteenth century's last years that gospel was actually born. Gospel music's first stage of development runs from the 1890s to the 1920s. (*Harp of Zion* by William Henry Sherwood published in 1893 was the first book of Black gospel music.) Therefore, Black gospel music is largely a twentieth-century phenomenon. Like blues and jazz, gospel was given birth by a populace in transit from rural towns to larger rural centers and most importantly to the burgeoning American cities of the early twentieth century.

Black evangelism arose quickly in these new urban settlements. Newcomers sought shelter from urban perils in churches. They needed spiritual resources for dealing with everyday life issues. They hungered for a feeling of collectivity. Ministers gave direction to these newcomers. And, these fragile Black communities were fertile ground for gospel music's emergence.

Pentecostal churches (known also as Holiness or Sanctified churches) and occasionally Baptist or Methodist churches were the birthplace of gospel. In these churches, congregational singing became prevalent. Soon, churchgoers experimented with tambourines, percussive instruments, the washtub bass, the piano, and drums. In these churches, the congregation sang with a driving intensity. The singing aimed at fostering the possession of members by the "spirit." Clapping, moaning, call-and-response preaching, and foot-stomping also added background to the gospel music.

All of this broke with the more formal Black churches, where services focused more on the sermon, music on standard hymns, and instrumentation was limited. In the gospel churches, the sermon, singing, and instrumentation were meshed. Everything was also more flexible. Classic hymns, for example, were altered on the spot.

Gospel was never just music. From its birth, gospel carried a spiritual message. Other Black music art forms are artistic statements in and of themselves, but gospel has always been a pathway to a state of spiritual transcendence. The preacher was essential to this transformation. The preacher was the appointed agent of his flock's transcendence. While gospel music acted as a musical pageant of the flock's travails and triumphs along the road to deliverance, the preacher guided the way. Moreover, the gospel singer and chorus were extensions of the preacher within the service and a witness for the assembled.

Epitomizing early gospel music's development was preacher-composer Charles Albert Tindley (1851–1933), a famous Philadelphia Methodist minister who hailed originally from Maryland. Tindley was first in a great chain of geniuses who laid gospel music's foundation. Beginning in 1902, he built a huge congregation, resulting in the ten-thousand-member Tindley Temple. His sermons attracted thousands (especially the annual "Heaven's Christmas Tree"), including droves of whites. To help members buy housing, he organized a church credit club. Feeding the poor and counseling the distressed, he served as a shepherd to an immense flock. Even with this busy life, he wrote a string of gospel classics: "Stand by Me," "Nothing Between," "We'll Understand It Better By and By," "Take Your Burden to the Lord," and "I'll Overcome Someday."

Mississippi-born Lucie E. Campbell (1885–1963) was another major influence in gospel's stirrings. From 1919 to her death, she wrote songs for presentation at the annual National Baptist Convention. Although college educated, she had no formal music training. Yet, her songs demonstrated a mastery over classical European music traditions, for she adapted these to gospel. A few of her songs were waltz gospels. Her songs were extremely popular: "Something Within," "The King's Highway," "In the Upper Room," and most of all, "He Understands; He'll Say, 'Well Done.'" "Miss Lucie," as called by her fans, served too as a stern, methodical teacher of gospel technique.

Most voluminous as a composer was Thomas Andrew Dorsey (1899–1993), a Georgian who grew up in a deeply religious country household. Early on, his family moved to Atlanta, where they sunk into poverty. Pressed to earn money, he went to work at eleven, stopping his schooling. His musical career started with his being a pianist in a movie theater. Later, he played in raffish quarters—for rent parties and even bordellos, where he specialized in smooth, soft sensuous blues, beloved by close dancers. By 1915, he was the city's leading entertainment pianist. By the 1920s, his blues pieces were performed by Joe "King" Oliver and he had accompanied Ma Rainey. He went by the moniker of "Georgia Tom."

A photo captured the Dorsey of those years—there he was, dressed in a spiffy coat, immaculate white shirt and bowtie, lighting a cigarette, wearing a racing cap cocked to one side, his face bathed in soft light, his eyes inviting. He was quintessentially a worldly man.

Slowly, through bouts of indecision and depression, Dorsey turned to gospel. He moved on up. From 1930 onward, he was one of the greatest American composers, writing hundreds of songs, including such favorites as "If You See My Savior," "Take My Hand, Precious Lord," "There Will Be Peace in the Valley for Me," and "The Lord Will Make a Way Somehow."

Now, gospel had the resources to surge forward. Significantly, gospel had—like other revered music traditions—a canon: genius moments, standards, musical techniques that practitioners knew and passed on. Once records and radio were exploited, gospel's audience swelled. Gospel moved beyond the churches where it was conceived and began to penetrate even the Black churches that had once rejected it.

Outstanding male gospel quartets appeared from the 1930s on, The Golden Gates, The Swan Silvertones, The Soul Stirrers (with

Sam Cooke), The Pilgrim Travelers, The Dixie Hummingbirds, The Five Blind Boys of Alabama. The Roberta Martin Singers, started in 1936 by Roberta Martin (1907–1969), set a new style of choral singing, especially suitable for women, and groomed a host of singers (Dinah Washington and James Cleveland). To members, her group was both a family and a "college." And then, there was William Herbert Brewster (c. 1897–1987): an advanced lyricist whose bluesy songs challenged the most talented singers to be more vocally ambitious.

Starting in the 1940s, it was this style that was raised to brilliance by The Clara Ward Sisters, Alex Bradford (1927–1978), and Mahalia Jackson (c. 1912–1972)—a chorus that began what is known as the "Classic Gospel Age." Brewster's "Move On Up to a Little Higher" was a million- seller hit for Jackson in 1947. His "Surely God Is Able" hit the million mark for the Ward Singers in 1950.

A great gospel performance is a wondrous event. No better proof of this exists than the singing of Marian Williams (1927–1994) in Bill Moyers's *Amazing Grace*, a 1990 film on the history of the famous hymn. Williams had tremendous standing among gospel connoisseurs. She had been in The Clara Ward singers. After she became a solo performer, she excelled in Langston Hughes's *Black Nativity* (1962–63) and in many splendid recordings. In Moyers's film, she was the last singer of "Amazing Grace." And, in only a few minutes, she transformed the song. Her singing glided from very soft, delicate phrasing to urgent statement and back to barely audible spoken lines. She never overpowered the words, but used them as her own. By the hymn's end, Williams had made the song her testimony to God's greatness. Moments like these keep gospel a force in American music and history.

### Why are Paul Laurence Dunbar's poems and novels important?

Early in this century, Dodd, Mead and Company published a beautiful multivolume edition of the poetry of Paul Laurence Dunbar (1872–1906). Up to that time, few Black authors had been honored with such lavish volumes. An artist designed the covers, photographs of rural Black life were added, and each poem had a decorative border.

When the books appeared, Dunbar was in his late twenties and already popular. He was the first Black American literary figure to achieve national recognition, appealing to both whites and Blacks.

His popularity was based in part on his ability to write in dialect, through which he attempted to capture Black folk speech. A passage from "The Deserted Plantation" illustrates his style:

*An' de banjo voice is silent in de qua'ters*
*D'ain't a hymn ner co'n-song ringin' in de air;*
*But de murmur of a branch's passin' waters*
*Is de only soun' dat breks de stillness dere.*

Another reason for Dunbar's popularity was that his work focused on heartwarming, amusing scenarios of Black life. The reader felt like he or she was strolling through a bucolic Black village. Some critics have argued that Dunbar's work seems superficial, but these poetic effects were difficult to achieve.

Dunbar was an exceptional portraitist of Black life. Through poetry, he was able to show the wonder of ordinary lives, from major concerns—like Black homes, families, churches, and the role of the old slaves in the community—to minor ones, like dinners, dress, dances, hunting, cooking, special foods, and finery. Dunbar's poems were verbal Polaroids of Black daily life, a point ignored in attacks against him.

Dunbar's most overlooked poetic achievement was his celebration of Black love. Many of his poems are Black Valentines, dedicated to falling in love, courtship, love letters, lovers' games, lovers' strolls, marriage, and, of course, sorrowful partings. He stood for Black love at a time when anti-Black writers questioned Black morality and the ability of Black men and women to have healthy relationships.

Dunbar's poems entered the oral culture of Blacks. In church meetings, speech competitions, and talent shows, his poems lived through spirited recitations. No writer before Dunbar was so taken to heart by the masses of Black people. Short, lively, and picture making, his poems were perfect for public performance. Some of the favorites were "When Malindy Sings," "Li'l Gal," "A Negro Love Song," "When de Colored Band Comes to Town," and "At Candle-Lightin' Time."

Dunbar was hurt that the public wanted him to confine his writing to dialect. Only his "jingles in a broken tongue," as he put it, were liked. Maybe this explains why, in the last years before his death, he wrote more outspoken poems. His "Ode to Ethiopia" said, "O Mother Race! to thee I bring this pledge of faith unwavering," and he urged "Be proud, my Race, in mind and soul." By 1902, Dunbar had published *The Sport of the Gods*, one of the first Black protest novels.

**Do those lawn jockeys, so detested by Blacks, actually represent sports heroes?**

It used to be a kind of game, in towns where there were Black colleges, for students to steal those little Black lawn jockeys from white neighborhoods and exhibit them proudly in their rooms. In Tiburon, California, in 1984, the town was forced to paint white its four jockeys on Main Street so as not to offend.

Now it seems that these figures should not be maligned as racist. Said Tiburon's Rev. Jerry Buckner, who is Black, in protest of whitening the jockeys, "All of the original first jockeys were Black, and most people aren't even aware of that fact." He went on to say that the first thirteen winners of the Kentucky Derby—beginning with Oliver Lewis in 1875—were Blacks who often owned the horses they rode.

Historical facts confirm this new interpretation. Many post–Civil War Black sports heroes were jockeys. In fact, the history of Black jockeys goes back to slavery. Slaveholders had used their slaves as jockeys in local races. Since the slaves were property and did not have to be paid, they were cheaper than hiring white jockeys. In the 1800s, a Black rider nicknamed "Monkey" Simon was the best jockey in the country and earned his master one hundred dollars per ride. Simon, aware of his power, used his fame to demand personal respect, becoming the first Black jockey celebrity.

Issac Murphy, in the late nineteenth century, was perhaps one of the greatest athletes—white or Black—of his day. He was the first jockey to win three Kentucky Derbys (1884, 1890, and 1891). Eventually he made a ten-thousand-dollar salary and was a horse owner in his own right. Murphy is considered by some to have been the greatest jockey ever, since he won more than six hundred out of fourteen hundred races. His high living—drinking champagne and eating heavily—slowed him down. But for a while, he rode like no other jockey before him and only a few after him.

Between the 1880s and 1905, Black jockeys like Murphy and Willie Simms, James "Soup" Perkins, and Jimmy Wakefield were the toast of the sport of kings. These jockeys were so successful that discrimination by white jockeys, the Ku Klux Klan, and owners forced them from the sport. By 1910, when segregation was in full swing, Black jockeys were no longer racing.

The Kentucky Derby Museum has mounted a small exhibit of photographs, Currier and Ives prints, and a thirteen-minute video honoring "African-Americans in Thoroughbred Racing." As for the lawn jockeys, they were symbols of a real history concerning Black jockeys. But for most people they will remain artifacts of racism.

## Was there ever a champion Black cyclist? Who was he?

It is little known that cycling was the first sport in which a national title was won by a native-born Black American. Marshall "Major" Taylor was his name, and he competed in an amazing number of cycling events. In his mammoth autobiography, *The Fastest Bicycle Rider in the World,* photos of his medals take up whole pages.

Born in Indianapolis in 1878, Taylor was attracted through a series of odd circumstances to bicycles, and he competed at thirteen in his first race. Eventually he met Louis D. "Birdie"

Munger—a white man who had been a great rider and was a manufacturer of bicycles (including lightweight ones)—who taught him about the sport of competition cycling.

In 1898, Taylor had won the most races for the year in America and was declared the national champion, despite the fact that as the *Philadelphia Press* admitted, Taylor had been harassed on the track: "Major Taylor is the greatest sprint rider in America and his white rivals all know it. . . . They have pocketed him [hemmed him in] at every opportunity, and ran him wide on the turns, and used other foul tactics in order to defeat him." But this did not undo him. The very next year Taylor competed in the World One-Mile Sprint Championship, won it, and later set a one-mile world record. He was now the national and world champion in his sport.

He raced for seventeen years, always attracting large crowds; "his presence doubles the receipts," it was said. Reflecting on his career, Taylor wrote, "I find that my name is now and then mentioned among that steadily increasing list of notables, placed in the immortal cycle hall of fame."

When he finally lost, he was—as he put it—"beaten fairly and squarely once at least by one who is destined eventually to trim even the best of us. I refer to the mightiest champion of them all, and one who needs no introduction—'Father Time.'"

True to his character, his autobiography ends modestly with "The Value of Good Habits and Clean Living." But even Taylor's modesty cannot conceal his glowing image: he was a lean, tough physical machine riding on the lightest bikes of the time, an unforgettable sight for Black and white America.

## What was lynching?

Lynching was one of the grisliest traditions in American history. It was murder by formula: an angry crowd would capture a person, take him or her to a remote spot, tie a rope around the victim's neck, and then string them up from a tree or post until they died. By far, the majority of lynching victims were men. Some whites were lynched, but most often Blacks were the victims. Photos of lynch scenes show the empty faces of milling whites as they observe—sometimes, smiling and joking—their handiwork.

Ultimately, lynching was a way of intimidating the masses of Blacks. It delivered an unmistakable message: Blacks had better not do anything that whites disapproved of. Even so, a person could be lynched for no real reason. The most trivial incident could set off the lynching mob. Historian W. Fitzhugh Brundage tells the story of Sandy Reeves, a seventeen-year-old Black of Pierce County, Georgia, who was lynched after snatching back his nickel from his white employer's three-year-old daughter. He had dropped it; she picked it up; in taking it back, he caused her to become enraged. That night in 1918, a mob lynched Reeves.

The 1890s to 1920s were the bloodiest years for Blacks, as lynching ran amok. In 1892, mobs killed some 71 whites and 155 Blacks. Never had the country seen so many lynchings. A white Methodist clergyman observed that Southern lynching was "not so extraordinary an occurrence to need explanation. . . . It has become so common that it no longer surprises."

Between 1880 and 1930 in the South, more than three thousand Black men and women were killed by lynch mobs, and over seven hundred whites! While the South led, the Southwest and Midwest shared in the tragedy. Mark Twain said the country should be called "the United States of Lyncherdom." By 1905, so ugly was the situation that a sociologist noted, "It has been said that our country's national crime is lynching."

During these years, no Black American escaped the fear of lynching. Everyone lived near a lynching, or had read or heard of one, or had a relative or friend who had been caught by a mob. The fear of lynching penetrated deep into the marrow of Black communities.

Identifying an exact cause of lynching is difficult because lies, tales, and rumors often triggered mob violence. Mobs were whipped up by people fomenting trouble against Blacks. What did these mobs hear? Many lynchings were carried out against Blacks who were accused of—but not tried for—murders, crimes, and rapes or advances against white women. Blacks could be taken from jails where they were being held for crimes.

Both Blacks and whites fought lynching. People joined hands across racial lines to denounce and work against it. But it took decades to get lynching to the top of the nation's agenda. Even Congress refused to enact laws against it. In the meantime, many a Black left the South fleeing the terror of the lynch mob.

## What was the *Afro-American Encyclopedia*?

How far had Blacks progressed in their first thirty years out of slavery? By the 1890s, Blacks had a great appetite for facts about their achievements. They hungered also for inspirational writing. Despite violence against them, Blacks were sparked with ambition. Moving the race forward was the widespread goal. Blacks loyal to racial advance were "race men" and "race women."

The *Afro-American Encyclopedia: Thoughts, Doings, and Sayings of the Race,* published in 1895, was the first popular survey of what Blacks had thought and done since the Civil War. It was sold door-to-door and written for the average person. Families bought a copy for the parlor. Churches sold copies. Leading citizens took out advertisements in it.

The goal of the encyclopedia's writers' was to build racial pride. "The Negro lacks confidence in himself and his race," they said. Their aim was to correct that, as one poem proclaimed:

*In farming, trade and literature,*
*A people enterprising!*
*Our churches, schools and home life pure,*
*Tell to the world we're rising!*

The encyclopedia was fabulous reading. Some of the subjects covered were "Afro-American Versus Negro," "The Causes of Color," "The Color of the First Man," "The Color of the First Woman," "Race Facts Worth Knowing," "The Colored Man in Medicine," "The Emancipation of Women," "The Greatest Negro Scholar in the World," "Virginia's First Woman Physician," and "Wealthy Colored New York Men."

Most wonderful was "The Seventeen Reasons Why the Negro

Should Be Proud of His Race." Some of the reasons were: "Because of its antiquity," Because of the literary achievements of . . . leading men and women," "Because of their kind disposition," "Because of its intelligence," and "It is willing to let bygones be bygones."

Black homes also invested in the encyclopedia for its abundant advice. Self-improvement was its mission. Readers got a sermon on "Ambition"; starting a bank account was explained, along with how to avoid the temptation of drinking. There was also advice on "Improving Negro Homes," "The Perils of "Extravagance," and "Trades for Your Children." Readers were even told why they should read a newspaper.

The advice was often quite specific. Black women were to "make it your highest aim to be good wives; the race needs you and must depend on you." Blacks should bank their money. The "walls of our homes should be adorned with the pictures of famous Black men and women." In "Have Courage," readers were admonished to "have the courage to wear the old suit of clothes, rather than go into debt for a new suit."

Two messages were repeated constantly: Blacks should cooperate to build unity, *and* Blacks should be self-reliant to forge their independence.

On a cold winter's evening, families gathered in front of the fire to be instructed and entertained by reading the encyclopedia. It was a companion to Blacks as they stood poised to enter the twentieth century.

## Who was Booker T. Washington? What was his 1895 Atlanta Exposition Speech?

Born in 1856, Booker T. Washington was to become the most important Black American of his time. He built the famous Tuskegee Institute, the nation's foremost vocational school, in Alabama's Black Belt. He was the leading advocate of Black self-help. An outstanding orator, he traveled the nation and overseas expounding his views on Black issues. And he was a national political force—often behind the scenes—always anxious to extend his influence.

In his own time, he was beloved by many Blacks, admired by whites, and opposed mightily by Black intellectuals. Decades after his death, he is still controversial. Yet few Black debates have escaped his imprint. Social critic Louis Lomax claimed, "We all, in a basic sense, are what we are because he was *there,* because he did what he did and said what he said."

His life started as a plantation slave in Franklin County, Virginia, where early experiences shaped his later beliefs. His work for a rich white family taught him the love of orderliness, and taking pride in even the lowliest work. The greatest opportunity came to young Washington when at sixteen, he enrolled in Hampton Normal and Agricultural Institute in Virginia, a school for Blacks. He walked the long distance to the school, sleeping under bridges and arriving penniless and ragged but determined.

Samuel Chapman Armstrong, a son of missionaries and former commander of Black Union troops, ran Hampton. His idea was to give Blacks a practical education and a strong sense of moral character. Washington was taught how to do fieldwork, care for farm animals, build furniture, keep himself clean, and have good manners. Washington particularly loved public speaking and debate. Hampton transformed him. Washington idolized Armstrong, later saying that he was "a great man—the noblest rarest human being it has ever been my privilege to meet." He graduated from Hampton in 1875.

The year 1881 was a landmark one for Washington. With Armstrong's backing, he went to head a new Black school in Macon County, Alabama. When Washington arrived there, he found a dismal situation. The school had no land, no buildings, and a very small budget. Washington went to work recruiting students, opening the first classes in a shanty, and borrowing money to buy land. Less than a decade later, Tuskegee sat on five hundred acres and offered classes in carpentry, printing, shoemaking, farming, and dairying. Women were trained in sewing, cooking, and canning.

Tuskegee was a beacon of light in the besieged Black Deep South. Thousands of Black farmers came for instruction. Graduates fanned out across the South. Tuskegee was gaining a regional, national, and international reputation. On this success, Washington raised money from white benefactors. His growing importance earned him an invitation to speak at the opening of Atlanta's Cotton States and International Exposition in September 1895. He was thirty-nine.

Both Blacks and whites packed Washington's audience. The speech was a masterpiece: coherent, clear, and brief. It merits a place alongside William Jennings Bryan's "Cross of gold" speech of 1896. It stirred white listeners to applaud and Blacks to weep, and catapulted Washington to the position of premier Black leader.

In fifteen minutes, he laid out his simple philosophy of Blacks' status and future advance: Blacks and whites were tied together in the South. Blacks were loyal to the habits of the region. "The most patient, faithful, law-abiding, and unresentful people that the world has ever seen would be friends to whites." In return, whites should be friends to Blacks. He urged the Blacks to seek an economic future in the South, to "cast down your bucket where you are."

All this seemed innocent enough. But Washington went on to proclaim that "in all things that are purely social we can be as separate as the fingers," and "the wisest among my race understand that the agitation of questions of social equality is the extremest folly . . . ." This was far different from Frederick Douglass's ideas about equal citizenship. Washington's speech came to be known as "the Atlanta Compromise." It bowed to white power. He had put more emphasis on Black responsibilities and less on Black rights.

Washington built a dazzling career out of this philosophy. As he became more famous, he attracted major teaching talent to Tuskegee (including plant scientist George Washington Carver); he was a friend of rich, powerful whites, including presidents; and he created organizations such as the National Negro Business League (in 1900). His influence penetrated every aspect of Black life, including prominent newspapers such as *New York Age.* By 1900, though, Black men—such as Harvard-educated newspaperman William Monroe Trotter—had grown restive under the doctrine of the famous man. By 1903, Washington was denounced as "deficient in fearlessness, self-assertion, . . . and heroic spirit."

Yet out of the poverty of Alabama, Washington had created an economic, educational, and political machine. He died in 1915. Today his reputation is being revived. Writer Nikki Giovanni sees him as trying to "empower Black folks."

### Who was W. E. B. Du Bois, and what was the significance of his *Souls of Black Folk*?

William Edward Burghardt Du Bois is one of the most distinctive names in Black history. His life matched his name.

Born in 1868 in Great Barrington, Massachusetts—a small, largely white town—Du Bois came from a family with Black, French, Dutch, and Native American roots. ("Thank God, no Anglo-Saxon," he once said.) From his early years, he decided to make his mark in the world through ideas—that is, through research, writing, and commentary.

For his education, he first went, at age seventeen, to Tennessee's Fisk University, one of the new colleges for Blacks (founded in 1860). Speedily completing his work there, he entered Harvard University in 1888. He went on to earn three degrees from Harvard, a B. A. in philosophy, an M. A. in history, and a doctorate in history in 1895. He was the first Black to achieve the Harvard doctorate. *Suppression of the Slave Trade to the United States of America, 1638–1870,* was the first volume of the Harvard Historical Series. In between his studies at Harvard, he spent a year studying in Berlin.

Du Bois was becoming a man of ideas but not an isolated, bookish man. Increasingly, he was aware of the link between his ideas and American society. In 1897, the University of Pennsylvania hired him to study Philadelphia's Blacks. A tireless worker, he interviewed five thousand people for study. Later, in Atlanta, he conducted more major studies. He thought solutions could be found to Black problems if more scientific knowledge about the group existed.

As Du Bois studied Blacks, he faced incidents that made him

wonder about being just a researcher. After learning that a Georgia Black man he was helping with a defense had been lynched, he wrote, "One could not be a calm, cool, and detached scientist while Negroes were lynched, murdered, and starved."

As the twentieth century began, Du Bois became a more public figure. He fought the Georgia legislature's plan to take Black voters off the rolls. He wrote for national magazines. In 1900, he went to London for the first worldwide gathering of Black leaders. The convention offered him a global perspective on Black problems, encompassing those of Africa and the West Indies, as well as the United States. Later, he became leader of the group working to stage future Black world conferences. He would become one of the great Pan-Africanists of the century, advocating self-determination for Blacks the world over.

A significant change came in Du Bois's life when he chose to support William Monroe Trotter, a cofounder of the *Boston Guardian* newspaper. Trotter bitterly opposed Booker T. Washington's ideas, which he thought were delivering Blacks to a bleak future under white supremacists. Together, Trotter and Du Bois fought for broader rights for Blacks.

In 1903, Du Bois published the landmark *The Souls of Black Folk*, easily one of the most important books by an American. It offered insight into American Blacks as they took stock at the turn of the century. The reader was invited to "listen to the striving in the souls of Black folk."

With its lyrical writing, the book almost sang. Within the book's first two pages, Du Bois created the most quoted sentences ever written by a Black: "It is a peculiar sensation, this double consciousness. . . . One ever feels his two-ness,—an American, a Negro; two souls, two thoughts, two unreconciled strivings; two warring ideals in one dark body." In the second chapter, he created yet another memorable phrase: "The problem of the twentieth century is the problem of the color-line—the relation of the darker to the lighter races of men."

Reviewers were not confident about Du Bois's ideas. But the book made history because he put forth new ideas about Black culture and the Black future. He wrote of his work among rural Southern Blacks as a teacher, thereby showing that he was not removed from the struggling Black masses. He also showed the profundity of Black culture—the spirituality of their songs and wisdom of their lives. The book went through six printings by 1905.

In the book's most famous section, he attacked Booker T. Washington for his idea that "the South is justified in its present attitude toward the Negro because of the Negro's degradation." To solve his race's problems, Du Bois proposed the creation of a "Talented Tenth"—a group of educated, committed, and dynamic Black people who would lead the race out of poverty, illiteracy, and disease. This group differed from Washington's Tuskegee graduates. Du Bois wanted an educated Black leadership; Washington, a leadership based on skilled farmers and tradespeople.

Du Bois was thirty-five when *Souls* appeared, and he already was well on his way to an amazingly diverse public life: as an editor of *Crisis* magazine (1910–32); as a prolific writer of history, essays, and novels; and as a political figure and Pan-Africanist. In August 1963 he died in the West African nation of Ghana.

## Who was George Washington Carver, and why was he called the "Wizard of Tuskegee"?

A wizard does wonders, is a person of amazing skill. When a person achieves wizardry in his line of work, we think magic has touched him. With justification, George Washington Carver (1864–1943) was called the "Wizard of Tuskegee."

When the annual Black history month arrives, George Washington Carver is given another round of applause. The standard recitation of his life includes these facts: born a slave, he went on to be a world-renowned scientist, taught at Tuskegee, and discovered a hundred ways for using the lowly peanut.

Carver was born in the last days of slavery near Diamond Grove, Missouri. Through the kindness of a white family, he enrolled in a tiny college but later switched to the State Agricultural College at Ames, Iowa, where he earned two degrees by 1896. In the same year, he went to Tuskegee to teach at the invitation of Booker T. Washington, the school's principal.

Carver had many motives for going to Tuskegee. He admired Washington, he wanted to do practical science, and he believed deeply that he could help the poor farmers of Alabama. He also wanted to honor his race with his work. As he said later, "the primary idea in all of my work was to help the farmer and fill the poor man's empty dinner pail. . . . My idea is to help the man farthest down." Carver was a highly motivated man, with a sense of spiritual calling about his work and life.

Tuskegee presented problems for Carver. At times, he was not accepted. He didn't have enough room for his equipment. The rest of the staff objected to him on color grounds (he was dark) and because he cared little about his dress. Despite these trivial problems, Carver flourished.

Carver's greatest achievement came in the laboratories, where he searched for products that could be extracted from the peanut and the sweet potato, common crops grown by Alabama farmers. Carver's aim was to create moneymakers out of Southern local produce. Soon Carver was issuing pamphlets on "How to Grow the Peanut" and "105 Ways of Preparing It for Human Consumption."

As director of Tuskegee's Agricultural Experimental Station, Carver issued many practical instruction booklets for farm families—"How to Build Up Worn Out Soils," "Saving the Sweet Potato Crop," "The Pickling and Curing of Meat in Hot Weather"—offering the kind of basic knowledge farmers needed. He also held meetings at Tuskegee devoted to farmers' issues. By 1912, his six-week "Short Course in Farming" was attracting 1,500 farmers.

Recently Carver's letters have been published, and they show us a much more complex person than we knew previously. He carefully cultivated his image as the humble scientist. He felt a close kinship to nature, believed in natural healing, and crocheted. He was also a truly inspiring teacher, a hard taskmaster who lightened the load with joking, teasing, and loyal friendship.

### Who was Ota Benga, and why was he a national sensation?

Ota Benga was a Congolese Pygmy brought to the United States in 1903 by Samuel Phillips Verner, an eccentric missionary and businessman, for public display at the Saint Louis World's Fair.

When Ota Benga arrived (along with nine other Pygmies), he was in his twenties, under five feet tall, and according to his society's customs, his front teeth were filed. According to Verner, he had snatched Ota Benga from the jaws of slavery in Africa, paying for him.

In 1904, when the Saint Louis World's Fair opened, Ota Benga and the other Congolese were given a premier place among the exhibits, displayed along with Philippine tribespeople, Eskimos, and Native Americans. Though officials claimed that their aims were scientific, their real goals were to make money and news. Dressed in loincloths, advertised as barbarians, and made to dance, Ota Benga and his fellows drew huge crowds. Newspaper headlines stirred public interest: "Pygmies Demand a Monkey Diet," "Cannibals Will Sing and Dance."

When the fair was over, the Pygmies were given gifts to take home, valued at $8.35. Ota Benga was taken to the Congo by Verner, but he came back, this time to a worse situation. His next stop was the Bronx's Zoological Park, where officials recognized a gold mine when they saw one. Ota Benga was installed in a cage in the Monkey House, with this sign:

The African Pygmy, "Ota Benga"
Age, 28 years. Height, 4 Feet 11 Inches
Weight 103 Pounds. Brought from the Kasai River,
Congo Free State, South Central Africa
by Samuel P. Verner
Exhibited each afternoon during September

Surrounded by bones strewn in the cage to indicate cannibalism and overwhelmed by the crowd's noise, Ota Benga had arrived in hell. Never before had the park exhibited a human. Several Black ministers led the outcry against "the degrading exhibition."

Ota Benga was rescued and sent to Virginia Seminary in Lynchburg, a Black college. Briefly, he had peace, freedom, and friends. He hunted in the woods. People took a genuine interest in his welfare. His great friend was the famous poet Anne Spencer (1882–1975), who became interested in his education, stories, and herbal cures. She came closest to knowing him as a person rather than an object.

It was too late. In March 1916, Ota Benga, using a gun no one knew he had, killed himself, ending his role in an American nightmare of racial obsessions.

### Where would Blacks and America be without the NAACP?

Today, the NAACP is passing through a period of troubles. Critics say it is obsolete, and a few predict that it will die the lumbering death of an aged noble elephant. This causes one to reflect on the NAACP's origins and its achievements.

In 1905 W. E. B. Du Bois launched a movement at Niagara Falls, Canada, because "today, we have no organization devoted to the general interests of the African race in America." Twenty-nine educated Blacks answered Du Bois's call. A manifesto was issued, attacking fifteen areas of discrimination against Blacks. Vowing to protest tirelessly, the group's members pledged to "assail the ears of America with the story of its shameful deeds toward us."

This organization was called the Niagara Movement. For a time, it was successful. But its membership did not grow quickly, and it lacked funds. The movement gradually died.

The NAACP began where the Niagara Movement left off. As Blacks made new lives for themselves, they still led a precarious existence, filled with threats, attacks against their rights, and little economic progress. The immediate stimulus for the NAACP was an anti-Negro outbreak in 1908 in Lincoln's hometown, Springfield, Illinois. After that, a newspaper editor wrote, "We are going nowhere, and the world we live in is a mean one, turned against us."

On February 12, 1909, Oswald Garrison Villard, a grandson of Boston abolitionist William Lloyd Garrison, called for a national meeting for "the renewal of the struggle for civil and political liberty." Joining Villard were fifty-three prominent Black and white Americans, representing a cross-section of concerned citizens; among them were political advocate Ida Wells Barnett, Mount Holyoke College president Mary E. Woolley, social worker and writer Mary White Ovington, Rabbi Stephen S. Wise, Bishop Alexander Walters of the African Methodist Episcopal Zion Church, and W. E. B. Du Bois.

A later meeting of "two or three hundred persons of all shades" started the NAACP. The early organization was ambitious. National in scope, it opened state and local branches, controlled by an interracial board of directors. Most far-reaching was the group's pledge "to make 11,000,000 Americans physically free from peonage, mentally free from ignorance, politically free from disfranchisement, and socially free from insult."

Just as Du Bois had been the Niagara Movement's intellectual leader, he once again took center stage in the formation of the NAACP. He was a perfect pick for the director of publications and research. Soon, Du Bois started *Crisis* magazine, providing him with a place to present his studies of Black life and his strong opinions about discrimination. *Crisis* was selling thirty thousand copies by 1914.

One of the NAACP's early targets was the federal government and President Woodrow Wilson. Under Wilson, federal departments like those of the post office and the treasury were segregating their restrooms and cafeterias and demoting their Black employees. People taking civil service exams had to present photos so that Blacks could be weeded out. Wilson supported these restrictions, even though during his campaign in 1912, he pledged that "justice [be] done to the colored people in every matter." The NAACP led the charge, collecting thousands of names on petitions, holding rallies, and actually presenting protests to Wilson in person. The president continued his policy, but the issue had been made more visible. The NAACP then took on tougher problems: judge and jury prejudice, worker protection, attacks on voters, Black-white labor tensions. NAACP workers traveled to any scene of heightened racial tension to collect information, which they then transmitted to the press. Without the NAACP, a strategy for dealing with these racial and reform issues would not have matured.

In 1915, the year that marked fifty years since emancipation but one that had seen many lynchings, the NAACP's lawyers won their first legal victory before the U.S. Supreme Court. The court declared unconstitutional the so-called "grandfather clause" that kept Black voters off the South's voting rolls. (The grandfather clause was a law that stipulated that only citizens voting on or before January 1, 1857—a date preceding the adoption of the Fifteenth Amendment—were eligible to vote. Obviously, this excluded Blacks.) The NAACP went on to win other Supreme Court cases: against segregated neighborhoods in Louisville, Kentucky, in 1917; and a new trial for a Black man convicted of murder in 1923. In 1921, the NAACP was responsible for getting the House of Representatives to pass an antilynching bill. It was defeated, however, in the Senate because of a filibuster.

The NAACP served as the foremost civil rights advocate for Black Americans in the twentieth century. It stood for Black rights when the country was awash in racial violence and prejudice.

### How did Matthew Henson come to be the first man to reach the North Pole?

Matthew Henson is famous in American history because few people imagine a Black explorer of the Arctic regions.

Born in Maryland in 1866, Henson became independent when, at eight, his parents died. He ran away from his foster home and worked as a dishwasher in Washington, D.C. Then, at thirteen, Henson got himself hired as a cabin boy on a ship bound for Hong Kong. Captain Childs, who hired Henson, became his second father and taught him reading, writing, navigation, and seamanship.

This trip opened the world to Henson. For the next five years, he sailed to Japan, China, the Philippines, Spain, France, and Russia. Henson learned the customs of other peoples and discovered that he was agile with languages. In Russia, for example, he learned to speak the language and to drive sleighs.

It was 1887 when Henson met Robert E. Peary, who was

already recognized as an important explorer. Peary offered Henson a job on a venture to Nicaragua, beginning a quarter-century partnership between the two men.

Their first Arctic trip was in 1891. Over the next twenty years, they returned six times. Peary and Henson faced hard times on these journeys. Temperatures dipped to thirty degrees below zero. Severe winds laced them. Food rations of biscuits, dried meat, and fruit were inadequate. Traveling by dog and sleigh was dangerous. Often they ate the dogs when food was exhausted. After each journey, they returned to New York, where Peary raised more funds while Henson worked as a porter.

On all their trips, Henson sought to create a special bond with the Arctic Eskimo, learning their language, building sleighs, following their customs. To them, he was Mai Paluk, "the dark-faced one," whose hair even was different from whites. Henson saw the Eskimos as a smart, skilled, resourceful people. Indeed, Henson's contacts helped Peary's team survive in crises.

Peary and Henson were obsessed with finding a way to the North Pole. In 1908, they left New York for the historic journey that achieved their goal. Once in Greenland, Peary, Henson, and several other parties set out for the pole. This was an incredibly tough drive. Only steadfast determination kept them going.

Henson's sleigh team consisted of him and four Eskimos. On April 6, Henson and his team arrived at the North Pole. Peary came minutes later. In a simple ceremony, Henson put the American flag at the pole—an exhilarating moment, captured in a famous photo. Henson had lived for years in, as he put it, "the savage, ice- and rock-bound country." He was to leave "with the added satisfaction of complete success." He added, with feeling, "I, a lowly member of my race, . . . had been chosen by fate to represent it, at this, almost the last of the world's great work."

Back home, Peary was instantly honored while Henson was ignored. He even found it hard to get a job. And a stingy Congress refused *four times* to vote him a pension, although they had given pensions to many other Arctic explorers. Eventually, he was given a gold medal by Congress, and in 1950 President Truman supported a ceremony that honored him. Five years later, Henson died.

In 1912, Henson told his breathtaking story in *A Negro Explorer at the North Pole.* Nothing in Black writing is as compelling as this record of trial and effort.

**Why was Madam C. J. Walker voted in 1992 into the Business Hall of Fame?**

Sarah Breedlove was her original name. Poverty-ridden Delta, Louisiana, was her birthplace in 1867. She began as a laundry woman, but by 1915, she was Madam C. J. Walker, head of a vast empire of women's hair-care products, hair salons, and sales agents. The journey from Sarah Breedlove to Madam C. J. Walker is one of the most impressive in American business history.

Walker explained, "I got myself a start by giving myself a start." Her first step was to move to Saint Louis in 1888, taking along her young daughter Lelia. (Her first husband, Moses McWilliams, was killed in an accident.) Saint Louis was a bustling town, with a large Black community and numerous upper-class, educated Blacks. Sarah was excited by the city's spirit and its business opportunities.

Her gold mine of an idea was to create and market products for Black women's special hair problems. Everywhere she noticed Black women with patchy, uneven hair or women having trouble growing long, healthy hair. Poor diets, bad care habits, stress, and dangerous hair potions contributed to these conditions. She had the same problems. For years she experimented at home with preparations, testing them on herself and others until she had three market-ready products with appealing names: Wonderful Hair Grower, Glossine, and Vegetable Shampoo.

Married now to C. J. Walker and giving herself the title of "Madam," like French beauty experts, she was prepared to spread her hair-care gospel. Her new name gave her enterprise a sense of flair, but her business planning made her a success. Although lacking any formal training, she was arguably the most astute marketing expert of her day.

A whirlwind of energy, she traveled constantly, giving product demonstrations to Black women. Using the Vegetable Shampoo, she washed a customer's hair, applied the Wonderful Hair Grower, and then, using Glossine, pressed the hair with a curling iron. While traveling, she began to hire women agents to sell door-to-door.

Walker put her photo on the products, figuring that Black women would identify with one of their own. Her husband, an advertising expert, ran dramatic advertisements in Black newspapers with "before" and "after" photos of women who used the products.

All lights were green after that. Madam Walker entered the mail-order business. She opened a training college for hairstylists in New York and named it Lelia College after her daughter. By 1912, her company had moved to Indianapolis, where it employed 1,600 agents and earned thousands of dollars a week. Crowning this success was a fabulous new beauty salon in New York City.

Madam Walker never forgot her origins and remained a staunch supporter of racial causes. Offended by petty discrimination against Blacks in Indianapolis, she built a blocklong complex in the Black neighborhood, boasting a factory, restaurant, and movie theater. She made donations to a host of Black causes and became a leader in antilynching crusades.

As her wealth grew, she wanted to live in New York and participate in the city's dazzling Black cultural life. In 1917, she built a spectacular mansion—Lewaro—at Irvington-on-Hudson. Its architect was Black, V. W. Tandy. Even the *New York Times* took note. Lewaro contained fine art and furniture and the most modern appliances; it became known for its dinners, intellectual fetes, and musical performances.

Madam Walker was elected in 1992 to the Business Hall of Fame. She was the first self-made millionairess in American history.

## Who was the "Black Apollo?"

In Greek mythology, Apollo was the god of medicine, poetry, and music—the higher arts—as well as of physical beauty and grace. "Black Apollo" is the name given recently to Ernest Everett Just (1883–1940) because he was a fine scientist, a diplomatic person, and incidentally, a man of classic good looks. His field was marine biology, and his research took him deep into the basic questions of life.

From Charleston, South Carolina, Just was educated at the country's leading institutions: New Hampshire's Dartmouth College, where he graduated first in his class, and the University of Chicago, from which he received a Ph.D. From 1907, he was on the faculty of Howard University.

For twenty summers, Just conducted research at the Woods Hole laboratory on Cape Cod, a great place for marine biology research. During his time there, he met some of the world's most distinguished scientists.

Just's main focus was the reproductive cells of marine animals. By studying how these cells developed normally, he hoped to discover how abnormal cell development took place in diseases like cancer, sickle-cell anemia, and leukemia.

Another Just specialty was devising new laboratory techniques. He pioneered safe ways to handle precious cell material and to conduct highly delicate experiments. Scientific journals carried his papers describing experimental procedures. In 1939, his tips to scientists were published in *Basic Methods for Experiments in Eggs of Marine Animals.*

Unfortunately, Just's brilliance and love of science did not protect him from discrimination. America was not ready for a Black scientist of his caliber. Few American research institutions invited him to work in their laboratories. In the 1930s, he went to work at the Kaiser Wilhelm Institute for Biology in Berlin, where he finally found what he longed for—intellectual friendship. When France fell to Nazi invaders on June 14, 1940, and foreigners were faced to leave Germany, Just returned home with a German wife. He was ill and depressed.

Just died on October 27, 1940, at age fifty-seven. This top-flight scientist was an important first for Black Americans.

## How did Blacks respond to the "War to End All Wars"?

More Blacks served in World War I (1914–18) than in any previous American war. Almost four hundred thousand Black men enlisted or were recruited into the military. When President Wilson appealed for volunteers, Blacks surprised the nation by showing up in great numbers. Said a leader, "They felt it was their patriotic duty to offer their services to their country." Two hundred thousand served in France. Black women served as well, some overseas but most as nurses to Black troops in stateside base hospitals.

Because of the numbers of Blacks serving in the war, many Black communities learned about Europe through the soldiers' eyes. Black newspapers, such as the *New York Age,* printed many articles on the soldiers' exploits. Soldiers' letters told of their sacrifices. When soldiers returned, the Blacks who had stayed at home hung on their descriptions of the war.

Living the war through stories was the best way. The war was a major conflict among the great powers, which used devastating chemicals and nerve gases. By the war's end, thousands of young men had died. Soldiers wrote of the fields of dead. World War I did not end war, but the world was justifiably shocked by its carnage.

On the battlefields of Europe, Black heroes were made. As was the custom, they served in all-Black divisions under white commanders. The Ninety-second and Ninety-third Infantry Divisions were the principal Black units.

Regiments from the Ninety-second Division served with French troops. Captain Hamilton Fish, one of the unit's officers, wrote, "Our regiment is the most envied American regiment in France. We are, to all intents and purposes, a part of the French Army." Back home in Black America, they were called the "Hellfighters." For 119 days, longer than any other American unit, they were under continuous shell fire. They pushed forward to the German front faster than other regiments. The Croix de Guerre and other honors were showered on them by the French.

Before and during the war, Black America asked, Will this war help our progress? A few Blacks felt that they should not fight for America against a Germany that was not their enemy. Everywhere, Blacks objected to the unfair treatment of their soldiers. Newspaper editors wondered: If Black soldiers secured democracy in Europe, would they be granted democracy on their return? Simultaneously, soldiers fought and communities debated.

On the blue, perfectly cloudless day of February 17, 1919, these weighty concerns evaporated. That was the day the "Hellfighters" came back to Harlem, some two thousand of them. First, they marched up Fifth Avenue, greeted by a crowd of New Yorkers. Proceeding north, they reached Lenox Avenue, one of Harlem's boulevards. James Europe's regimental band was ready with its famous syncopation. The Black crowd went wild. A reporter wrote, "The Hellfighters marched between two howling walls of humanity."

The war was over. But the irony of Black soldiers fighting so valiantly for other people's democracy and having so few rights themselves underscored issues that would eventually rise to the surface.

### Why was Oscar Micheaux called the "dean of Black filmmakers"?

Before becoming a filmmaker, Oscar Micheaux (1884–1951) was a successful homesteader in South Dakota; the author of *Homesteader,* a story of hard life on the frontier; a railway porter; and the owner of a book business. But he dreamed of making movies; in fact, it was an obsession. *The Homesteader,* his first film, was made in Sioux City, Iowa, in 1919. To finance it, he sold shares in his virtually nonexistent company to local farmers and small businessmen. Then he went from city to city with the film and its posters under his arm, asking theaters to show it. He even asked theater owners to invest in his next film, promising them the first option on showing it.

Micheaux's films began to draw crowds. In the 1920s Blacks were so anxious to see themselves defined by a Black in the new medium of the movies that they made Micheaux a success, encouraging him to work like a fanatic.

Over ten years, Micheaux made thirty films. He filmed fast. He did not pause to reshoot scenes or perfect the mood of a dramatic moment, and occasionally he did not edit carefully. But his films had a more important quality: besides being entertaining, they commented on society and made people think. Fluffy musicals, comedies, gunslinging Westerns, and gangster films didn't interest him. Instead he brought to the screen the real issues of struggling city Blacks: the lowlife element of Black prostitutes, gamblers, and thieves; racial mixing, interracial marriage, and Blacks passing for white; and injustice against Blacks.

For all of his productivity, few of his films survive. For years it was thought that his 1925 film *Body and Soul* was the only one that still existed. However, luck struck in the 1980s. A copy of his 1919 film *Within Our Gates* was found. It has been restored and shown, with a new score composed by violinist India Cooke and pianist Mary Watkins. The main story line revolves around a woman—played by Evelyn Preer—who has a violent past and who seeks a benefactor for a Southern school for Black children. The film is surprisingly long, has high-quality camera work and good lighting, good acting, and some great scenes. It is Micheaux's answer to D. W. Griffith's *Birth of a Nation* (1915), a very popular film among whites that depicted Blacks as demons. *Within Our Gates* confronted racism in a bold way, daring to show the taboo subjects of lynching and of white men raping Black women.

Micheaux's film company survived the Great Depression to continue to make films in the 1940s; his last film came out in 1948. When he died in 1951, he was not a celebrity. Current recognition has come from film studies and the Black Filmmakers Hall of Fame. In 1987, Hollywood's Avenue of the Stars gave him a star.

## What was the "Great Migration"?

From 1915 to the 1930s, 1.8 million Southern Blacks migrated to industrial cities in the North and Midwest.

Called "a stealing away from the South," this was a tremendous population shift for America. At its peak, sixteen thousand blacks a month left the South. The layout of Blacks on the American landscape changed forever—hence the term the "Great Migration." Black relations with other Americans were fundamentally altered.

In 1900, there were nine million Black Americans, nearly 12 percent of the nation's population of over seventy-five million. Over 90 percent of Blacks lived in the South.

The South had been the American homeland of Blacks

from slavery's earliest days. In 1900, they were one-third—eight million—of the South's total population. No one could have predicted that a people so fixed within a region would leave in such numbers.

The Black South cracked first during World War I, from 1914 to 1918. Almost five hundred thousand left in those few years. During the 1920s, another eight hundred thousand moved. The Depression in the 1930s slowed Black migration a bit. Still, even under these bad conditions, four hundred thousand Blacks fled, most leaving Florida, Georgia, South Carolina, and Virginia.

Causes of the migration varied. At first, Blacks came North to get wartime factory jobs; factory work paid well compared to farm wages. Violence also pushed people North. In many places in the South, Blacks lived poorly, were paid the lowest wages, had the least nutritious diets, and suffered bad health. Fifty years after the Civil War, Blacks were virtually prisoners in the South's rural backwaters.

When Southern whites came to their senses, they were thunderstruck. Migrating Blacks challenged the basic Southern assumption that Blacks were passive, dependent, and unambitious. Still, some whites approved of the migrating Blacks, figuring that this freed the South of "educating and civilizing" Blacks. Blacks were considered "the white man's burden." New Orleans's *Times Picayune* cheerfully declared, "As the North grows blacker, the South grows whiter." Whites believed that Blacks would return in "a backswing of Negroes" once "many darkies" found "the promise of high wages and social equality" illusory. When the North's welcome turned chilly, the Blacks would come scampering back.

But Blacks did not return. They stayed in the North, the Midwest, and the Far West, familiarizing themselves with their new homes and new ways. In time, prominent white Southerners realized that Blacks had real gripes about their oppression. The realization came too late.

## Was Marcus Garvey serious about going "Back to Africa"?

When Marcus Garvey arrived in Harlem in 1916 from Jamaica, he brought a powerful dream of a Black empire run by

Blacks. He had traveled to South America, Central America, and Europe, and everywhere Blacks were at the bottom of society, exploited and mistreated.

To reverse this, Garvey created the United Negro Improvement Association (UNIA), aimed at restoring Black independence. Soon he had acquired the *Negro World,* a newspaper whose pages could be used to broadcast his message. Central to his dream was a Black return to Africa, because "the Negro peoples of the world should concentrate upon the object of building . . . a great nation in Africa." His motto was "Africa for the Africans."

Garvey's idealism became the property of Black millions—not only in the United States but also in Africa and the Caribbean. He explained his vision in this way: "Gradually . . . the Negro peoples of the world will have to . . . go forward to the point of destiny as laid out by themselves, or must sit quiescently and be exterminated ultimately by the strong hand of prejudice." And "nationhood is the strongest security of any people and it is for that the Universal Negro Improvement Association strives at this time."

Not immediately successful, Garvey persisted, and eventually he flourished. "Broad, short, always in a three-piece suit," he was a brilliant mass organizer, a superb speaker, and a smart self-promoter. Secretly, Black leaders—like W. E. B. Du Bois— probably envied his abilities while disparaging his ideas. The bad racial situation in America after World War I also helped Garvey. Blacks saw a wall of intolerance being built around them. Northern cities were not paradises, and the South was still hellish. Soldiers had fought to make a "world safe for democracy" but had not secured it at home.

Beyond this, the Garvey movement often presented Harlem with a visual feast. Annual UNIA parades were holidays; brass bands, Black Cross nurses, and the uniformed African Army marched while Garvey led the parade in a carriage, wearing a plumed hat. He was the "provisional president of Africa." Shiny banners carried his messages. Waving in the air were his movement's proud flag, of red (representing the one blood of Blacks around the world), black (for the world's Blacks), and green (for Mother Africa). In 1921, Garvey was at the height of his power.

Garvey never settled a Black colony in Africa. He formed a steamship company, the Black Star Line, to take Africans back to Africa and raised millions for it. Then New York officials investigated his accounts, and soon the federal government charged him with mail fraud. In 1925, he was given a five-year sentence, with two served in an Atlanta federal prison before being deported to Jamaica. In 1940, he died in London.

Garvey definitely jolted the Black masses with his strong Black nationalism. He shook American Blacks out of their narrow view of the world, furthering their identification with Africa and the West Indies. Leaders such as W. E. B. Du Bois were not so thrilled, regarding his plans as fantasy. But this did not prevent his message from reaching Africa, where it influenced the struggle against colonialism. Black nationalism was refreshed by this Jamaican immigrant and his vision.

### Where did the Blues come from?

The Blues are a Black American musical form that grew into distinctive shape at the turn of the twentieth century.

They came from a rather small territory within Black America. Yet, they became the basis for nearly all of today's popular music—for sure, rhythm and blues. Rhythm and blues is just a few evolutionary musical steps from blues. Obviously too, rock and roll has profited immensely from blues. Many a blues lyric, yelp, moan, and guitar lick have entered rock. British rockers of the 1960s were mentored long-distance by listening to blues records. The Blues are also the soulful backbone of that other great indigenous American art form—jazz.

The Blues' influence goes even further: it has helped form twentieth-century identity. Any definition of modern identity has to wrestle with the blues legacy. It is for this reason that when Ralph Ellison wrote *Invisible Man* (1953)—that celebrated anatomy of post-World War II individuality—his main character spoke of his condition in the blues idiom. Ellison's character ponders: "What did I do/To feel so black/And blue?"

The first published and recorded blues were not actually

"blues." They were derivations of the "coon songs" of Black minstrel shows that had toured throughout the South since the days of Reconstruction and into the 1910s. In 1912, W.C. Handy (1873–1958) published an instrumental composition called "The Memphis Blues." In 1914 came his "The St. Louis Blues" and still later, "Yellow Dog Blues." Each of these were written for bands or orchestras, and while Handy promoted himself as "Father of the Blues," these were rather distant from the music that today is considered the blues.

When first Okeh and then Paramount Records began recording blues in the 1920s and found a public hungry for the music, it was primarily female singers rhyming scads of double entendres that were heard, which came originally from traveling orchestras. This was not blues either as it came to be known. Within this bevy of female performers, the incomparable Bessie Smith (1894–1937) arose. A 1925 Paramount poster boasts other female songsters—"Paramount Stars"—such as Alberta Hunter (1895–1984), Gertrude "Ma" Rainey (1886–1939), and Ida Cox (1896–1967).

Initially, these female singers' songs borrowed heavily from pre-blues forms like the jump-up and the cakewalk. However, they were soon making records of their own compositions. Soon, they were also introducing more rural folk blues into their work. Male blues singers were coming along too—Sylvester Weaver (1897–1960), Papa Charlie Jackson (? – 1938), Lonnie Johnson (1889–1970). They too drew upon more of the country flavor. Blind Lemon Jefferson (1897–1929) from Texas entered the record market in 1926 with his deep-South blues and caused a dramatic upsurge in demand for down-home singing and playing. During the 1920s–1930s, blues came into their own. They became "down home."

What are the blues? Most basically, the blues are a form, a structure. Although early blues ranged from 8 bar to 32 bar patterns the traditional form is what is referred to as "a 12 bar blues." The verses are made up of three lines each four bars long. The chord progression moves across the 12 bars like so: the first four bars on the tonic chord, the next two bars on the sub-dominant chord, with a return to the tonic chord for the last two bars of the line, and for the third and final line the progression moves for two bars to the dominant seventh to then return for the last two bars to the tonic (I-IV-V-IV-I). As the old bluesmen say, the verses are "rhymed up."

*I got ramblin'*
*I got ramblin' on my mind*
*I got ramblin'*
*I got ramblin' on my mind*
*Hate to leave you baby*
*but you treats me so unkind.*

Secondly, the blues are a feeling. Blues are a strongly contoured use of "floating choruses" in a personal statement that generally is not "political" but details an individual's confrontation with some troubled situation—a mean lover, a worried mind, debts, misfortune. In getting away from these troubles, blues often express deep human yearnings. Songs tell of the need to be with someone beautiful and kind, to be loved, to be someplace good, to be free of dignity-stripping law, to be decent in work, and just as essential, to have some fun out of life.

Yearning in blues is understandable. Blues drew upon the songs of hardluck Blacks—the hollers of fieldworkers, the plaintive appeals of prison road gangs, and the songs of work crews. These songs influenced not only the blues' phrasing and rhythms, but gave them an unmistakable feeling that has allowed the music to speak across cultures. When Willie Brown sang "Can't tell my future, I can't tell my past, And it seems like every minute sure gonna be my last," he seems to speak for many. When Ma Rainey says despairingly in "Bo-Weevil Blues" that "I went down town, I got me a hat, I brought it back home, I laid it on the shelf, I looked at my bed, I'm gettin' tired sleepin' by myself," she appears to personify a common predicament. Blues did not hold back on feelings.

Third, blues is geography and economy. Country blues came from the Mississippi Delta, East Texas, and the Piedmont. But, the Delta—with its thousands of Blacks (four to every one white), its vast poverty of sharecropping, its back-breaking field labor in cotton—surpassed the other regions in creativity. In the Delta were tiny towns and juke joints (often owned by big landowners)—with live blues, dancing, cooked food. Few Blacks made money from sharecropping. Even fewer ever bought their own land. Many were cheated by their landlords. But, on payday, with their few dollars in hand, workers streamed to hear the blues. The juke joints were oases.

In blues, the world has the greatest example of a Black art form called into being by the most outcast, lowly class in a race that was at the bottom of a society at the turn of the century in America. The instrumentation of the single guitar fretted by hand and often with a bottleneck or kitchen knife to create a slide sound and the self-conscious vocal documentation of the particulars of Black life is the most perfectly sealed aesthetic world in Black music and American music history. And it was this music that Robert Johnson (1891–1938), Son House (1902–1988), Charlie Patton (1891–1934), and Furry Lewis (1893–1981) produced. Tough-singing women, like Memphis Minnie (Lizzie Douglas, 1896–1973) also added their special narratives to blues' development.

From the Mississippi Delta, blues traveled up Route 61, the blues highway within the World War II era Black migration from Mississippi to Chicago. Their first stop was often Maxwell Street, a social and business center catering to the newcomers. In the late 1940s Chicago, Big Bill Broonzy (1893–1958) was playing the blues of the Delta for transplants. But it was Muddy Waters (1915–1983), a son of Clarksdale, Mississippi, a disciple of Robert Johnson, who in Chicago in the early 50s transformed the blues of the Delta into the electrified "Chicago Blues." By 1955 with the rise of Soul Music, the blues would never again top the Black charts as they had in the early 1950s. Generally, Black audiences turned away from the Blues, considering them "that old slave music."

British rockers' adulation of blues caused a 1960s revival. Likewise, the 1990s returned the focus to blues as many historic folk blues compilations were released—as in 1991, with the Grammy-winning *Robert Johnson: the Complete Recordings,* which sold a half million copies. A club circuit and festivals support present-day blues performers—such as Riley "B.B." King (1925– ),

Cora "Koko" Taylor (1928– ), Robert Cray (1953– ). But in relation to a Black audience the modern blues performer is as much a marginal as the wandering blues man of the 1920s.

### Who was Alain Locke, and what was his concept of "the New Negro"?

Alain Locke (1886–1954) was, first of all, a thinker, a man infatuated with ideas. In his professional life, he was known as a "philosopher," a person who earned his living by thinking about things. Up to that point, this was a rarity in Black life.

From the start, he was successful. From his elite Philadelphia family and excellent high school record, he went to Harvard, where he was elected to Phi Beta Kappa and graduated magna cum laude in 1907. After Harvard, Locke went to Britain's Oxford University (1907–10) as a Rhodes Scholar—the first (and only, until 1962) Black American to receive this honor. Studies in the intellectual capitals of Paris and Berlin followed. Later, he acquired a doctorate from Harvard. When Locke became the chair of Howard University's philosophy department in 1918, he was easily one of the most educated people in America.

Locke was a keen observer of the Black American scene. He saw Blacks changing outwardly and inwardly—not only were they wearing more stylish clothes but they were also thinking different thoughts. Everywhere, he saw a more forceful, more expressive Black people. Black arts, literature, and music were areas where this energy expressed itself most intensely.

In the 1920s, Locke wrote about "the New Negro" and "a spiritual emancipation" that was occurring. The "Negro community" was entering "a new dynamic phase," he explained. "The younger generation is vibrant with a new psychology; the new spirit is awake in the masses." He quoted a poet who captured these sentiments:

*We have tomorrow*
*Bright before us*
*Like a flame.*

*Yesterday, a night-gone thing*
*A sun-down name.*

*And dawn today*
*Broad arch above the road we came.*
*We march!*

He went on to say, "The day of the 'aunties,' 'uncles,' and 'mammies' is equally gone. Uncle Tom and Sambo have passed on." Blacks were holding their heads high.

Harlem—in Locke's words, "the Mecca of the New Negro"—was his prime example of this revolution in Black consciousness. It was the epicenter of Black newness. In the 1920s, Harlem was the largest Black urban community in the world. Blacks came from all over America, the West Indies, and even Africa—and from every class—to participate in Harlem's scramble for modernity. By 1930, more than 164,000 people lived there. Its collective life in the 1920s became a testament to the fact that Blacks were shedding their old selves.

Locke was also a cultural impresario for 1920s Black arts and writing. He sought new talent and found wealthy patrons to help them. Many were introduced to the public by Locke: writers Rudolph Fisher, Zora Neale Hurston, Langston Hughes, Claude McKay, sculptor Richmond Barthe, and muralist Aaron Douglas. Locke edited the first collection of Black plays in 1927. This cascade of talent became the Harlem Renaissance.

By the mid 1930s, Locke was critical of the movement he had promoted so zealously. Economic and social improvement was desperately needed by Blacks. But Locke was still clear that ideas, as presented in arts and literature, should be pivotal to any Black progress.

## What was the Harlem Renaissance?

A renaissance is a new birth, rebirth, or revival. In the 1920s, Harlem was the center of a Black rebirth. All of a sudden, a small bit of territory in Manhattan—from 130th to 155th Streets—began to acquire a magical reputation.

Blacks—and many whites—responded to its legend and came running to it, seeking their fortune. By the mid 1920s, Harlem was bursting at the seams with newly arrived Blacks and Black talent. It was a mixture that generated a spectacular life.

The most prominent Harlem Renaissance product was literature—poetry, novels, plays, reminiscences. Today, the period's literature is considered some of the world's greatest. As early as 1917, signs were clear that new literature would be coming from Harlem. In that year, a time of intense anti-Black violence, Claude McKay (1889–1948), a West Indian living in Harlem, published these militant lines:

*If we must die, let it not be like hogs*
*Hunted and penned in an inglorious spot. . . .*
*What though before lies the open grave?*
*Like men, we'll face the murderous cowardly pack,*
*Pressed to the wall, dying, but fighting back!*

McKay would go on to write *Home to Harlem*, a distinctive novel of the Black underworld. Countee Cullen (1903–46) produced collections of fine poetry, decorated with art deco drawings, the most important being *Color* and *Copper Sun*. Wallace Thurman (1902–34) wrote a major novel, *The Blacker the Berry*, which took a hard look at the problem of color prejudice among Blacks themselves.

Just as important was the poet Langston Hughes (1902–67), an active writer and a celebrator of mass Black society. Hughes's poem *The Weary Blues* was a literary triumph. Deftly using folk music and dialect, this and his other poems explored the common Black urban experience, the world of the average people—workers, church folk, street people, and artists. One of Hughes's greatest contributions was "The Negro Speaks of Rivers," a poem tracing Blacks' history through the rivers important to their lives, from the Nile to the Mississippi.

Attuned also to Black folklife was writer Zora Neale Hurston. Her greatest writing came after the Renaissance, though the significant stories "Spunk" and "Drenched in the Light" appeared in the 1920s. These stories foreshadowed her later novels, like *Their Eyes Were Watching God.* Her great strengths were her command of Black folklore and her ability to celebrate the vibrancy and creativity of uneducated rural Black people.

Nella Larson's (1893–1963) *Quicksand* and *Passing* dealt with the middle classes, featuring Black characters who passed for whites. Rudolph Fisher (1897–1934) wrote interestingly of the divisions and deceptions of Harlem life in *The Walls of Jericho* and *The Conjure Man Dies,* a detective novel.

Harlem also gave rise to gifted visual artists. Augusta Savage (1900–62) made plaster, stone, and bronze sculptures highlighting the Black body and face. Her *Gamin,* a sculpture of an attractive street-smart boy, was a popular piece. Aaron Douglas (1899–1979) gave the mural new life as an art form, making large, multihued, geometric murals of Black history.

The Harlem Renaissance attracted white artists too. Carl Van Vechten's novel *Nigger Heaven* allowed voyeuristic whites to experience Harlem's "unleashed passions." Van Vechten even included a glossary of "Harlem's almost secret language." The book sold one hundred thousand copies, but Black critics resented its title. Less provocative were Charles Cullen and Winold Reiss, white artists whose illustrations turned Harlem Renaissance books into handsome collectibles.

Reflecting on the era, Alain Locke called it "our little renaissance," a modest appraisal. The range of expression and quality of work made this a very big rebirth of Black cultural expression.

### What were some great places to go in Renaissance Harlem?

Harlem attracted Blacks from every area of the world, as well as white Americans and Europeans. Every class of people came: from the fur-wearing, silk-clad, Bentley-driving white aristocrats to "bohemians" from the art and literary worlds.

Harlem did not disappoint. A spectacular social and political life could be had there: clubs, dances, private parties, bars, political rallies, street-corner orators, literary workshops and salons, great bands, spiritual seances, pimps and gangsters, and, when needed, powerful churches with great preachers. Above all, Harlem could be glamorous. Some Harlemites had it good: Savile Row suits for the men, furs and five-hundred-dollar dresses for the women, fancy cars, diamonds, slick hairstyles, and the latest hats. Celebrities ran from Lillian "Pig-Foot Mary" Harris, who made a fortune selling pig-feet meals, to Walter White, who was a novelist and a spy on anti-Black Southern groups.

If you were on a tour of Renaissance Harlem, these would be some "must-stops":

- *The Cotton Club:* This upscale nightclub, largely frequented by whites, featured performers such as Duke Ellington, Cab Calloway, Edith Wilson, and a talented group of women ("bronze beauties") known as the Cotton Club Dancers.

- *Seventh Avenue between 127th and 134th Streets*: Called "Black Broadway," this was a wide "promenade" frequented by the stylish during the roaring twenties. Churches, cabarets, speakeasies, and restaurants lined the street.

- *Dunbar Apartments:* Named for poet Paul Laurence Dunbar, this huge apartment complex housed many of the best-known Harlemites, from writer and political leader W. E. B. Du Bois to bandleader Fletcher Henderson.

- *Lafayette Theater:* This large auditorium admitted racially mixed audiences for popular shows given by major performers and groups.

- *The 132nd Street "Tree of Hope":* The bark of this aging elm tree had been rubbed smooth by passersby seeking good luck.

- *Karamu House:* This theater and neighborhood association put on plays by Black playwrights.

- *The Hobby Horse:* This bookshop in Harlem attracted young Black writers and artists.

- *Marcus Garvey's African Orthodox Church:* This was a place where you could hear Garvey's message, buy his newspaper, and join one of his auxiliary groups. A highly decorated stage included Garvey's "throne-seat."

- *Edmond's Cellar:* A small place, seating 150 to 200 people "jammed close together around a handkerchief-size dance floor," the cellar was a haven for Harlem's lowlifes—drug dealers, prostitutes, and underworld figures.

- *Savoy Ballroom:* This huge dance hall, opened in 1926, could hold four thousand people and featured every major Black performer in Harlem at some point.

If you had any time left over, you might try out "Jungle Alley" (a district packed with cabarets and nightclubs), Saint Philip's Episcopal Church (a prominent social center and church for the wealthy), or Small's Paradise (a cabaret featuring mostly Blacks).

### What was the "house rent party"?

Poor Blacks moving into apartments and houses in Northern cities often found it hard to meet the rent, especially since landlords bilked the newcomers of every cent they could get.

The "rent party" originated in the South and was brought north to places like Harlem. Soon, much to the dismay of upper-class people, house rent parties caught on. Often masked by names such as "parlor social" or "social whist party," they were thrown in hopes of collecting enough money to avoid eviction.

A traveling printer created rent party "tickets"—actually little advertisements with lines from songs, slang, or a joke. "Come and Get It Fixed." "If You Can't Hold Your Man, Don't Cry After He's Gone, Just Find Another." "Old Uncle Joe, the Jelly Roll King Is Back in Town and Is Shaking That Thing." "If you can't do the Charleston or do the pigeon wing, you sure can shake that thing at a Social Party." These tickets announced the party's location and date, and let it be known that "good music" and "refreshments" would be available.

Partygoers were received by a hostess—who did not know them and would not remember them—to whom they paid twenty-five to fifty cents admission. In the corner was a little band, occasionally featuring a celebrity pianist, a drummer, and maybe a saxophonist. The evening's first music was slow, but as the night went on, the pace quickened. At some point, the hostess would announce over the din that the "eats" of the evening were available for purchase; pig's feet, chili, chitterlings, okra gumbo, chicken, hog maws, black-eyed peas, biscuits, sweet potato pone, and sweet potato pie were the usual offerings.

Harlem's rent parties were very popular for Saturday night entertainment. They attracted the ordinary people, though some high and mighty folk might also drop by. Writer Langston Hughes and photographer Roy De Carava published photos of Harlem rent parties in *Sweet Flypaper of Life.*

**Was February made Black history month because it is the year's shortest month?**

A pernicious little rumor has it that a conspiracy of unnamed people packed Black history month into February because it shortened the observance of Black historical achievements. Usually skeptical and right-on syndicated columnist Cynthia Tucker of the *Atlanta Constitution* recently bought into the idea: "The contributions of Blacks are trivialized by the act of picking a month, just one month, to celebrate them. Is it a mere coincidence that they are celebrated during February, the shortest month of the year?"

She is right that Black history month should really be Black history year. But nothing is fishy about February being the home of Black history month. It was chosen by Blacks.

Black history month was once limited to "Negro History Week." Both celebrations have always been in February. Carter G. Woodson, "the father of Black history," established the week in 1926 as an annual event. He chose the second week of February to commemorate the births of two people who had great impact on Black history—Frederick Douglass (born on February 7, 1817) and Abraham Lincoln (born on February 12, 1809).

Woodson wanted to increase the knowledge of Black history within Black communities, but he was just as anxious to spread it to interested others. He was convinced of the therapeutic power of Black history: Blacks would be seen as integral to the country if their contributions were known. Explaining his plan, Woodson wrote, "If a race has no history, if it has no worth-while tradition, it becomes a negligible factor in the thought of the world, and it stands in danger of being exterminated."

Easier said than done. Months before the first week's celebration, he sent out pamphlets and other materials to elementary and secondary schools, colleges, women's associations, the newspapers and journals of the Black community, and state departments of education. This literature explained why the week was necessary and gave instructions on how to celebrate it. Each year, Woodson improved this material, offering new books to groups. One was a valuable "Table of 152 Important Events and Dates in Negro History."

Once the word spread, Negro History Week took off. In his diary, Lorenzo J. Greene—a research assistant for Woodson—reports on February 10, 1930: "When I stopped by his office [in Washington, D.C.], I found Woodson was busy with people coming from different parts of the country for the Negro History Week celebration." After a banquet for four hundred people, a public program drew a crowd that "finally numbered between 5,000 and 6,000." The same fervor was present throughout the country.

Woodson was an odd combination of workaholic and visionary. Born in 1875 in Buckingham County, Virginia, where there is now a monument to him, he struggled to get an education, starting school when he was twenty. By 1912, Woodson had been to Berea College in Ohio and the University of Chicago, and he graduated from Harvard University with a Ph.D. He was the first Black of slave parentage to earn a doctorate in the United States.

In 1915, he was the guiding force behind the Association for the Study of Negro Life and History, today the most influential organization promoting Black history. Besides nurturing the

Black history movement, Woodson wrote in rapid succession many books, including *The Negro in Our History* (1922), still an important reference book.

In 1940, W. E. B. Du Bois declared that Negro History Week was one of the great 1920s cultural innovations. The week helped Blacks to overcome the sense of inferiority that was the inheritance of slavery and Jim Crow. Much of what the public knows today about past Blacks comes from Woodson's plan for a February celebration.

## RESOURCES

Adero, Malaika, ed. *Up South: Stories, Studies, and Letters of This Century's African-American Migration.* New York: New Press, 1983.

Anderson, Jervis. *This Was Harlem, 1900–1950.* New York: Farrar, Straus, Giroux, 1982.

Aptheker, Herbert. *A Documentary History of the Negro People in the United States.* Vol. 3. New York: Citadel Books, 1973.

W. Fitzhugh Brindage, *Lynching in the New South: Georgia and Virginia, 1880-1930.* Urbana and Chicago: University of Illinois Press, 1993.

Brodie, James Michael. *Created Equal: The Lives and Ideas of Black American Innovators.* New York: William Morrow, 1993.

Davis, Francis. *The History of the Blues.* New York: Hyperion, 1995.

Greene, Lorenzo J. *Working with Carter G. Woodson, the Father of Black History: A Diary, 1928–1930.* Baton Rouge: Louisiana State University, 1989.

Haskins, Jim. *The Cotton Club.* New York: Hippocrene Books, 1977. Text with photos.

Hazzard-Gordon, Katrina. *Jookin: The Rise of Social Dance Formations in African-American Culture.* Philadelphia: Temple University Press, 1990.

Higginbotham, Evelyn. *Righteous Discontent: The Women's Movement in the Black Baptist Church, 1880–1920.* Cambridge: Harvard University Press, 1993.

Lewis, David Levering, ed. *Harlem Renaissance Reader.* New York: Viking Books, 1994.

———, ed. *W. E. B. Du Bois: Biography of a Race.* New York: Henry Holt, 1993.

———, ed. *W. E. B. Du Bois: A Reader.* New York: Henry Holt, 1995.
———, ed. *When Harlem Was in Vogue.* New York: Alfred A. Knopf, 1981.
Lott, Eric. Love and Theft: *Blackface Minstrelsy and the American Working Class.* New York: Oxford University Press, 1993.
Neverdon-Morton, Cynthia. *Afro-American Women of the South and the Advancement of the Race, 1895–1925.* Knoxville: University of Tennessee Press, 1989.
Philadelphia Museum of Art. *Henry Ossawa Tanner.* Philadelphia: Philadelphia Museum of Art, 1991. Text with artwork.
Reagan, Bernice. *We'll Understand It Better By and By.* Washington, D.C.: Smithsonian Institution Press, 1992.
Studio Museum in Harlem. *Harlem Renaissance Art of Black America.* New York: Harry N. Abrams, 1987. Text with photos.
*Three Negro Classics.* New York: Avon Classics, 1965.
Willis-Thomas, Deborah. *VanDerZee: Photographer, 1886–1983.* Washington, D.C.: Smithsonian Institute, 1994. Text with photos.

A TEXT FOR YOUTH

Freedman, Suzanne. *Ida B. Wells-Barnett and the Antilynching Crusade.* Brookfield, Conn.: Millbrook Press, 1994.

# 7

# Earnest Black America

## The 1930s–1950s

1930s: The Swing Era commences, with Duke Ellington, Count Basie, and Jimmie Lunceford as leaders

1931: Scottsboro trial of nine Black males, ages thirteen to nineteen, starts in Alabama

1931: J. A. Rogers publishes *World's Greatest Men of African Descent*

1932: Franklin D. Roosevelt is elected president, promising a "New Deal"

1933: Mary McLeod Bethune joins Roosevelt's "Black cabinet"

1937: Zora Neale Hurston publishes *Their Eyes Are Watching God,* a classic novel

1938: Joe Louis, in his second try, defeats German Max Schmeling

1939: Marian Anderson sings before seventy-five thousand at the Lincoln Memorial

1941: America enters World War II

1942: Baseball star pitcher Leroy "Satchel" Paige's Kansas City Monarchs win the Negro World Series

1945: Rev. Adam Clayton Powell Jr. goes to Congress from Harlem

1945: *Ebony* magazine is started by twenty-seven-year-old John H. Johnson

1945: World War II ends

1947: Historian John Hope Franklin publishes *From Slavery to Freedom*

1948: President Truman signs Executive Order 9981, urging an end to segregation and discrimination in the armed forces

Late 1940s: Manhattan's West Fifty-second Street becomes home to jazz jam sessions

1949: Jackie Robinson is the National League's Most Valuable Player

1950: Korean War begins; it ends in 1953

1954: Supreme Court rules, in *Brown v. Board of Education,* that school segregation is unconstitutional; Thurgood Marshall leads the brilliant legal team

1955: In Montgomery, Alabama, Rosa Parks is arrested for not giving up her bus seat; a major Black bus boycott begins

1956: Rev. Martin Luther King Jr. begins to emerge as a significant leader

1957: Ghana becomes the first independent state in Black Africa

1957: Congress passes the Civil Rights Act, protecting all citizens' right to vote

1957: Althea Gibson, tennis player, wins the singles and doubles titles at Wimbledon, England

1957: President Eisenhower sends U.S. Army paratroops to Little Rock to enforce school desegregation

1959: Malcolm X visits Africa, meets Egypt's Gamal Abdel Nasser

It is downright peculiar: some prominent Blacks are making the late 1930s to 1950s into their favorite decades. They are bypassing the usual choices of the Harlem Renaissance in the 1920s and the civil rights movement of the 1960s. Could this time become the new "Age of Black Progress"?

That prospect was clear on August 28, 1993, when Charlayne Hunter-Gault, on the *MacNeil-Lehrer NewsHour,* interviewed prominent Blacks on the March on Washington's thirtieth anniversary. Her question: Where are Blacks today?

Cultural critic Stanley Crouch answered by praising "the power of the Afro-American communities . . . in the middle fifties that supported education, . . . discouraged criminal behavior, that did not take a cowardly or whining position about opposition." Novelist Charles Johnson agreed: "We have to look back to the communities we had in the fifties and forties." In a later broadcast, Crouch declared, "In the middle fifties, Afro-American communities had as high a quality civilized life as any in the Western world."

For many Blacks, these decades still stand for racial problems and bitter memories: Blacks being denied World War II rations, Jackie Robinson being booed, the existence of separate but unequal Black schools.

Yet Crouch and Johnson are right. Once the Great Depression ended, Blacks entered a period of tremendous gains—in the fight against prejudice, in World War II acclaim, in stellar individual achievement, and in the nation's economic life. In 1930, there were almost twelve million Blacks; in 1950, over fifteen million, 10 percent of the nation. Overcoming hurdles, much as the spirited runner Jesse Owens did, Blacks went on to record tremendous gains. "We were on the move then," as Johnson says.

As we look at these few decades, it will be clear why today's Blacks are increasingly admiring them.

## How did jazz—that internationally acclaimed music—develop?

Jazz, to use a nature metaphor, is a mighty river, eddying off into smaller crucial streams, flowing undaunted from turn-of-the-century New Orleans through the twentieth century, dynamically changing its contour decade by decade, and rolling on into the future. It is one of the great American art forms.

The river's point of origin: New Orleans. By 1900 New Orleans contained African, Cuban, American, Parisian, Martinican, and Iberian musical influences. New Orleans was a sonic carnival of the likes rarely seen before in human history. The essential elements and tensions that loosed jazz were the French and Spanish elements in this Southern American city; the cultural clash of the two Black populations, the Creoles and other Blacks; and the explosion of all these elements in Storyville, the city's fabled red-light district.

It is important to note that before Jazz there was Ragtime. Ragtime, beginning around 1890, was composed piano music, written in the style of nineteenth century piano works—waltzes, polkas, marches, Lizst and Chopin— that was spiced with swing. It was, in short, European piano music tradition propelled and reconstructed by Black rhythmic dynamism. Scott Joplin (1868–1917) was the foremost ragtime composer and instrumentalist. Of his over 600 compositions, Joplin's most famous pieces were "Maple Leaf Rag" (1899), "The Entertainer" (1904), and the ragtime opera, "Tremonisha" (1915). Although ragtime was extremely popular, it lacked an essential of jazz: improvisation.

Which brings us back to New Orleans. In 1860, there were several hundred thousand Black Creoles in Louisiana. They were house servants, educated, possessing property and sometimes slaves. They were a landed Black French gentry of soldiers, poets, print journalists, and layman aesthetes. The Creoles heated up jazz. Their music was not simply a Black development from earlier sources, but the sound of their diverse New Orleans, both elegant and crude. Eventually, the Creole style combined with other Black musics. Together, this new music began a most elegant rupture from rural folk music and old surviving African traditions. The late great jazz drummer Art Blakey is famed for his discovery while traveling in Africa that jazz is so distinctly American, a wild coming-together and cross-fertilization of influences.

The earliest jazz dates from the first two decades of the century, and 1917 gives us the first recorded jazz cut, "Dixie Jass

Band One-Step/Livery Stable Blues" by the Original Dixieland Jazz Band. It was a "hot" record, intense with aggressive phrasing. The trumpet, trombone and clarinet played "free counterpoint." The trumpet led, underpinned by trombone and the rhythm section, with the string bass, banjo or guitar and piano adding percussive heft. The 1910s were the era of Dixieland. Many white New Orleans bands—such as the Original Dixieland Jazz Band and the New Orleans Rhythm Kings—held court. In many of these Dixieland bands, there were often black musicians.

One should not think of early Dixieland as "white-washed jazz," for it embodied the European musical traditions that gave jazz its tonal system, its outward form and four-beat measure. Indeed, European musical influence is not limited to the Dixieland era. It stretches throughout jazz's history as various Black American virtuosos bend it to their will. Ellington's harmonic sense echoed the French composer Ravel (1875–1937).

Storyville was closed down by Naval decree in 1917 as New Orleans became a naval port during World War I. The paid revelers and musicians shut out of work, out of their paradise, went to Chicago. In 1922 in Chicago, Louis Armstrong joined King Oliver's Creole Jazz Band. In the following years the first jazz masterpieces ensued: New Orleans-style in Chicago. From 1925 to 1928, Armstrong generated highly original recordings—featuring trumpet experiments, scat singing, casual delivery, unique solo accenting, and his exuberance.

For some, this marked the birth of the "Jazz Age." More important, Armstrong's vast sharing of music insights influenced future jazz musicians in several fields—from singers like Mildred Bailey (1907–1951) to Nat "King" Cole (1919–1965); from trumpeters such as Hot Lips Page (1908–1954) to Roy Eldridge (1911–1989); from pianist Earl Hines (1904–1983) to Art Tatum (1910–1956).

Jazz is like a river. From jazz's widening and majestic development in Chicago in the 1920s, many streams developed:

- Also in 1920s Chicago was a young group of white players, most notably the delicate and lyrical cornettist Bix Beiderbecke (1903–1931). They played "Chicago Style." In their music, solo and parallel voicings ruled. With World War I and the war's great decentering of the self, individualism reached new potentialities in jazz. The saxophone player would become a major instrumentalist.

- Jazz's second migration was from Chicago to New York in the late 1920s. Parts of this migration ended in Harlem in the late 20s. With that migration the "Swing Era" was born. Most jazz to this point had been two-beat jazz (emphasis on two beats of the four in the measure). Swing was four-beat jazz. With its rise came the rise of big bands. As with the "Jazz Age," the "Swing Era" was mostly a commercial tag to hook young white consumers.

- But, the heroes of swing—Benny Goodman (1909–1986), known as "the King of Swing," Artie Shaw (1910– ) and Glen Miller (1904–1944)–were outstanding musicians. Fats Waller (1903–1943) added his musical girth to this swinging time, with his impulsively happy piano playing and tunes, "Ain't Nobody Business If I Do," "Honeysuckle Rose," and "Ain't Misbehaving."

- Edward Kennedy "Duke" Ellington (1899–1974) began his most ambitious, challenging work as a pianist-composer in the late 1920s–1930s. Ellington went on to create the most distinguished corpus in jazz's history. He wrote thousands of compositions, many of which were played by his own orchestra—compositions whose titles stirred the imagination, "Black and Tan Fantasy," "Creole Love Call," and "Mood Indigo." As well, Billy Strayhorn (1915–1967) wrote famous works for Ellington's band, "Lush Life" and "Take the A Train." And, as Ellington's band gained renown, he attracted star performers, such as Cootie Williams (1903–1983) and Johnny Hodges (1906–1970). With such talent, the Ellington orchestral sounds became like a painter's palette, which he could pick and choose for the most exquisite effects.

- Around this time, the big band founded by William "Count" Basie (1904–84), a Kansas City product, hit its stride. Jimmie

Lunceford (1902–1947) also created and led several famous bands. In Kansas City, the "riff style" developed, a musical variant of the Black call-and-response patterns. As bands got bigger, soloing paradoxically became more distinct and important. Saxophonists Lester Young (1909–1950) and Coleman Hawkins (1904–1969) were the greatest of the age. Hawkins' "Body and Soul" was a historic moment in jazz.

- The 1930s marks the height of jazz's commercial popularity (not in regards to sales figures but everyday familiarity) and hence often a period of its commercial debasement. Radio, magazines, movies, and records sold swing as the sound of the coolest and as a lifestyle. At the end of the 1930s, Billie Holiday (1915?–1959) and her haunting interpretation of love songs demonstrated that jazz still could aspire to an artistic pinnacle. She set standards for interpretation, a softer singing style, and ability to play off of musicians.

- When a great popular art is being debased, artists usually rise up to re-invent it. So in the 1940s bebop was born. Some tremors were heard in Kansas City, but it was in Minton's Playhouse in Harlem that the bebop revolution started. The birth of be-bop, modern jazz, wasn't guesswork. Great musicians conceived ways to modulate, accelerate, and explode what had come before. With World War II, this music would be sent overseas with the GIs as marching music and nostalgic mementos of home. At Minton's the musical meeting of the minds was as deep as the minds who met at Los Alamos. Dizzy Gillespie (1917–1993) was on trumpet, Thelonious Monk (1917–1982) on piano, Charlie Parker (1920–1955) on alto-sax, Charlie Christian (1916–1942) on guitar, Kenny Clark (1914– ) or Max Roach (1925– ) on drums.

- They stripped Swing down to its harmonic essentials and turned it inside out. Charlie "Bird" Parker and Dizzy Gillespie would begin with unison riffs, a statement, often a melodic fragment of some Swing standard and run with it into the unimagined musical future. They expanded the musical vocabulary of the West by introducing notes that would be standardly thought of as "wrong" into the mix. It sounds revolutionary and it was, as revolutionary as cubism or surrealism or high modernism. Then, it was a new beginning when jazz wasn't even half a century old.

- The 50s brought the birth of the cool. Miles Davis (1926–1991) mellowed jazz out. Meanwhile, young boppers—notably John Coltrane (1926–1967) and Sonny Rollins (1930– )—intensified bebop's manic intensity and density. Miles Davis proved himself the greatest modern jazz bandleader by placing Coltrane in his band.

- The 1960s brought "Free Jazz," in which free tonality ruled. Meter and beat were unstable elements. Jazz moved into the realm of pure noise. Charlie Mingus (1922–1979), John Coltrane, Archie Shepp (1937– ), and Ornette Coleman (1930– ) are prime movers. The 1970s brought the rise of fusion in the wake of Miles Davis's electric jazz rock masterpiece "Bitches Brew."

- The 1980s–1990s saw a grand trend toward diversification on all levels. The Marsalis Brothers—Wynton (1961– ) and Branford (1960– ) of the New Orleans music dynasty—popularized and restated the now classic elements of 40s and 50s Jazz. Meanwhile Sonny Rollins, Ornette Coleman, Sun Ra (1914– ), and Don Cherry (1936– ) continue to push the boundaries, throwing away conventions. Jazz is continuing to push forward artistically.

### Who were the "Scottsboro boys"?

On March 25, 1931, two groups of boys—one white, one Black—stole a ride on a freight train headed across northern Alabama. Hopping trains was what thousands of Americans, mostly hobos, did in those days. It was the route to new place and, perhaps, better times.

Fighting broke out. The Blacks chased the whites off the train. Angry, the white boys ran to the next town with their story of mean Blacks. Forty miles down the track, at Paint Rock, Alabama, a crowd of gun-toting white men greeted the train.

Nine boys ages thirteen to nineteen were captured, taken on the back of a flatbed truck to the Scottsboro jail, then charged with "assault and attempt to murder." They were Olen Montgomery, Clarence Norris, Haywood Patterson, Ozzie Powell, Willie Roberson, Charley Weems, Roy White, Eugene Williams, and Andy Wright. Hours later, they were told that two white women—Victoria Price (age twenty-one) and Ruby Bates (age seventeen)—had accused them of rape. Later, Norris recalled, "I knew if a white woman accused a Black man of rape, he was as good as dead."

They became the famous "Scottsboro boys." Their trial came quickly. By April 9—only two weeks after the incident—eight were condemned to death and one was given a life sentence. But this was just the beginning. With appeals and retrials, twenty court cases followed, continuing into 1938.

The case was a cause célèbre, known throughout the nation and overseas. The boys' parents chose the International Labor Defense, an arm of the Communist party, to defend them. Samuel Leibowitz, Jewish and from New York, took the case pro bono.

During 1932, things changed for the defendants. First, the Supreme Court reversed the convictions in Powell v. Alabama, because the defendants had not been permitted proper counsel. Then Ruby Bates, in a letter, denied that she had been raped. Doctors had not found evidence of violence against the two women.

In 1933, they were reconvicted, but the judge, James Horton, made a bold move and dismissed the death penalty verdicts against them because of the shoddy evidence. (Being an elected official, he was defeated during the next election.) In 1934, they were tried, convicted and given the death penalty again. The Supreme Court, in 1935, overturned these convictions in Norris v. Alabama because of improper selection of the jury. Owing to this ruling, one Black man was put on the next grand jury: he voted to reindict the defendants.

During 1937, an agreement was reached between the defense team and the prosecution that charges against four men would be dropped. Five would serve terms—of one year or less—and then be paroled. But, Alabama officials did not uphold their part of the arrangement. They kept them in jail longer. In 1943, three men were paroled. Later, one of these was reimprisoned. In 1946, one was paroled. In 1948, one escaped to Michigan. After he was found by the FBI, the liberal governor refused in 1950 to extradite him to Alabama. In 1950, the last is released, the one who had been imprisoned. Finally, nineteen years after the original charges, the last Scottsboro Boy was freed.

Long-lasting changes came from the case. Blacks realized that the world could be awakened to their suffering, that there was a court of world opinion. Blacks and whites rediscovered cooperation on cases of racial injustice. A glimmer of white Southern liberalism was seen here and there during the trial. In Norris v. Alabama, the Supreme Court had ruled against exclusion of Blacks from juries. These were small victories in an otherwise tragic story.

### In what way was James Weldon Johnson a "Renaissance man"?

Had James Weldon Johnson (1871–1938) accomplished only two things—the coauthoring of "Lift Every Voice and Sing" (1900) and the authoring of *God's Trombones: Seven Negro Sermons in Verse*—he would have been voted into the pantheon of Black immortals. But he did much more.

"Lift Every Voice and Sing," written for Florida's Black schoolchildren, attained such popularity that by the 1940s it was known as the "Negro national anthem." Johnson's *God's Trombones* poems were recited all over Black America, especially the first poem-sermon, "The Creation." Its inspiration came from a mighty Black minister Johnson had heard in Kansas City. Both the song and poem are still immensely popular.

Born into Florida's Black middle class, Johnson graduated from Atlanta University in 1894. From the moment he got a degree, he speedily embarked on a succession of careers and major creative projects. After years as a Jacksonville school principal, he became a lawyer, the first Black admitted to the Florida bar. With his brother, J. Rosamond Johnson (1873–1954), he wrote some of America's hit songs, among them "The Maiden with the Dreamy Eyes," "Congo Love Song," and "Since You Went Away."

By the time he was in his early thirties, he already had been successful in three fields. Then President Theodore Roosevelt in 1906 appointed Johnson U.S. consul to Venezuela and in 1909 consul to Nicaragua.

Returning home, James Weldon Johnson launched his next career as a writer. His novel *The Autobiography of an Ex-Colored Man* (1912) allowed readers into "the inner life of the Negro in America, into the freemasonry of the race." Johnson's book showed the diversity of the race—the manual laborers, porters, servants, rural farmers, the professional classes. But the real revelation was the story of an unnamed light-complexioned man, deeply ambivalent about his own race, who slipped into white life. He "passed."

Johnson was also a fighter against prejudice. In 1916, he became the NAACP's field secretary. Tremendously successful in the post, he helped increase the association's branches from sixty to almost four hundred. Crisscrossing the country, he delivered hundreds of speeches. Tall, scholarly looking, and charming to a fault, Johnson made a striking platform figure. No grand orator, still his quiet, well-documented speeches educated all who heard them.

In 1920, he became the NAACP's first Black secretary. Using this position, he lobbied hard for the Dyer Antilynching Bill in 1922. It was a great disappointment that after the House passed the bill, the Senate failed to. But Johnson thought the bill "made of the floors of Congress a forum in which the facts [about lynching] were discussed and brought home to the American people as they had never been before."

Johnson continued to write, one of his greatest books being *Black Manhattan* (1930). Johnson had the instincts of a social historian: he wrote not just of the mighty and educated classes but also of the mass culture—dances, songs, restaurants, parades, celebrities, weddings, finery, street-corner oratory. He loved

Black social vitality in all its forms. In 1938, a car accident ended Johnson's life.

When Aretha Franklin sang "Lift Every Voice and Sing" at the White House in 1994, she acknowledged Johnson, truly a multitalented "Renaissance man."

**Why did the famous writer Alice Walker go looking for Zora Neale Hurston's grave?**

Alice Walker's *In Search of My Mother's Gardens* tells of her journey to Eatonville, Florida, to find the grave of Zora Neale Hurston (1901–60). "'Zora,' I call again. 'I am here. Are you?'" As a Black woman writer, Walker wanted to restore Hurston and her work to their proper place in the literary pantheon. She found that Hurston lay in an unmarked grave.

This was a surprise, given Hurston's stature in the 1930s when she published several great books. Hurston's unconventional vision of Black life made her famous. And she enjoyed her fame to the hilt: she loved parties and dancing; she wore colorful scarves or dashing hats and dressed in pants and boots; she enjoyed telling stories to the rich and famous and being a public figure. Photos show her to have been a jaunty figure.

Every element in her early development contributed to her writer's viewpoint. Born in Eatonville, an all-Black town, she absorbed its lore and music. Her secure, stable family of eight children headed by a warm mother and minister father offered a shelter for her imagination. These years gave her unshakable confidence in the creative power of Black culture and its wisdom.

When her mother died and her father remarried, Hurston was forced to wander, living with relatives, being a maid for a touring Gilbert and Sullivan troupe, and gaining a high school degree in Baltimore while doing housework. She learned to survive and protect her vision that life is an adventure and that gain comes through risk—motifs that appear in her writing.

Hurston was a prominent figure in the Harlem Renaissance. Her first stories appeared in *Opportunity* and *Fire!* magazines. She attended Barnard College (graduating in 1928) and later

Columbia University, where she studied the ideas of anthropologist Franz Boas. Boas believed that races were not different, not superior or inferior to each other, and that every culture had its own logic.

Just as the Great Depression appeared, Hurston was making her mark as a writer with *Jonah's Gourd Vine* (1934) and *Mules and Men* (1935). The latter was (and still is) truly pioneering, for it gave many examples of Black folklore—everything from stories, sexual boasts, songs, aphorisms, and language play to religious rites and healing formulas. Just as Boas had taught, Hurston collected what she called this "bright, new material" herself.

Wrapped in a gold cover, *Their Eyes Were Watching God* arrived in bookstores in 1937. Janie, the main character, is animated by an idea of freedom, but she has found it hard to achieve. Her marriages to Logan Killicks, aged and well-off, and to Jody Starks, ambitious to be a "big man," made her the hostage of male designs. Finally, she runs away with Teacake for a life of uncertainty and changes but full of joy. This work is now a classic.

Hurston's productive years included trips to Haiti and Jamaica, where she collected more folklore. Out of these trips came *Voodoo Gods* (1938) and *Tell My Horse* (1938). Once again, she presented experiences that few people had understood before.

In the 1930s, her talent shone brightly, but her last years were spent in poverty and obscurity. Walker's rediscovery of Hurston began the restoration of her fame. And her grave now has an ebony mist stone and Walker's inscription:

Zora Neale Hurston
A Genius of the South
Novelist, Folklorist, Anthropologist, 1901–1960.

### What was the Great Depression's impact on Blacks?

The Great Depression lasted from 1929 to 1941. Unemployment increased dramatically, industrial production fell, and prices fell. By the early 1930s, many workers were idle.

The Depression was especially devastating for Blacks. By 1934, 17 percent of whites and 38 percent of Blacks were unemployed. Other impacts on Blacks included the following:

- In some urban centers, the number of unemployed Blacks grew to over 40 percent. In Atlanta, 65 percent of the Blacks needed relief.

- Many Blacks couldn't afford even basic food or housing.

- Discrimination operated in the public relief organizations. Blacks had a hard time getting their share.

- Black protests increased. The "Jobs for Negroes" movement, involving pickets and boycotts of businesses that refused to hire Blacks, was launched in cities. The aim was to pressure businesses, particularly those in Black neighborhoods, to hire more Blacks.

    In New York's Harlem, the campaign was spearheaded by Adam Clayton Powell Jr. Harlem businesses and public utilities were the targets. Powell used this movement as a stepping-stone to a city council position and later to Congress.

- The Depression brought a significant change in Blacks' political loyalties. Prior to the 1930s, most Blacks had been Republicans, as Abraham Lincoln had been. Black Democrats were few in number; as one politician put it, they "were as rare as a five-dollar bill in the middle of Broadway. . . . It was a form of heresy practically unknown and unpracticed. Being a Black Democrat was like announcing one had typhoid."

    Blacks started drifting away from the Republican party in the 1928 presidential election, when Democrat Al Smith gained votes in Northern cities. Yet even in 1932 Blacks did not flock to Franklin Delano Roosevelt in his contest against Republican Herbert Hoover. The choice was, as the *Chicago Defender* said, "between the devil and the deep blue sea." Roosevelt was distrusted as a rich man and because he chose

Southerner John Nance Garner as his vice president. Not until 1936 did a great bloc of Black votes go to the Democrats.

The Great Depression was a disaster for Blacks, sweeping away their meager economic gains. To resist this devastation, Black labor and electoral politics were transformed.

### How did a Detroit women's group help Blacks during the Depression?

The Great Depression was no mere set of statistics for Black women. Since they managed their families' budgets, they saw the economic downturn in their daily lives: it meant more bread-and-gravy meals, more hungry nights for children; it meant clothes wearing thin and shoes with paper soles. Often they were jobless. Even jobs as domestics were scarce, as many whites couldn't afford to hire them.

During the 1930s, Black women in Detroit took an innovative approach to holding the line against the Great Depression by forming the grass roots Housewives League of Detroit (HWLD). As their founding statement said, "the great loss of employment upon the part of our people has reduced us to the place where we have not sustenance of our bodies, clothes to wear, homes to live in, books to read." Reacting to these conditions, the organization stressed the necessity of mutual aid among Blacks, the need for racial pride expressed through the support of Black businesses, and the idea of Black independent action.

Just fifty Black women started the organization. Fannie B. Peck—wife of Rev. William H. Peck, pastor of Detroit's large Bethel African Methodist Episcopal Church—was the originator of the idea. Peck's idea found natural supporters among the city's women. They had come from strong traditions, mostly Southern and rural, of building community. By 1934, five years into the depression, the HWLD had grown astronomically to around ten thousand members.

HWLD meant business, Black business: "We emphasize and

declare it to be the most desirable to own our own business and manage it ourselves." Their commitment to Black businesses was a form of economic nationalism. Not only did members pledge to spend their money with Black businesses but they purchased products made by Blacks and used Black lawyers, doctors, and dentists from their neighborhoods.

HWLD literature exhorted housewives, "Think! Think! Think! before you give your business to any place. . . . Can't you see the need of keeping as much as possible [of] your little money in the hands of your own race?" HWLD wielded women's dollars with clear objectives. "It is our duty," one statement said, "as women controlling 85 percent of the family budget to unlock, through concentrated spending, closed doors."

The HWLD was responsible for helping many Black businesses and professionals survive during the Depression. It also promoted self-help, teaching Blacks to rely on themselves for economic advance and, specifically, teaching women techniques of organization. It lasted through the 1930s to the 1960s, spreading to many cities such as Washington, D.C., Chicago, New York City, and Baltimore. Historian Darlene Hine has observed that the group's decline in the 1960s was ironic: "Just as the Black Power movement gathered momentum in the late 1960s, the Housewives' League of Detroit faded from view."

### What did the New Deal mean to Black Americans?

When Franklin Delano Roosevelt accepted the Democratic party's nomination for president in 1932, he pledged "a new deal for the American people."

Relations between Roosevelt's New Deal and Blacks were slow to develop. When Roosevelt first ran for president, the majority of Blacks supported Republican Herbert Hoover. In his first term, Hoover had offended Blacks by cutting Black federal appointees, keeping the capital's cafeterias and rest rooms segregated, and refusing to condemn lynching. Most sinister, he wanted to put an openly anti-Black North Carolina judge, John J. Parker, on the Supreme Court. (Parker was defeated, largely by the NAACP.) Yet Blacks still loyal to the party of Lincoln voted for Hoover.

Roosevelt won anyway. His first term was mixed at best. He failed to endorse federal antilynching legislation. He took little action against white Southern officials who prevented Blacks from voting. When the NAACP asked for a meeting, he refused. New Deal programs often failed to benefit Blacks. The Tennessee Valley Authority, which ran many flood control and soil preservation projects, hired Blacks only as manual laborers and did not let them into training programs.

Roosevelt's first term did see a few Black gains. There were Black federal appointees, and relief was provided for many jobless people, among whom Blacks were numerous. At Howard University, he told a Black audience, "Among the American people there should be no forgotten men and no forgotten races."

It was Eleanor Roosevelt who was the early force pulling Blacks toward Roosevelt and the Democrats. She could not create policies, but she could, and did, generate powerful, positive racial symbolism. Visiting Black churches, colleges, and housing projects, she sent the message that the first lady was not afraid of mingling with Blacks. Commenting on her appearance at a mass meeting, the *Afro-American* was astonished: she "seemed as much at ease . . . as in her own home." Countless times, she was photographed with Blacks, breaking a long-standing taboo and antagonizing the South. As the film *The Tuskegee Airmen* (1995) showed, Eleanor Roosevelt later gave decisive support to Black air units. She used her influence to raise funds to build an airstrip at Tuskegee, deflected southern white criticism of the units, and boosted their image by going on a flight with a Black pilot.

The human approach worked. As Rayford Logan has said, "Negroes had been so depressed, so frustrated, almost having given up hope, . . . the outlook was so bleak . . . that little things counted a great deal. . . . Treating Negroes as human beings was a very significant factor." In 1936, on Roosevelt's second presidential bid, he won a majority of Black votes. After his sizable win, the *Pittsburgh Courier* said, "My friends, go turn Lincoln's picture to the wall. That debt has been paid in full."

During Roosevelt's second term, Blacks began to profit from relief programs. The Works Progress Administration (WPA) employed almost four hundred thousand Blacks. By 1939, over one million Black families had income from the WPA. Some five thousand Black teachers were hired to teach Blacks to read and write. Government aid helped thirty-five thousand Blacks to go to high school and college. The Public Works Administration built many Black schools. The Civilian Conservation Corps (CCC) also provided thousands of jobs for Blacks.

Blacks often criticized these programs. In a letter to Eleanor Roosevelt, Roy Wilkins, an NAACP official, wrote, "There is hardly a phase of the New Deal program which has not brought some hardship and disillusionment to colored people." Most Blacks, however, agreed with the relief participant who said, "The CCC may have been segregated, but Blacks could get into CCC camps. The WPA may have been discriminatory, but before that, Blacks had no bread."

By 1936, so many Black advisers—twenty-seven men and three women—were in the Roosevelt administration that they were nicknamed the "Black cabinet." Not since Reconstruction had the capital seen this many highly placed Blacks.

The New Deal was a watershed event in Black relations with the federal government. Wedded to Southern whites, Roosevelt did not attack racial inequality. Nor did he open as much opportunity as possible. But his administration helped Blacks survive the depression, put Blacks in federal agencies, and promoted a new interracial symbolism.

## Why was Mary McLeod Bethune called "the Amazon of God"?

When Mary McLeod Bethune (1875–1955) was a child, her mother thought she was special. The fifteenth child of seventeen, she did not play with the other children. Even as a child, she had serious thoughts. When she did play, she organized her playmates. "They accepted my leadership because I was always striving to set up something going in the opposite direction from the mass of things." Early on, she was also deeply religious.

A disturbing childhood encounter made her desperately want an education. One day she followed her mother to her work at a rich white family's home. Mary wandered into the children's playroom and picked up a book, prompting the family's daughter to scream, "Put that down. You can't read." Later Mary recalled, "When that nice little white girl said that . . . I thought, 'Maybe the difference between white folks and colored is just this matter of reading and writing.' I made up my mind I would know my letters before I ever visited the big house again."

A week later, Mary was picking cotton when a family friend told her about a new school. For six years, Mary walked five miles each way to that school, where a Miss Emma Wilson ("the first Negro I ever heard called 'Miss'") taught her to read and do arithmetic. Assuming the role of mentor, Wilson arranged for Mary to go to North Carolina's Scotia Seminary in 1888, where a "balanced education" included the training of head, heart, and hand. The day Mary left for school, "all the neighbors stopped work that afternoon."

Finished with college, she worked in South Carolina, Georgia, and Florida as a teacher. During these years, she married Alburtus Bethune and bore a son, but she thought "this married life was not intended to impede things I had in mind to do." Burning in her heart was the desire to start her own school.

In 1904, with one dollar and sixty-five cents in her pocket, she chose the site of the Daytona Educational and Industrial Schools for Negro Girls. Her first students were five girls. Building the school required that she do everything. Just when she was about ready to call it quits, supplies would miraculously appear; for this reason, her first building was named "Faith Hall." Education there was practical and vocational, "for living in everyday America." In 1923 the school merged with the Cookman Institute to become the co-ed Bethune-Cookman College.

After two years in office, President Roosevelt summoned Bethune to Washington, D.C., to direct the Black section of the National Youth Administration (NYA). She was honored but initially doubted his commitment. When he explained his plans to her, she rose, wagged her finger in his face, and said, "Mr. President, you've got to do better than that for me." They were fast friends from that time on.

Bethune's NYA work (1936–43) demonstrated her ability to organize national resources and get them to needy people. Thousands of Black youth were given educational opportunities through NYA. In these years, she was photographed hundreds of times for newspapers. Soon, the statuesque "Mrs. Bethune"—very black, wearing a hat and standing tall, the epitome of rectitude—had become a national symbol.

In 1935, she was made president of the National Council of Negro Women. She took a skeleton organization, begun by thirty women on a winter's day in Harlem, and converted it into a group embracing twenty-two national women's organizations. She served until 1949.

"Amazon of God" was the 1940s label for her. Her physical person, combined with her rugged purpose, made her an Amazon to her followers. When Bethune died in 1955, she left a last will and testament to Blacks, in which she gave advice for "her people's future and success." Her nine legacies were: "I leave you love." "I leave you hope." "I leave you the challenge of developing confidence in one another." "I leave you a thirst for education." "I leave you a respect for the uses of power." "I leave you faith." "I leave you racial dignity." "I leave you a desire to live harmoniously with your fellow men." "I leave you finally a responsibility to our young people." Today, her will seems written for our times.

### What Black athlete upset German dictator Adolf Hitler?

In the summer of 1936, Jesse Owens (1913–80) ran right into world stardom with his performance at the Olympics in Berlin. Born in Danville, Alabama, Owens had been an outstanding runner since his high school days in Cleveland, Ohio.

Ohio State was the college he chose. Black papers criticized him for enrolling there: the school's administration was known for its poor treatment of Blacks. Owens became the star of the track team, but when traveling, he and other Black members were relegated to the YMCA while white members stayed in whites-only hotels.

Still, he showed his stuff! On a single afternoon, May 25, 1935, he broke three world records (220-yard dash, 220-yard low hurdles, long jump) and tied one (100-yard dash) in the space of an hour. His record in the long jump—26 feet 8¼ inches—remained unbroken for twenty-five years. William J. Baker, his biographer, says that knowledgeable track people still look at his performance as the single greatest day in the history of track and field.

A year later, he was in Berlin. His appearance in the Olympics was greatly awaited not only because of his reputation but because he would be challenging white runners, including Germans. In 1936, Nazis controlled Germany, and with them came the racist notion of Aryan physical and mental superiority. According to German racial theories, Owens could not beat a European athlete. Hitler had called Blacks "an inferior race, . . . a degraded people."

Owens upset past records and racist theories. He won four gold medals. Newspaper headlines yelled, "Jesse Owens Star Performer in Berlin." He tied the world record for the 100-meter dash. He set an Olympic broad-jump record, jumping over twenty-six feet. He set a 200-meter dash record. He was the first runner in the 400-meter relay, assisting his team to a record.

When Owens completed his triumphal performance by winning the 200-meter race in a record 20.7 seconds, the stadium's spectators rose in unison to pay tribute to him. The applause could be heard for miles.

Hitler, in attendance, was visibly shaken by the Blacks' stellar performances: they had won seven gold medals. He congratulated German winners but refused to do the same for any Blacks. Instead, he left the stadium hastily, citing bad weather and other engagements as the reasons.

Jesse Owens's many victories were captured on film and shown globally in theaters. Magazines ran whole series of photos. Newsreels in Black theaters showing Owens wearing his gold medals were greeted with cheers. Owens returned home to parades in Cleveland and Columbus. For all Americans, Jesse Owens's swift legs had brought them a triumph in a distant land.

### Why were Paul and Eslanda Robeson one of the world's most dynamic couples?

He was a highly educated man, a versatile star athlete, a self-taught scholar, a connoisseur of languages, a world-famous singer and actor, an uncompromising advocate for Black justice, and a supporter of great international causes. She had an early career in the medical sciences, was trained as an anthropologist, was a writer, a global goodwill messenger, and a famous speaker and promoter of important issues. Paul and Eslanda Robeson were one of the most talented couples of their time.

Paul Robeson's brilliance in so many fields has often overshadowed Eslanda Robeson's achievements. In most history books, she goes unmentioned. But she had her own distinctive life and may have been responsible for spurring him on to ever-greater heights of achievement. They grew into a perfect symbiosis as a married couple. They were America's equivalent to England's Sydney and Beatrice Webb, a couple who became cultural and political actors of extraordinary influence. No history of the twentieth century would be complete without them.

Robeson was born in Princeton, New Jersey, into an upright Black household—his father was a preacher—in 1898. Robeson was always a good student, but he really shone at Rutgers University. On graduating from Rutgers, he had earned his class's highest grade point average; had been elected to the prestigious Phi Beta Kappa honor society; had been voted a member of the All-American football team; and had earned letters in four sports. He headed for New York's Columbia University law school and years of living in Harlem—years that changed his life.

Arriving in 1920s New York, he was just in time to sample the Harlem Renaissance and, most important, to meet Eslanda Cardoza Goode, who was born in 1896. Her mother was from the famous Cardozas of Charleston, South Carolina, a family with both white and Black ancestry. Eslanda had been trained as a chemist and was working as a clinical pathologist at New York hospitals—unusual studies and work for a Black woman of her day.

Undoubtedly, Eslanda was charmed by Robeson's intellectual vitality, wit, and gallant manners and attracted by his immense physical magnetism. She saw his great potential as a public figure. He was enthralled with her graciousness, accomplishments, and her striking beauty. They married in 1921. Eslanda knew that the arcane technicalities of the law were not his strength. Instead, she steered him toward theater. This was characteristic of her risk-taking. And she insisted that he work extremely hard to make the risk pay off.

The year 1925 was a turning point in Paul Robeson's life. Before a packed house in New York, he sang, as one critic said, "the first concert in this country made entirely of Negro music." His bass voice would become his signature from this moment on. That year, he also appeared in Eugene O'Neill's *Emperor Jones* in London, his first major dramatic role. Robeson brought a brooding intensity to the main character, Brutus Jones.

The next years were a whirlwind of travel, performance, and acclaim. Robeson triumphed in Jerome Kern's *Showboat* (1928), where his bass-baritone singing was met with wild applause. In 1930, he played Othello in London. When it was announced he would play opposite a white woman, a row ensued. He was breaking a social taboo. But he was so commanding as Othello that the racial issue was overcome. This production of *Othello* eventually traveled to New York's Broadway in 1943, where it played for an astonishing 296 performances. The photo on the program of Robeson in costume revealed him in his true dramatic glory.

Living mostly in Europe from 1928 to 1939, the Robesons broadened their activities. (Their son, Paul Jr., was born in 1929.) Using London as a base, Robeson acted and sang all over Europe. Films—*Emperor Jones* (1933) and *Sanders of the River* (1935)—were also an outlet for his talent. Both he and Eslanda were drawn increasingly to international issues and to leftist politics. They visited Spain, where Paul sang in the camps of anti-Fascist forces. In 1935, he went to Soviet Russia to see the socialist revolution at work.

Eslanda Robeson pursued anthropology studies at the London School of Economics. In 1936, she took an extensive trip through Africa, resulting in *African Journey*, published in 1945. Though a travel diary, it was one of the most incisive books on Africa to be published at that time and was particularly insightful about colonialism and African cultural traditions. One of her main points was that the Black world was larger than just the Black American experience.

During the 1940s, Paul and Eslanda Robeson returned to live in America and brought a new militancy to the Black struggle. Paul met with President Truman on the issue of lynching. The Council on African Affairs, a group lobbying against colonialism in Africa, was founded with the Robesons' assistance. Eslanda represented the council at the first United Nations convention in 1945.

The 1950s were hard times for the Robesons. They were often ostracized as Communists. But they did not bow under the condemnations nor recant their opinions. In 1950, the State Department revoked Paul's passport, thus cutting off his large income from overseas performances. (The Supreme Court restored it in 1958.) Both Robesons were called before the House Un-American Activities Committee (HUAC) to testify. They refused to cooperate with the witch-hunting activities of this group. Paul Robeson told HUAC, "My father was a slave, and my people died to build this country. . . . I am going to stay here and have a part of it just like you."

Fleeing American persecution, the couple moved to Great Britain in 1958, occasionally living in the Soviet Union. Returning to the United States in 1963, they began campaigning immediately against the Vietnam War. But their strength was waning. Eslanda Robeson died in 1965. Paul Robeson lived in seclusion until his death in 1976. Once, though, they had been a radiant presence on the world's stage.

## What singer had "the voice of the century"?

In her second tour of Europe in 1935, Marian Anderson sang at the Hotel de l'Europe, one of Salzburg's leading hotels. Attending her concert was the world-famous conductor Arturo Toscanini. In her autobiography, *My Lord, What a Morning*, she recalled, "I was as nervous as a beginner. . . . I was in a state by the time I got out onstage."

At intermission, Toscanini came backstage to congratulate her, but Anderson was too nervous to hear him. Afterward she was told he said, "Yours is a voice such as one hears once in a hundred years."

"The voice of the century" was just one of many accolades Anderson received during her career. But Toscanini's assessment carried extraordinary weight. It was as if Anderson, in her midthirties, had been awarded a Nobel Prize for singing.

Anderson was born in Philadelphia in 1902 into a frugal, loving household in a racially mixed neighborhood. Union Baptist Church, with its great music program, allowed her her first opportunities to sing. Occasionally she was on the church's program with Roland Hayes (1887–1977), the celebrated Black tenor. Rev. W. G. Parks often asked for special collections for "our Marian." Her first voice teacher was Mary Sanders Patterson, a Black soprano, who donated the lessons.

Looking for advanced training, Anderson was rejected by a local school, where the admissions officer said, "We don't take colored." She went on to study with Giuseppe Boghetti in Philadelphia, who shaped her voice for the concert stage. In 1925, she won the Philadelphia Orchestra's annual vocal contest, beating out hundreds of competitors. Later that year, she sailed for Europe for her first tour of European capitals.

Europe proclaimed her one of the world's greatest contraltos. Her voice always astonished audiences because of its richness, depth, and range. Almost unanimously, music critics felt her voice was a magnificent instrument.

But more was at stake in an Anderson performance. Touring Europe in the 1930s, she stood for the excellence of a race that Nazi ideas assaulted. She also stood for her people's musical history. Each of her programs included songs by Schubert ("the composer I love the best"), French songs, Spanish and Italian selections, as well as spirituals, placed on par with the European works.

When Anderson returned to the American concert stage, she encountered acclaim along with racial slights. Traveling by train exposed her to segregated cars. Once a Los Angeles hotel would not serve her in its public dining room. Going to the South brought more difficulties; in Memphis, for example, newspapers would call her anything but "Miss Anderson."

In 1939, at the height of her fame, she was denied the concert use of Washington, D.C.'s Constitution Hall, owned by the Daughters of the American Revolution. Reports of the slight traveled across the country like lightning. Eleanor Roosevelt resigned from the DAR in protest and arranged for Anderson to sing instead at the Lincoln Memorial on Easter Sunday. The issue and the crowd of seventy-five thousand made this one of the most important events of the 1930s.

For all her skill and experience, it was not until 1955 that Anderson appeared at the Metropolitan Opera, as the old sorceress, Ulrica, in Verdi's *Un Ballo in Maschera.* Television news programs showed her taking her bows. She was the first Black to become a permanent member of the Metropolitan Opera and also the first Black to sing at the White House.

When not singing, she represented America on many occasions. Awards were heaped on her. No singer of the twentieth century made more of an impact on America and the world.

**What Black athlete caused street celebrations in Black American neighborhoods?**

Joe Louis was born Joe Louis Barrow on May 13, 1914, in Chambers County, Alabama, the seventh of eight children in a poor sharecropper's family. After his father became mentally ill, Joe's mother and all her children went to live with Pat Brooks, who migrated north to Detroit in 1926, got a job working for Ford, and then sent for his new family. When the family settled in a rough neighborhood, Joe learned to fight to protect himself.

In his first amateur fight, using the name Joe Louis, he was knocked down seven times in two rounds. He gave up boxing for six months, but his mother encouraged him to try again. In the next year, he won fifty out of fifty-four amateur fights, forty-three by knockout. He won merchandise checks as prizes, with which he bought groceries for his family.

Louis turned pro in 1934, and immediate success was his.

Still, racial barriers existed in boxing, and it was hard for him to book big-time fights. Promoters since the era of Jack Johnson—the first Black world heavyweight champion in 1908—had rationalized the color line by claiming that interracial fights would encourage racial antagonisms. The leading white fighters such as Jack Dempsey avoided fighting Blacks. But a white promoter saw Louis as a moneymaker and began to arrange major fights for him.

In 1935, Louis fought and defeated Primo Carnera, the Italian former heavyweight champion. By then Louis was widely seen as a symbol of Blacks' ability. After the Carnera fight, Joe Louis went to church with his mother and heard Rev. J. H. Maston talk about "how God gave certain people gifts," and he realized that "through my fighting I was to uplift the spirit of my race."

In 1936, Louis's stock fell when he was knocked out by the German heavyweight, Max Schmeling. The German fighter received hundreds of congratulatory messages from American whites and a telegram from Adolf Hitler.

Later that year, Louis redeemed himself, winning the world heavyweight title from American James J. Braddock. A newspaper reported, "With a head-jarring right to the jaw, the 'Brown Bomber' from Detroit ended Braddock's two-year reign and emerged as the first Negro heavyweight champion in twenty-two years. . . . Battered, bruised, and bleeding, Braddock slumped to the canvas." He held the world heavyweight title for a record twelve years and made twenty-five defenses—another record.

Still, beating Schmeling had become a private mission for Louis, and in 1938 he got his chance at a rematch. The second fight had even more important implications than the first. News of German concentration camps had spread, and Nazi propaganda now consisted of a steady stream of anti-Black and anti-Jewish lies. The result was swift: "Joe Louis Floors Max in the First Round" proclaimed a morning paper's headlines.

Each of Louis's major fights was a matter of intense interest among Black Americans. He was also a symbol to those Americans who did not believe in notions of a "master race." The night Louis lost to Schmeling, songstress Lena Horne said, "Until that night, I had no idea of the strength of my identification with Joe Louis. We had the radio on behind the bandstand, and during breaks, we crowded around it to hear the fight. I was near hysteria . . . when he was being so badly beaten. . . . Some of men in the band were crying."

His victories caused Black pandemonium. Roi Ottley, a Black reporter, described Harlem following Louis's beating of Schmeling: "Tens of thousands marched through the streets. . . . There was shouting, clapping, laughing, and even crying. Youngsters who should have long been in bed were on the streets pounding dishpans and yelling. The din was deafening. . . . Much whisky was guzzled;. . . music of jukeboxes . . . blared forth. . . . Young couples went into furious lindy hops and Suzy Q's." Malcolm X called these scenes "the greatest celebration of race pride our generation had ever known."

Even in the rural areas of Arkansas, Louis's victories were great events, as Maya Angelou tells us: "Then even the old Christian ladies . . . would buy soft drinks, and if the Brown Bomber's victory was a particularly bloody one, they would order peanut patties and Baby Ruths also. Champion of the world. A Black boy. Some Black mother's son. He was the strongest man in the world."

## What was "passing?"

As the word suggests, when people "passed," they "passed" from one race to another, from one social territory to another, like a refugee stealing from one country to another. Like illegal refugees, those who passed probably lived an anxious life, fearing discovery.

Blacks who passed as whites were acting out a role, pretending to be something they were not. It was risky: a passing person could be detected, exposed, and possibly punished. Yet for all its problems, some Blacks accepted the challenge and passed. It was their choice—and in some ways, it was a very American choice. Americans have always loved forgetting their origins and reinventing themselves.

For Blacks to pass, they had to possess at least two physical traits: light skin color and less curly hair. It was thought that race could be detected by skin and hair. Actually, passing worked because of centuries of racial mixing—in the big houses, slave cabins, woods, speakeasies, brothels, even luxury apartments.

Despite state laws barring racial mixing, it has occurred throughout the whole of American history. By 1850, the federal census counted almost 250,000 mulatto slaves, almost 8 percent of the slave population. Ten years later, there were over 400,000 mulattos, or more than 10 percent. Many free people had mixed origins. Some authorities say that 75 percent of all Blacks have at least one white ancestor, with 15 percent having predominantly white ancestral lines. Most whites have Black blood lines too.

It is impossible to say when passing started or to estimate how many Blacks passed. But in Black society there have always been abundant tales of passing. And it is no coincidence that twentieth-century stories appear in greatest profusion. Passing was more likely in the post–World War I period, when Blacks migrated out of the South. As they settled anonymously in Northern cities, greater numbers passed. In general, American cities have always allowed people to abandon their identities, to change their roles.

In the 1930s, Nella Larson wrote two novels, *Quicksand* and *Passing,* which reflected this situation. Her leading women, engaged in passing, were city dwellers. Previously, the Black writers Charles Chesnutt (1854–1932) and Frances Ellen Watkins Harper (1825–1911) had explored analogous themes. That Black writers have kept coming back to tales of passing shows the evocative power of the experience.

Walter White (1893–1955) was probably the twentieth century's most famous example of passing. Born and educated in Atlanta, White was hired in 1920 by the NAACP to go South to conduct undercover investigations of lynchings and race riots. White could be a spy behind the lines because he was light skinned, blond-haired, and blue-eyed—"a white colored man."

To White's amusement, white Southerners boasted while looking him straight in the face that they could tell a Black no matter how light he was. Lynchers talked easily with him about who instigated the mobs. White became famous for his sensational reports that exposed lynching.

Passing has waned, but the fascination with it has not, as proved by the reception to Shirley Taylor Haizlip's *The Sweeter the Juice* (1994), the story of a family with Black and white sides.

### What bomb did writer Richard Wright drop in 1940?

Richard Wright was born before World War I, in 1908, in Roxie, Mississippi—a hard place in the Deep South. It was there, living in a sharecropper's family, that he first glimpsed the meanness of life for Blacks. For his first ten years, owing to his mother's illnesses, he moved around a lot and lived with relatives. During his teenage years, he went to live with his grandmother, who ran a stern, frugal household.

Wright could see that even under these tight conditions, Blacks were trying to counter racism and improve their lot. For instance, teachers at his segregated high school were imbued with a sense of mission: "They realized that this was all the schooling the colored kids of Jackson [Mississippi] were likely to get. So, they gave it all they had."

In 1925, Wright went to live on his own in Memphis. His desire to write began to grow. To write, Wright figured he had to question, think, and, most of all, to read. Reading allowed him to travel, through books, beyond the cage of his own existence. By pretending he was seeking books for a white employer, he was able to get the white librarians in Memphis to lend him books, which he devoured in his boardinghouse room. H. L. Mencken (1880–1956), the Baltimore-born editor and satirist, became one of his favorite authors. His truth telling, especially about the South, attracted Wright.

By the winter of 1927–28, Wright had moved to Chicago. He was only nineteen but had a man's experience of life. The city enthralled him: its skyward-reaching buildings, its crude industrial life, its polyglot working classes, and its Black quarter congested with wide-eyed masses who had migrated from the Mississippi Valley. Later, when introducing the sociologist St. Clair

Drake's *Black Metropolis* (1945), Wright described his feeling for Chicago: "I felt those extremes of possibility, death, and hope, while I lived half hungry and afraid in a city to which I had fled with the dumb yearning to write, to tell my story."

Just as Wright arrived and got a post office job, the depression descended. Wright saw Black families evicted, living on the streets. But he also saw poor whites as beaten down as Blacks. The depression put him out of work until he landed a job with the Writers' Project of the Works Progress Administration (WPA). Through the project, he learned much about the craft of writing. And on his own, he worked tirelessly to continue to learn the basics. It was also in Chicago that Wright participated in the radical, pro-Communist John Reed Club, where the focus was on society's class issues. But Chicago could not hold him, and he moved to New York in 1937. By the time he left, though, his first stories had been bought by a publisher and were released as *Uncle Tom's Children* in 1938.

In 1940, Wright published the book most often associated with his name, *Native Son.* It was as much a cultural bombshell as a literary event. How Black life was represented, especially in literature, was changed forever. Writing a year later in *New Masses* magazine, writer Ralph Ellison voiced his deep admiration for the novel: "*Native Son,* examined against past Negro fiction, represents the takeoff in a leap which . . . marks the merger of the imaginative depiction of American Negro life into the broad stream of American literature." Wright had helped to "create the consciousness of his oppressed nation."

No quaint folklore or sentimental scenes graced *Native Son.* Wright's goal was to write a novel that "no one would weep over," for "it would be so hard and deep that they would have to face it without the consolation of tears." Readers could see the causes and extent of Black rage against white society. Bigger Thomas was the main character. Wright made him into a man created by the racial environment—poor, living in a rat-infested single room, trapped, and fearful of whites. Bigger had a job as a chauffeur for the wealthy white Dalton family.

Accidentally, he killed Mary Dalton, his employer's daughter, decapitated her, and put her body into a furnace. He also had to kill his girlfriend, Bessie Mears, to whom he had confessed Mary's murder. After each murder, Bigger grows more fearless—in fact seems to take on an identity. He is caught, tried, defended by a Communist lawyer, and sentenced to the electric chair. Near the novel's end, Bigger says, "But what I killed for, I am," and "I didn't know I was really alive in this world until I felt things hard enough to kill for 'em." Although a stark story, *Native Son* sold several hundred thousand copies, making Wright a new star on the literary scene.

Wright was prolific. In 1941, he published a collection of black-and-white photos and text, *Twelve Million Black Voices.* The sheer realism of the photography made the book into a remarkable documentary of the "qualitative and abiding in the Negro experience." In 1945, his autobiography, *Black Boy,* was issued. Finally, the public was able to understand the formation of this intricate intellectual. Again, he rendered tough judgments, even on himself, establishing a high standard for candor in autobiography and changing the tone of future Black reminiscence.

When World War II ended, he left for France, where he lived until his death in 1960. There, his writing never ceased, even amid charges that he was out of touch with the American experience.

### Did union leader A. Philip Randolph really make President Franklin Roosevelt cave in?

Before World War I, Asa Philip Randolph (1889–1979) came to New York City to try out this bustling place. He waited tables, carried luggage, ran an elevator. He read voraciously, especially Shakespeare. At one point, he hired a speech tutor who transformed his Florida accent into high-sounding, upper-class Oxford English.

Eventually, he became a socialist, believing that capitalism's greedy instincts had to be tamed. For a while, he diagnosed the American illness as a budding soapbox orator on Harlem's street corners. Using his swift, clear logic, deep voice, and impressive accent, Randolph could draw a good crowd. He was now "A. Philip Randolph" and a labor agitator.

With writer Chandler Owen, "a facile and acidulous writer," Randolph started *The Messenger* magazine in 1917, whose cover boasted, "The Only Radical Negro Magazine in America." They proclaimed, "We do not accept the doctrine of the old, reactionary Negroes that the Negro is satisfied to be himself, because of our recognition that the principle of social equality is the only guarantee of social justice." He began to doubt socialism.

Randolph's first effort was to organize elevator operators in New York City. Then, in the 1920s, Randolph began to identify with the plight of the badly treated Black porters and maids of the railway system—the Pullman Company workers. Although not a porter, he knew about this laboring class and their desperate plight; these workers put in up to eighty hours per week, were responsible for big expenses for their uniforms and meals, and were harassed by the company every time they tried to organize. Out of Randolph's 1925 effort to build a union, the Brotherhood of Sleeping Car Porters and Maids (BSCPM) was born, which eventually became the most significant Black labor group until the 1960s.

Initially, the union had its ups and downs. In 1935, its strapped membership raised half a million dollars for a campaign to gain company recognition. In 1937, the Pullman Company signed a contract with the brotherhood. After this, Randolph, on behalf of the union, gained pay increases, shorter work weeks, and a cut in the miles its workers had to travel.

By 1941, Randolph was the gray-haired patriarch of the Black labor movement. He was in the nation's headlines because he proposed a march on Washington by more than one hundred thousand Blacks to protest discrimination in the defense industries.

As America readied itself for war, industries were hiring large numbers of workers. Thousands of jobs were being given to whites while Black workers were being ignored. The government was using Black and white taxpayers' money to give contracts to industries, but this had little impact on Black employment.

Randolph's march—scheduled for July 1, 1941—was designed to dramatize this outrageous situation. Black newspapers took up the cause. Black churches, women's clubs, Masonic groups, and even children's clubs raised money. It appeared that Washington would be packed with Blacks.

Official Washington began to worry. Southern members of Congress were predicting trouble, so President Roosevelt called Randolph and the other leaders to Washington. Randolph did not budge. He demanded that the president issue an "executive order to effectuate the speediest possible abolition of discrimination in war industries and the armed service."

On June 25, just a week before the march was to occur, Roosevelt issued Executive Order 8802, decreeing that "there shall be no discrimination in the employment of the workers in the defense industries and in Government because of race, creed, color, or national origin." The march on Washington was called off. Roosevelt had caved in.

Randolph built a long record of achievement. In the 1950s and 1960s, he was an important civil rights leader. The 1963 March on Washington was based on his 1941 plan.

### Why does the world owe Dr. Charles Drew a debt?

When Andrew Spurlock, a famous Black photographer, arrived to take a photo of Charles Drew (1904–50), he found the doctor wearing a white laboratory coat and seated beside a microscope. He wanted to be seen as a man of science. It was 1941, and Drew had already pioneered the preservation of blood for transfusions and was revered as an expert on the banking of blood.

Drew was born in Washington, D.C., went to the competitive Dunbar High School, and entered Massachusetts's Amherst College, gaining a reputation for both academic and sports excellence. Upon graduation in 1926, Amherst awarded him the coveted Messman Trophy. At McGill Medical College in Montreal, he was attracted to the growing field of blood research. While there, he was inspired by seeing a man's life saved by a transfusion.

By 1935, Drew had returned to his beloved hometown and

joined the faculty of Howard University's Freedman's Hospital, where he began intensive blood research. Immediately, he got good results. He created a procedure for separating plasma from whole blood, making it easier to keep blood—a first step toward the establishment of blood banks. Prior to Drew's procedure, blood spoiled quickly and could not be kept more than two days.

Drew also discovered that if blood was handled gently, it did not deteriorate so quickly. He then discovered that many patients could be given only plasma and not whole blood. Since plasma does not have a "type," this makes transfusion easier and safer.

Columbia Presbyterian Hospital in New York City offered Drew a blood research fellowship in 1938, which led to a doctor of medical science degree for his thesis, *Banked Blood: A Study in Blood Preservation.* His thesis surveyed all the world's work in this field, even that published in Russian. This second degree put Drew at the pinnacle of scientific achievement.

It was logical in 1940 for the American Red Cross to seek out Dr. Drew as the project director of its blood donor program. Soon, though, Drew found himself embroiled in a big controversy. The U.S. War Department announced a continuation of the policy of segregating blood by races, for "psychologically important" reasons.

Drew was furious: this offended him as a scientist and as a human being. He argued at a news conference that "there is absolutely no scientific basis to indicate any difference in human blood from race to race." Individuals' blood differed only by "blood groupings." Fears of blood mixing was a social paranoia and was not based on scientific fact. Resigning his post, he went back to work at Howard University.

In the prime of his career, Drew was killed in a 1950 automobile accident in North Carolina while headed to a meeting at Tuskegee. He was taken to a local hospital, where he was given good medical treatment. Out of his tragic death came a powerful rumor: that the white hospital refused to give him a transfusion—the very procedure he had researched. In fact, his doctors tried to give him a transfusion, but his injuries were too severe. Yet the rumor continues to make its own history.

## How did World War II affect Black America?

The United States entered World War II after the bombing of Pearl Harbor in 1941. With that, both Black and white Americans were galvanized.

Nearly a million Blacks entered the U.S. Army and Navy, including thousands of women in the Women's Auxiliary Army Corps (WAAC). Half of these—five hundred thousand—were sent overseas. All served in segregated units, although toward the war's end, in December 1944, some Blacks and whites fought side by side in the Battle of the Bulge.

Even with segregation, some barriers weakened. Blacks were admitted to officer training, and by 1944, the army had some seven thousand Black officers. At first the Navy permitted Blacks only as stewards and mess attendants, but in the war's last year, the navy commissioned its first Black officers. Also just before the war ended, Blacks were allowed into the marines and the coast guard and went on to be officers.

Advances did not ensure fair treatment. Base housing, recreation clubs, and transportation were poor for Black soldiers. White officers, especially in the South, insulted them, and clashes between white and Black soldiers were commonplace. Black soldiers told how German prisoners of war got better treatment than they did.

Despite this, Black combat soldiers went overseas to do a job, and they did it valiantly. Twenty-two Black units fought in Europe. Particularly distinguished were the 761st Tank Battalion, which fought six months straight; the 614th Tank Destroyer Battalion; and 593rd Field Artillery. Within the air force, the 99th and 332nd fighter squadrons were "hell on the enemy." The Tuskegee Airmen, the first Blacks to be trained as combat fighters, won medals for their outstanding service in Italy and Germany. Black units such as the 93rd Combat Division—the largest unit in the Pacific theater—also fought well.

Gains for Blacks resulted from the war, but they had to be fought for. One million Blacks entered the civilian labor force in the war years. In the defense industry, five thousand worked in aircraft factories, fourteen thousand in shipyards, and over one hundred thousand were in the iron and steel industry. Over four hundred Black domestics left for other jobs, many in factories. Yet, in Robin D. G. Kelly's brilliant *Race Rebels* (1995), he captures how many Blacks "were still faced with job discrimination, and employment opportunties for Blacks tended to be [in] low-wage, menial positions."

This did not deter thousands of Blacks.

Seeking jobs, a second wave of Black migrants struck out for the North and West. Between 1940 and 1945, the Black population of Los Angeles doubled, going from 75,000 to 150,000. The new Black migrants faced white resistance, which proved to be a volatile mixture; in the horrifying Detroit riots of 1943, seventeen Blacks and nine whites were killed.

Blacks fought and worked for American victory. "The Double V" was their vision of the war, meaning that there should be two victories—one over the Axis powers and a second over American prejudice. In fact, during the war, Blacks grew more militant, with the NAACP increasing to five hundred thousand members. New evidence, collected by historian Robert Hill, demonstrates that Blacks were highly active in protesting bad treatment, on a scale that was unparalleled. Many urban Black youth—wearers of the zoot suit, dancers of the lindy hop, speakers of hepcat lingo—like Malcolm Little (later Malcolm X) stayed clear of the war altogether.

When the war ended in 1945, Black life had been changed. Blacks had been exposed to new experiences—the factory and life at the front. New knowledge rushed into Black neighborhoods. Black war participation had been high and positive. President Harry Truman, Roosevelt's successor, capitalized on this, lobbying Congress on civil rights and job discrimination and issuing an executive order toward ending segregation in the military.

The changes did not stop there. In 1946, when Charles and Medgar Evers returned from overseas military service to Mississippi and attempted to register to vote, the angry white registrar asked, "Who you niggers think you are?" They replied, "We've grown up here. We fought for this country, and we should register." In the war's finale lay the civil rights movement's opening act.

### Why was Jackie Robinson a hero?

A simple sentence on April 10, 1947, announced the revolutionary event: "The Brooklyn Dodgers today purchased the contract of Jackie Robinson from Montreal." A Black man was going to play major league baseball, ballyhooed as the American pastime, and not for just any team but for the "Bums" located in that ethnic hotbed of fan loyalty, Brooklyn.

In 1947, America was still relishing the victory over Hitler, was prosperous and at the top in world influence and admiration. If ever there was a chance for Dodger executive Branch Rickey's decision to work, it was now. It was clear too, as sportswriter Gordon Macker said, if a Black "is good enough to stop a bullet in France, he's good enough to stop a line drive" in a baseball park.

Baseball had been rigidly segregated throughout its history. In 1889, Moses "Fleetwood" Walker played briefly for a major league team but was forced out. Since then, talented Blacks had appeared on college teams. In 1920, Andrew "Rube" Foster—a brilliant pitcher, promoter, and team manager—had assembled the first National Negro Baseball League. During the 1930s, three more leagues were founded, together becoming "the Negro leagues."

Most famous of the league's teams were the Homestead Grays, American Giants, Kansas City Monarchs, and the Saint Louis Stars. Black fans loved these teams: their games were remembered as superb baseball and great entertainment. When one of the hot teams arrived in town, Black communities would come alive; everyone from the professional folk to the lowlifes joined in the fun. Director John Badham's 1976 film *The Bingo Long-Traveling All-Stars and Motor Kings,* starring Billy Dee Williams, James Earl Jones, and Richard Pryor, was a loving tribute to the fun and sport generated by such teams.

Stars like James "Cool Papa" Bell, Walter Buck Leonard, John Henry Lund, and Satchel Paige caused tickets to go like mad. Before Robinson went to the Dodgers' Montreal farm team, he played for the Kansas City Monarchs. While in Kansas City, he was scouted by Rickey.

Although born in Georgia in 1919, Robinson's mother moved her family to Los Angeles, where he attended UCLA. There, he became one of the most impressive college athletes of all time: an All-American as the nation's foremost running back, a championship broad jumper, a leading scorer in basketball, and an excellent baseball player. Besides athletic versatility, Robinson had a college education and had served in the army; he was discharged in 1943 as a lieutenant.

When Robinson was first interviewed by Rickey, he asked, "Do you want a ballplayer who's afraid to fight back?" To this, Rickey answered, "I want a ballplayer with guts enough not to fight back! You've got to do the job with base hits and stolen bases and fielding balls, Jackie. Nothing else."

Robinson was a perfect choice for Rickey's strategy. In his first year, he hit .297 and stole twenty-nine bases, helping put the Dodgers in the World Series. He was voted Rookie of the Year. In 1949, he was named the league's Most Valuable Player, hitting .342 and stealing thirty-seven bases. Although there were insults and threats against Robinson, he never lost his temper—not even when the Phillys' manager, Ben Chapman, sprayed him with racial venom about fat lips, flat noses, and extra thick skulls that restricted brain growth.

Robinson's calm disciplined performance was an example to the country of grace under pressure. Even during his first season, Robinson's heroic quality was becoming apparent: he was the nobleman of the ballpark. In 1962, he was voted into the Hall of Fame. After him would come greats like Roy Campanella and Don Newcombe.

## What was the impact of *Amos 'n' Andy?*

When CBS-TV bought *Amos 'n' Andy* in 1948, it was taking a tremendously popular radio show and bringing it to a new medium. CBS paid the creators of the show the unheard-of sum of $2.5 million for television rights. But it ran as a television show only from 1951 to 1953.

Chicago was the show's birthplace in 1926, where it went by

the name of *Sam 'n' Henry.* NBC radio picked it up for nationwide distribution three years later. The show was created by Freeman Gosden and Charles Correll, two white entertainers who used prominently what was assumed to be Black dialect. At first, Gosden and Correll spoke all the roles themselves; later, they began hiring Blacks.

During the 1930s, the show's popularity reached its peak, with an audience of nearly forty million. The show focused on Andrew Brown, a rather slow-witted character prone to infatuations with numerous women, and Amos Jones, his level-headed partner in the ownership of the Fresh Air Taxi Company. Brown and Jones were members of the Mystic Knights of the Sea fraternity, a lodge whose members included George "Kingfish" Stevens, a con man and smooth operator who was forever embarking on schemes to make money, especially at Andy's expense. The show's most important women characters were Sapphire, the Kingfish's wife; Mama, his mother-in-law; and Madame Queen, who was always trying to get romantically involved with Andy.

Americans took to the show because its characters expressed issues that were important to the depression and did so in a comic way. Money was forever on the characters' minds, and jobs were hard to find. Get-rich schemes were constantly being tried. Audiences were also amused by the show's Black folk humor tradition. The show's creators actually had some knowledge of the language and attitudes implicit in Black humor. However, it is questionable whether they understood the functions of Black humor and its subtleties; their version was too obvious and broadsiding, rather than delicate and insinuating.

From 7:00 to 7:15 P.M., six evenings a week, America came to a screeching halt. Stores, bars, and restaurants had their radios turned up so that customers could hear the episodes. Presidents Truman and Eisenhower were fans. American language was influenced by the program. Popular phrases came from it: "I'se regusted," "Check and double check," "Now ain't dat sumptin," and "Holy mackerel."

When the show came to television in the 1950s, CBS was surprised by the heightened political consciousness of the post–World War II Black community. The NAACP adopted a resolution at its 1951 convention condemning the show because "it tends to strengthen the conclusion among uninformed and prejudiced people that Negroes are inferior, lazy, dumb, and dishonest." Eventually, CBS responded to pressure from the NAACP and other groups, dropping the show in 1953. It remained in syndication, however, until 1966.

One of the most damaging aspects of the show was its stereotyping of Black women through the character of Sapphire, Kingfish's domineering wife. She was the embodiment of many stereotypes—aggressive, razor-tongued, oppressive, and overwhelming of her spouse. She was always on the attack, especially at Kingfish for his laziness and deceptions. The name "Sapphire" lived on after the show to symbolize the domineering Black woman.

The show was an important landmark in American popular culture, but history changed while the show did not. That caused its demise.

## What was "the new meritocracy" of post–World War II Black America?

After World War II, a door onto the larger world opened—slightly—for Black Americans to walk through. In 1944, Gunnar Myrdal's *An American Dilemma*—researched by a Swedish economist and Black American social scientists such as Charles S. Johnson and St. Clair Drake—had told America that to solve "the Negro problem," opportunity had to come.

When it did, Blacks were ready; it seems as if they had been studying, almost rehearsing for the day when a door would open and they would march forth to earn their renown. When America asked, "What do you want?," Blacks replied, "All we want is a chance, a chance to prove our worth, our merit."

In those years, every Black schoolchild was drilled in one axiom of success, an eleventh commandment of Black life: "To succeed, you must be twice as good as a white person." From this deep resolve came a "new meritocracy"—a highly skilled, intelligent, powerful class of Black superachievers, whose contributions to different fields proved that America needed their abilities.

Perhaps the country was learning a lesson from the earlier fate of three of America's greatest Black talents: Josephine Baker (1906–75), Paul Robeson (1898–1976), and Richard Wright (1906–60). All had been leaders in their fields and had created cultural works that the world applauded, but all had left America because they could not tolerate its racism and its limitations on their careers. This Black self-exiling was an indictment. It robbed America of Black talent. With the door ajar after the war, Black talent surged through, ending up in unexpected places.

Leontyne Price (born 1927), a Mississippian educated at New York's Juilliard School of Music, appeared in 1952 in Virgil Thomson's opera *Four Saints in Three Acts*, in 1954 in Gershwin's *Porgy and Bess*, and again in 1954 as the lead in Puccini's *Tosca.* Katherine Dunham (born 1910) had studied dance in the Caribbean and presented this form for the first time on American stages. In 1945, she opened the Dunham School of Dance. Dorothy Dandridge (1924–65), the first Black actress marketed as a sex symbol, was nominated for an Academy Award for her tender performance in *Carmen Jones* (1954), a modern film adaptation of Bizet's *Carmen.*

In the world of global geopolitics, Ralph Bunche (1904–71)—a graduate of UCLA with a Harvard University Ph.D., later a leading drafter of the United Nations charter—negotiated a peace agreement between Israel and Egypt in 1949. At forty-five, his diplomacy was recognized with the Nobel Prize for peace.

Another place the new meritocracy worked was in the world of ideas. Black authors' books hit the public in rapid succession. Gwendolyn Brooks (born 1919) published two smashing volumes of poetry, *Street in Bronzeville* and *Annie Allen*, the latter winning the Pulitzer Prize. No doubt the judges were impressed with her sharp images, unusual word usage, and gentle look at human foibles. Ann Petry (born 1908) startled over a million readers with *The Street*, which offered a deep Freudian look at the novel's characters.

In 1952, Ralph Ellison (1914–94) added to this creative surge with *Invisible Man.* Drawing on literary conventions from Russian, nineteenth-century American, and European modernist traditions, Ellison's was a grand novel with a magnificent historical sweep from Reconstruction to post–World War II Black urban life. The novel was a virtual encyclopedia of Black culture: myths, icons, sayings, music, oratory, dress styles. In 1953, his work received the National Book Award.

As the fifties progressed, a new mood in Black writing surfaced. Authors were more introspective, looking inside Black life, often criticizing and even reproaching it. James Baldwin (1924–89) made a habit of this. *Go Tell It on the Mountain* is a skeptical look at the Grimes family's Pentecostal religious beliefs. *Notes of a Native Son,* razor-sharp essays, demonstrated Baldwin's capacity to turn the tables on America and interrogate its racial history. In *Black Bourgeoisie,* E. Franklin Frazier (1894–1962), a Howard University sociologist acclaimed for his studies of the Black family, assailed the Black middle class for its materialism and its obsession with social forms over substance.

Hovering over this scene were the new sounds of jazz from Miles Davis, Sonny Rollins, and John Coltrane; the gospel singing of Mahalia Jackson; and the banshee screaming, erotic lyrics, and wild instrumentalism of early rock and roll with Little Richard and Chuck Berry. Blacks had earned their place in the new kingdom of post-1945 American culture, undoing the ideas behind segregation.

## What are the characteristics of Black rock and roll?

It would be misleading to characterize Black rock and roll and white rock and roll as separate entities. Chuck Berry (1926– ), Little Richard (1935– ), Elvis Presley (1935–1977), Bo Diddley (1928– ), Fats Domino (1928– ), and Jerry Lee Lewis (1935– ) were not creatively on different sides of the fence. Together, they kicked the fence down and didn't look back. Though there is some validity in the general idea that Elvis Presley was a white boy who sang Black, rock and roll as an art form sprang from Black American Southern culture and the hillbilly culture of that region. Rock and roll is part of our national transracial heritage.

From the beginning, it was about exceeding limits, boundaries. Besides jazz, most African-American musical forms of the twentieth century showcase the artist's negotiation of a form's limits with grace, elegance, and crudity. Rock and roll is about driving the emotion beyond the form so that the formal boundaries are seemingly destroyed. Rock and roll is a progression from the blues in the way that a smoggy superhighway represents progress to a naturally wooded dirt pathway.

If there is a defining characteristic of black rock and roll, it is the descending pentatonic riff of Chuck Berry's "Johnny B. Goode." But, that trademark sound is also the sonic foundation of rock and roll just as the song is rock and roll's founding myth. Beatle John Lennon (1940–1980) said once of the genre: "If you tried to give rock and roll another name, you might call it 'Chuck Berry.'"

Berry grew up Black and middle class in the 1930s and 1940s in St. Louis. By the time he wrote the songs that dreamed the rock and roll nation into being, he was in his late twenties. Chuck Berry is the first poet laureate, the first to give witty lyrics to rock, and is considered by many to be the father of rock and roll. "Johnny B. Goode"—with its playful sentence name title— is as good a point as any to analyze the notion of Black rock and roll. When composing it Berry first wrote: "Way down Louisiana close to New Orleans/Way back up in the woods among the evergreens/ There stood a little cabin made of earth and wood/ there lived a colored boy named Johnny B. Goode." He changed "colored boy" to "country boy" before anyone else had ever heard it. With that change he made the song not only his story but Elvis's and all who would follow Elvis.

Berry also made theatrics central to the rock performance. It was his duckwalk—like Elvis's swiveling hips and Jerry Lee Lewis's humping the piano or Fats Domino pushing the piano across the stage with his belly while belting "Ain't That a Shame"—that defined rock and roll stagecraft as alien ferocity and gimmickry beamed by television into the homes of 1950s teen America. As for stagecraft, exhilaration, and uncompromising flamboyance, Little Richard (actually Richard Wayne Penniman, from Macon, Georgia) had no peer. If his continuing boast that he is the "architect of rock and roll" is a tad overstated, he was one of the first Black artists to cross over to the white pop charts, and he did so with the blasts of wildness that would come to define the rock and roll spirit. "Tutti Frutti," "Rip it Up," "Good Golly Miss Molly," and "Long Tall Sally" took gospel vocalizing way beyond the secular arena of love and into the country of lust and irreverence and anarchy. His banshee shrieks, trills, no-holds-barred piano thumping, shiny pompadour are rock and roll archetypes.

Many critics have said the British Invasion of the 1960s that brought the Beatles and the Rolling Stones to the top of the American pop charts could be called a Chuck Berry and Little Richard revival. There is much to that. As foreigners playing American music, these British musicians were wowed by the blues greats who inspired them but more so by the sound of Black American rock and rollers, who had been locked out of the American Dream, singing and performing as if no restrictions existed.

Of Black rock and rollers, from Chuck Berry to Little Richard to Jimi Hendrix to the Bad Brains and Fishbone of today, it is indisputable that if there is one characteristic that sets them apart from their white peers it is a drive to annihilate boundaries of race and class. It is with Jimi Hendrix that this statement is most dynamically proven. John Lennon once said "after Hendrix everything was different."

A single day gave proof of this. John Seabury and Charles Murray have described Jimi Hendrix's appearance at the Monterey Pop Festival on Sunday, June 18, 1967, as the day "the electric guitar was, to all practical purposes, reinvented." Using the unfashionable Fender's Stratocaster guitar, Hendrix invaded the upper reaches of rock performance—usually reserved for whites—with a music that ranged so widely over blues and rock genres that the audience was overwhelmed. "No one in America had ever heard anything like it." Hendrix became a 1960s futuristic attraction—and today, twenty-five years after his untimely death at twenty-seven, other Seattlites—notably Pearl Jam—pursue his dream.

In Cleveland's new Rock and Roll Hall of Fame is a list of the most influential songs that have shaped the music. Of the 500 songs, 184 were performed by Blacks. Casting a wide net, the list begins with Afrika Bambaataa's "Planet Rock," runs to Lavern Baker's "Jim Dandy" to Little Eva's "The Loco-Motion" to The Ronettes' "Be My Baby," and ends with Stevie Wonder's "Uptight." For many, the Black share in this list is not enough. But, it is a start in documenting Blacks' place in this century's most popular music.

### How did a young girl's case and a big-time lawyer's crusade team up to tear down an old racial wall?

When Linda Brown (born 1943) went to elementary school in Topeka, Kansas, in the 1950s, she walked to a segregated one a mile, and over an hour and half, from her home, across a dangerous railway yard.

Most bothersome was the fact that she passed the all-white Sumner School only seven blocks from her house. Oliver Brown, her father, a welder and lay minister, was upset every day as he waved good-bye to Linda. In September 1950, he tried to enroll her in Sumner. When Linda was refused, her father filed suit to win admission for her as part of an NAACP test case. Her case would become the famous *Brown v. Board of Education* when it ended up before the United States Supreme Court.

In the 1950s, segregation was a way of life in schools, hotels, restaurants, sports, theaters, bathrooms, and even at drinking fountains all across the South and in some states outside the South, like Kansas. Beginning in the 1930s, the NAACP challenged the "separate but equal" doctrine established by the Supreme Court in 1896 in *Plessy v. Ferguson.* The NAACP focused on segregation in education because it was indispensable to legal equality and economic opportunity for Blacks.

The NAACP's first victory came in a Missouri case. In 1936, Lloyd Gaines had been rejected for admission to the University of Missouri Law School. He sued—and lost—in the state courts, but the Supreme Court ruled in 1938 that a state must provide education for all citizens. Missouri responded by building a separate school for Blacks. After World War II, the NAACP won similar victories, such as the Supreme Court's 1950 order that Texas admit Herman Sweatt to the all-white state law school, even though Texas had one for Blacks. The latter was judged not equal to the law school for whites.

In the 1930s and 1940s, a tall, handsome, brilliant, folksy lawyer was making a legal name for himself—this was Thurgood Marshall (1908–92). Marshall was from a middle-class Baltimore family, his mother a schoolteacher and his father a Pullman steward on the railway. In 1926, he attended Lincoln University in Pennsylvania, and instead of becoming a dentist as his mother wanted, he applied to the all-white University of Maryland's law school and was rejected. When he met Linda Brown, he must have known how she felt.

Howard University's law school was his next choice, a decision that profoundly altered his life. There he came under the charismatic influence of Charles Hamilton Houston (1895–1950), who was instructing young Black law students in the strategy of using existing laws to undermine racial discrimination. Houston was the architect of the legal undoing of school segregation.

Marshall graduated valedictorian of his class and returned to Baltimore, where after some false starts, he became a lawyer for the NAACP. In 1935, his first case and victory forced the University of Maryland to enroll a Black law student—this was the same law school that had turned Marshall down five years earlier. A Baltimore lawyer described the decision's impact: "The colored people . . . were on fire when Thurgood did that. . . . He brought us the Constitution as a document like Moses brought his people the Ten Commandments."

Marshall went to work at NAACP headquarters in New York, eventually becoming its leading lawyer. Over the next twenty years, Marshall and his legal staff compiled a most remarkable record of success. In 1952, Marshall and his cohorts initiated simultaneous lawsuits in four states and the District of Columbia challenging the constitutionality of segregated education. The Supreme Court grouped all these cases together under *Brown v. Board of Education of Topeka, Kansas.*

On May 17, 1954, Chief Justice Earl Warren ruled that "separate educational facilities were inherently unequal." The ruling stated, "To separate them [Negro children] from others of similar age and qualifications solely because of their race generates a feeling of inferiority . . . that may affect their hearts and minds." Linda Brown's rights were upheld, and even though she did not attend Sumner Elementary School, she did go to an integrated junior high school.

The reaction to the *Brown* decision in the South was twofold: a sense of relief among Blacks and rebel yells from many whites. Unfortunately, the 1954 decision left one matter unresolved: how and when racial barriers would drop. In 1955, in a decision called *Brown II*, the Court said that desegregation "would be gradual and would be influenced by local conditions"; this allowed for much Southern white resistance.

Marshall continued to work for the NAACP. President Kennedy nominated him to be a justice for the Second Circuit Court of Appeals, but because of intense Southern opposition, it took nearly a year before he was confirmed in 1962. In 1965, President Lyndon Johnson nominated him to be solicitor general of the United States, and two years later, he nominated him to the Supreme Court.

In 1990, when Marshall went to speak at a Washington, D.C., hotel, Black bellhops, maids, doormen, and waiters crowded the corridor by which he entered, some crying, others waving.

### Who was Rosa Parks and how did she triumph over the Montgomery, Alabama, segregated bus system?

On the evening of December 1, 1955, seamstress Rosa Parks (1913– ) was just too tired to give up her seat on the bus to a white man. She was also sick and tired of the law that required her to do so. When she refused to move to the Black section of the bus, she was arrested and put in jail.

In these minutes, a historic moment had occurred. Crystallized in this episode were two major opposing forces—a determined segregationist society colliding with Black resistance. On its surface, what happened on the bus did not appear revolutionary. But, in the history of defeating segregation, it was the same as if Rosa Parks had stormed the Bastille. Over twenty years later, historian Lerone Bennett would define this as one of the "Great Moments in Black History"—"The Day the Black Revolution Began." Later, Rosa Parks would be christened "the Mother of the Civil Rights Movement."

In addition to being a seamstress, Rosa Parks had been secretary of the NAACP and was now running the office of E.D. Nixon, president of the state NAACP. In the summer prior to her courageous act, she had gone—using a scholarship—for a summer workshop to the Highlander Folk School in Monteagle, Tennessee. Highlander had a long history of training labor organizers. In the 1950s, it turned to providing instruction and support for the new civil rights activists from all over the South. Training was offered in literacy education, voter registration tactics, forming cooperatives, and understanding one's rights under the Constitution. Before her Highlander experience, Parks had had a history of passive resistance to bus drivers. Although whites had not expected her civil disobedience, it was no surprise to those who knew her when she decided to take her stand.

After Parks's arrest, a local committee of women—the Women's Political Council (WPC)—called for action. By December 5, Black leaders had formed the Montgomery Improvement Association, with the then little-known Rev. Martin Luther King Jr. as president. They announced that as long as city buses remained segregated, Blacks would walk. For three hundred and eighty-one days, buses were virtually empty as the city's seventeen thousand Blacks (75 percent of the bus-riding population) boycotted local transportation.

Finally, in May 1956, NAACP lawyers argued the case before the Federal District Court, which declared segregated buses unconstitutional. Later in the year, the U.S. Supreme Court upheld the ruling.

Many thousands of people—including some few whites, such as activist Virginia Durr—contributed to the success of the Montgomery Bus Boycott. But it is generally accepted that Rosa Parks provided the movement's stimulus, which launched Rev. Martin Luther King Jr.'s career as a civil rights activist. And, her act was the inspiration for Blacks throughout the South.

## RESOURCES

Aptheker, Herbert. *A Documentary History of the Negro People in the United States.* Vols. 4 and 5. New York: Citadel Publishing Group, 1993.

Berry, Mary Frances. *Black Resistance, White Law: A History of Constitutional Racism in America.* New York: Penguin Press, 1994.

Duberman, Martin. *Paul Robeson: A Biography.* New York: Alfred A. Knopf, 1989.

Egerton, John. *Speak Now Against the Day: The Generation Before the Civil Rights Movement in the South.* New York: Alfred A. Knopf, 1994.

Etter-Lewis, Gwendolyn. *My Soul Is My Own: Oral Narratives of African-American Women in the Professions.* New York: Routledge, 1993.

Falkner, David. *Great Times Coming: The Life of Jackie from Baseball to Birmingham.* New York: Simon & Schuster, 1995.

Goodman, James. *Stories of Scottsboro: The Rape Case That Shocked 1930s America and Revived the Struggle for Equality.* New York: Pantheon Books, 1994.

Harris, Middleton. *The Black Book.* New York: Random House, 1974.

Hill, Robert A. *The FBI's RACON: Racial Conditions in the United States During World War II.* Boston: Northeastern University Press, 1995.

Hutcheson, Earl Ofari. *Blacks and Reds: Race and Class Conflict, 1919–1990.* East Lansing: Michigan State University Press, 1995.

Johnson, Thomas L., and Philip C. Dunn. *True Likeness: The Black South of Richard Samuel Roberts, 1920–1936.* Chapel Hill, N.C.: Algonquin Books of Chapel Hill, 1986. Text with Photos.

Kluger, Richard. *Simple Justice: The History of Brown v. Board of Education.* New York: Alfred A. Knopf, 1976.

Meier, August. *Black History and the Historical Profession, 1915–1980.* Urbana: University of Illinois Press, 1986.

Robeson, Susan. *The Whole World in His Hands: A Pictorial Biography of Paul Robeson.* New York: Citadel Press, 1981. Text with photos.

Rowan, Carl. *Dream Walkers, Dream Breakers.* Boston: Little, Brown, 1993.

Sandler, Stanley. *Segregated Skies: All-Black Combat Squadrons of World War II.* Washington, D.C. and London: Smithsonian Institution Press, 1992.

Taulbert, Clifton L. *Once Upon a Time When We Were Colored.* Tulsa, Okla.: Council Oak Books, 1989.

Washington, Linn. *Black Judges on Justice.* New York: New Press, 1994.

Watson, Denton. *Lion in the Lobby: Clarence Mitchell's Struggle for the Passage of the Civil Rights Laws.* New York: Morrow Books, 1990.

TEXTS FOR YOUTH

Clayton, Ed. *Martin Luther King: The Peaceful Warrior.* New York: Pocket Books, 1968. Text with illustrations.

Lee, George L. *Interesting People: Black American History Makers.* New York: Ballantine Books, 1989. Text with illustrations.

Marsalis, Wynton, *Marsalis on Music.* New York: W.W. Norton & Company, Inc., 1995. With special audio CD.

Patillo, Melba. *Warriors Don't Cry: A Searching Memoir of the Battle to Integrate Little Rock's Central High.* Abridged ed. New York: Beals, 1994.

Sampayo, Carlos and Jose Muñoz, *Billie Holiday.* Seattle: Fantagraphics Books, 1993.

# 8

# The Civil Rights Decade

## The 1960s

1960: "Sit-in" movement begins: four Black students from North Carolina's A & T College sit in at a Woolworth's "whites-only" counter and are denied service

1960: The Student Nonviolent Coordinating Committee (SNCC) comes into being at Shaw University, Raleigh, North Carolina

1960: Black runners—especially Wilma Rudolph—electrify the Rome Olympics

1961: John F. Kennedy is inaugurated as the thirty-fifth president of the United States

1961: Congress of Racial Equality sends Freedom Riders on bus rides through the South to test compliance with federal laws integrating bus stations

1962: Rev. Martin Luther King Jr. conducts a massive sit-in and marching campaign against segregation in Albany, Georgia

1962: Leroy Johnson becomes the first Black legislator in the Georgia state senate since Reconstruction

1963: Birmingham, Alabama, civil rights demonstrators, led by Rev. Martin Luther King Jr., are attacked by police headed by Eugene "Bull" Connor

1963: Medgar Evers, NAACP field secretary and civil rights leader, is assassinated June 12 at his house in Jackson, Mississippi

1963: Over 250,000 participate in the March on Washington, the largest protest gathering in U.S. history; Rev. Martin Luther King Jr. delivers his "I Have a Dream" speech at the Lincoln Memorial

1963: John F. Kennedy is assassinated and Lyndon B. Johnson is sworn in as president

1964: President Johnson presses Congress to pass the Civil Rights Bill, banning discrimination in public accommodations, education, and employment

1964: Cassius Clay is heavyweight champion and assumes "Muhammad Ali" as his name

1964: Malcolm X breaks formally with Elijah Muhammad and the Black Muslim movement; establishes the Organization of Afro-American Unity

1964: The Council of Federated Organizations sponsors the Mississippi Summer Project—"Freedom Summer"—bringing over a thousand Black and white volunteers into the state; violence—burnings, bombings, and murders—marks the summer

1965: Malcolm X is assassinated at a meeting called by the Organization of Afro-American Unity, Audubon Ballroom, New York

1965: Martin Luther King Jr. begins campaign to register Black voters in Selma, Alabama; more civil rights advocates are killed

1965: President Johnson signs the Voting Rights Act of 1965

1965: Watts, a largely Black district in Los Angeles, erupts

1965: President Johnson announces a major increase in U.S. Armed Forces in Vietnam

1966: Robert C. Weaver becomes cabinet officer and Constance Baker Motley becomes first Black woman federal judge

1966: President Lyndon Johnson announces the Great Society programs

1966: Rev. Martin Luther King Jr. starts campaign in Chicago

1966: The Black Panther Party is founded in Oakland, California, by Huey P. Newton and Bobby Seale

1967: Martin Luther King Jr. leads thousands of marchers protesting the Vietnam War in New York City

1967: President Johnson nominates Thurgood Marshall to the Supreme Court

1967: A "long, hot summer" of racial eruptions in major cities and towns

1967: U.S. Supreme Court rules state laws against interracial marriages are unconstitutional

1968: Martin Luther King Jr. is assassinated in Memphis, Tennessee; the nation mourns and cities are engulfed in rioting

1968: *New York Times* poll reports that 59 percent of Blacks prefer to be called "Afro-American" or "Black" rather than "Negro"

1969: Richard M. Nixon is inaugurated as president

1969: Chicago police raid headquarters of the Black Panther Party, killing two leaders in their beds and wounding four others

When the film *Freedom on My Mind* appeared in 1994, it became the latest award-winning film devoted to Black history of the 1960s.

Created by Connie Field and Marilyn Mulford, the film told the story of the 1964 Mississippi Freedom Summer. Interspersing vignettes of segregated life with footage of registration efforts, the film portrayed the people who worked to change the state. Already, *Eyes on the Prize* (1986) and *Eyes on the Prize II* (1989) had given America a rare panoramic history of the civil rights movement. Now audiences got to see a more personal story.

More films, television shows, and videos exist on the 1960s than any other period in twentieth-century Black history.

America can see this era, hear its words, be swept up in its songs. When the period's photography is added, our connection to this decade is made even stronger.

Rallies, Rosa Parks's gentle face, the expectant eyes of young protesters, the Fruit of Islam, scenes from LeRoi Jones's *The Dutchman,* the 1964 Mississippi Freedom Democratic party protests, Malcolm X in Harlem, bandolier-wearing Black Panthers, Motown acts, Sly and the Family Stone and Aretha Franklin on Ed Sullivan—so many sharp images have been preserved from this time. Even Hollywood sought to exploit the Black 1960s in *Mississippi Burning* (1988), a cinematic excursion turning real history into fiction.

Why this decade became the focus of so much attention is understandable. The 1950s brought Black America to the brink of major change. But its preoccupation with legal remedies and integration as solutions to Black problems was too limited. Segregated life in the South had improved only slightly. In other areas, Black poverty cast a pall over any political and social progress.

Blacks in the 1960s were ambitious. They took it all on, in one bold move after another, moving the whole country in their direction. During the first half of the decade, the nearly nineteen million Blacks, 10.5 percent of the nation, seemed to act as one. After that time, disagreements appeared within Black America. Still, a sense of solidarity was maintained to the decade's end.

**In general, what was the civil rights movement about?**

Civil rights—the very words are synonymous with the 1960s. During much of the decade, the civil rights movement dominated the national consciousness. Twentieth-century America had never witnessed anything quite like it. The movement focused mainly on changing the South's racist society. But racial backwardness was challenged all over the country as the movement's impact rippled to the north and west.

When the 1960s dawned, millions of Blacks still lived in some of the world's most oppressive societies, as in the Deep South states of Alabama, Georgia, Mississippi, and South Carolina. Millions lived in an American gulag, where both laws and custom dogged Blacks' every step. Even with oppression, Blacks had built many striving, sharing, interesting communities, as Clifford L. Taulbert's loving memoir of Glen Allan, Mississippi's "colored town," confirms.

The movement's basic goal was to secure for Black Americans their full rights as guaranteed by the Constitution. After Blacks were freed from slavery, they were made citizens by the Fourteenth Amendment in 1868. Moreover, they were granted "equal protection of the law." But Blacks were citizens in name only.

A tragic example of Black oppression was in the area of voting. Blacks found it next to impossible to participate in this basic American process. Taxes, property qualifications, special tests, literacy exams, and interrogations were all used to thwart potential Black voters.

This was in direct violation of the Constitution's Fifteenth Amendment, adopted in 1870, which assured that "the rights of citizens . . . to vote shall not be denied . . . on account of race, color, of previous condition of servitude." In many districts in the South, Blacks were the majority of the population yet only a tiny minority of the voters. Even with these restrictions, John Kennedy's 1960 election was possible because of Black votes in Michigan, South Carolina, and Illinois.

When the civil rights movement came, Blacks became its chief participants. Every Black family probably had someone in the movement. Young Blacks were especially crucial as the movement's day-to-day soldiers, organizing rallies, boycotts, marches, sit-ins, jail-ins, voter drives, and labor support actions. Blacks of all ages and ranks transformed themselves into a great moral crusade. This was essential. Nothing less than their full intensity and power would have destroyed segregation.

They lifted the centuries-old siege of denial. But they did not act alone. The civil rights protests gave birth to a new interracial family of supporters and workers. White students increased in the movement until 1966, when their role was questioned. Labor groups sent organizers, money, and supplies to the South; churches and synagogues also sent emissaries.

Blacks and whites died together. In 1989 in Montgomery, Alabama, a monument designed by Chinese-American architect Maya Lin was dedicated to the movement's martyrs. Forty people—the majority Blacks and several whites—were honored. Not since the abolitionist movement of the 1830s to 1860s had there been a cross-racial campaign for Black freedom.

Civil rights groups orchestrated the vast mass war against segregation. Older groups, such as the NAACP and the National Urban League, were crucial in preparing the way for the 1960s movements. The NAACP's brilliant legal work in the 1950s proved that the daunting edifice of segregation could be undermined. Segregation was not invincible.

The 1960s saw proud new groups come to the struggle's foreground. The Congress of Racial Equality (CORE), founded in 1942 in Chicago, had a national presence in the 1960s. Its first leaders—four white men—were imaginative and aggressive in their tactics. In the 1960s, James Farmer, a Black, became its head.

The Southern Christian Leadership Conference (SCLC), which grew out of the Montgomery bus boycott in 1955, entered the fray under the leadership of Rev. Martin Luther King Jr. In the early 1960s, King was developing into an electrifying presence. He was unequaled in oratory and as a social philosopher and had an ability to focus national attention on racial injustice.

The Student Nonviolent Coordinating Committee (SNCC) was established in 1960 by Black college students committed to direct-action campaigns throughout the South. SNCC was one of the most spirited and effective civil rights organizations of the 1960s.

The civil rights movement can be divided into three phases. The first started with the Montgomery bus boycott on December 1, 1955. The movement's early tactics emerged in this period. On February 1, 1960, a second phase started with student sit-ins by four freshmen from North Carolina Agricultural and Technical College at segregated lunch counters in Greensboro. SNCC was formed based on this event.

The third phase began with the March on Washington in August 1963—one of the largest demonstrations in American history. Three hundred thousand people came to the capital. After that, the movement was clearly of national import. In the next years, the movement confronted the hostile racist Southern communities. Rivalries within the movement sprang up too.

Two of the movement's great achievements were the Civil Rights Bill of 1964 and the Voting Rights Act of 1965. The former law prohibited discrimination in public accommodations and forbade discrimination in federal programs. The latter struck down literacy tests in voting registration and gave powers to the U.S. attorney general to send federal agents to register Blacks in places where they encountered obstruction. More than 250,000 Negroes in the South registered that year.

But the civil rights movement's contributions surpassed specific legislation. It was a great mass movement that changed the South by eliminating legal segregation, engaged Black and white participants, and also changed America by breathing new life into the Constitution's promises.

### Who was Rev. Martin Luther King Jr.?

Just as the depression was setting in, Martin Luther King Jr. was born in Atlanta, Georgia, in 1929. He was named after his father who had changed his name to Martin Luther, after the German monk and heretic who started the Reformation.

Hard times faced the King family in Martin's early years, but they were able to provide for him. His security was enhanced because of his family's strong religious beliefs and their tradition in the ministry. His father, Martin Luther King Sr., was a Baptist minister. His mother's father, Rev. Alfred Daniel Williams, had founded Atlanta's Ebenezer Baptist Church in 1895.

Martin Luther King Jr.'s father had been a pioneer in the field of Black rights in Atlanta. Although it was dangerous in those days to belong to the NAACP, he was one of the Atlanta chapter's first members. In the 1920s, he led a campaign to force the city to build a public high school for Blacks. He organized a boycott of a white newspaper that had ridiculed Blacks' desire for equal education. He frequently spoke from his pulpit against racial injustice.

Atlanta had one of the country's largest Black urban populations during the years in which Martin Luther King Jr. was growing up. Discrimination took its toll on the community, but Blacks had built important churches, colleges, women's and men's clubs, newspapers, and businesses. During King's adolescence, this community had two great ambitions: to improve itself and to overthrow segregation.

King attended Booker T. Washington High School, the first public high school for Blacks. At the age of only fifteen, by skipping two grades, he entered Atlanta's prestigious Morehouse College, a Black men's school, whose president was the charismatic Benjamin Mays. Mays's speaking ability, memorable chapel lectures, scholarship on Black religion, high standards, and personal caring made a profound impact on King.

Aided by George D. Kelsey, a professor of religion, Mays persuaded King to enter the ministry. Each of these men was a Socrates to King—a friendly but probing teacher. A year before King graduated from Morehouse, he was ordained to be a minister at a ceremony in his father's church. After he graduated in 1948, he left Atlanta to further his education.

By 1955, King had graduated from Boston University's School of Theology with a Ph.D. In his years in Boston, he met Coretta Scott, from Marion, Alabama, a voice student at Boston's New England Conservatory of Music. Over the objections of his parents, who wanted him to marry someone from Atlanta, they were married on June 18, 1953.

After Boston, he returned to the South where he became the pastor of Dexter Avenue Baptist Church in Montgomery, Alabama. He succeeded the famous Vernon Johns, a brilliant minister who was uncompromising in his opposition to segregation.

During his years away from Atlanta, Martin Luther King Jr. was influenced by several important ideas, which he fashioned into a life's philosophy. Henry David Thoreau's politics and his essay *Civil Disobedience* were very instructive. Later, King recalled, "I read the work several times. This was my first intellectual contact with the theory of nonviolence and resistance." Another influence on King was theologian Walter Rauschenbush, who believed that Christianity was to be used in everyday life to address the present's issues, that it should be a "social gospel."

The greatest influence on King came from Mohandas Gandhi of India, who led his country to independence from the British empire. King devoured Gandhi's writings and learned from them that nonviolence could work effectively for social reform. Gandhi also reinforced King's concept of using love for one's enemies as a tool for change.

From 1955 to the mid 1960s, King went from being an unknown preacher to a great presence in American public life. His campaigns turned an irate, sullen people into a glorious, righteous multitude. He was crucial in a string of highly dramatic assaults on the racist South—as head of the Montgomery bus boycott in 1955; in the Albany, Georgia, protests in 1962; in the Birmingham, Alabama, demonstrations in 1963; and in the Selma-to-Montgomery march of 1965. Each campaign became etched in the nation's vision of the times.

David Halberstam, in a recent speech on the American 1950s, said that King was "the nation's first tele-evangelist," meaning that he was keenly aware of the new power of television and skillfully used it to focus attention on civil rights campaigns. The television images of brutality directed at peacefully assembling and marching Blacks sickened much of the country.

King was in danger many times during these campaigns. He was beaten and jailed, his marches were attacked, and his life was threatened continually. Early in the movement his home was bombed. None of this deterred him.

King was a man of prodigious talent. His writings were superb syntheses of religious thought, social philosophy, and statements of the Black cause. The "Letter from a Birmingham Jail" (1963) is regarded as one of the greatest American essays. From his famous "Give Us the Ballot" speech in 1957, he earned a reputation as perhaps the greatest American orator of his day. His "I Have a Dream" speech during the March on Washington in 1963 is one of the most significant speeches of the twentieth century.

In 1964, King's talents were recognized by the world: he was selected by *Time* magazine as Man of the Year and awarded the Nobel Prize for peace.

On April 4, 1968, King was assassinated in Memphis, Tennessee, while supporting the Memphis garbage workers in a strike. His death ignited more than a week of rioting in over 125 cities across America.

Today, on the third Monday of January, a federal holiday honors the birth date of Dr. Martin Luther King Jr. Each year, the holiday is marked by marches, assemblies, pageants, oratory, and readings of his writing.

### What were the "sit-ins"?

On February 1, 1960, Ezell Blair Jr., Franklin McCain, Joseph McNeil, and David Richmond—four Black students from North Carolina Agricultural and Technical College in Greensboro—began the sit-in movement, a tidal wave of student protest. Other sit-ins had been initiated by Blacks in Chicago in the late 1940s, in Wichita in 1953, and Oklahoma City in 1958, but the Greensboro sit-in was the most earthshaking.

The students' protest was a simple act that today does not appear at all revolutionary. Downtown Greensboro had a Woolworth's variety store with a lunch counter at which Blacks could not eat. On this day, they walked into the store, sat down at the counter, and ordered coffee and doughnuts. The response was "We don't serve colored here." Although refused service, they remained seated all day until closing.

That evening, they formed the Student Executive Committee for Justice. They pledged to follow the tactics used by Martin Luther King Jr. during the Montgomery bus boycott of 1955. No matter what they faced, they would be strictly nonviolent. No response would be made to name-calling, threats, or violence.

The next day, they returned with twenty-five male and four female classmates. To signify their dignity, the men wore coats and ties and the women wore dresses. White opposition also showed up.

By demanding service at a public place, they had started an attack on one of the South's fundamental racial rules. In 1960, Blacks could not have equal service at restaurants, bars, or hotels anywhere in the South. They could not sit side by side with whites. Restaurants permitted Blacks to eat only in small, separate cubbyholes, often dingy and hot, with entrances marked "Colored," "Colored Entrance," or "Colored Side."

Thulani Davis's recent novel *1959* tells the story of a sit-in transforming the Black community in the little town of Turner, Virginia. Suddenly, the whole community was brought into the protests by the young students and the support of larger, more established organizations. This is just what happened in Greensboro. The local NAACP, headed by Dr. George Simpkins, came to the aid of the sit-in. Soon, the Congress of Racial Equality (CORE) in New York City sent aid too. Len Holt came to Greensboro to give instruction to students in nonviolent protest tactics.

Two weeks after the Greensboro protests, forty students from Fisk University in Nashville sat in at the local Woolworth's. After a few days of protests, many students were jailed and Blacks boycotted downtown stores. The local NAACP leader's house was bombed. By spring, Nashville yielded. After negotiations between city officials and the movement's leaders, several stores desegregated their eating places.

Within months of the Greensboro protests, sit-ins spread to fifty-four cities in nine states. Thousands of students joined sit-ins. Unfortunately, change did not come peacefully or quickly in many places. Student protesters were beaten by irate whites; they were jailed and held in extreme conditions or expelled from college. A race riot occurred after the Jacksonville, Florida, sit-ins.

In April of 1960, Ella Baker, executive secretary of the Southern Christian Leadership Conference, organized a conference at Shaw University in Raleigh to channel the energy of the sit-in movement. Two hundred students attended, more than half from the South. This was a turning point. Out of this conference came the Student Nonviolent Coordinating Committee (SNCC). It was based on the ideas of nonviolent direct action and pacifism. Under Baker's guidance, it was shaped into grass roots campaigns all across the rural South.

The symbol of a young Black man or woman, dressed smartly, often carrying college books, and seated at a lunch counter was a powerful image to the nation. It said that Black youth were going be a vanguard in challenging racial barriers everywhere.

### What were the "Freedom Rides"?

One of the most famous photos of the civil rights movement was of a motionless Greyhound bus outside Anniston, Alabama, with fire and smoke pouring out of its windows. It was Mother's Day, May 14, 1961.

Before the bus was set ablaze, its tires had been shot out and windows broken. The passengers, many young Blacks, had to flee into a hostile white mob. Hank Thomas, who was on that bus, remembered, "I couldn't have survived that burning bus. . . . But I was so afraid of the mob that I was gonna stay on that bus."

The Freedom Rides campaign, of which this bus was a part, was started in 1961 by the Congress of Racial Equality. The first rides had a basic formula: Blacks and whites would leave Washington, D.C., traveling South on Greyhound or Trailways buses, deliberately challenging the segregated seating on the buses, and at each rest stop along the way they would enter the segregated facilities. Strict adherence to nonviolence was practiced. They understood that their actions would lead to harassment and arrests. Preparing for what faced them, the Freedom Riders were trained in coping with tense situations.

A main goal of the Freedom Rides was to test the Supreme Court's recent *Boynton v. Virginia* decision that extended the prohibition against segregation in interstate travel. Segregation in interstate bus travel had already been prohibited by the Court in 1946, but the South had openly ignored the rulings. Blacks were still arrested if they sat in the front of buses, where whites normally sat. They were thrown out of terminal restaurants reserved for whites. Defiantly, signs separating the races still hung at bus stops.

James Farmer, the head of CORE, reported that traveling "through Virginia, we had no problem. . . . Greyhound and Trailways had taken down the For Colored and For Whites signs. The same was true in North Carolina."

South Carolina, Alabama, and Mississippi were a different story. "Nigger, you can't come in here" greeted John Lewis, a tenacious protester, in South Carolina. His white compatriot, Albert Bigelow, was beaten. Birmingham police knew that a mob of white men armed with baseball bats and chains was going to be "the welcoming party" for Freedom Riders. They allowed the mob to beat the demonstrators for fifteen minutes before they arrived to break it up.

The Freedom Rides movement grew during May 1961. The violence galvanized the civil rights movement; more volunteers came forward. During that summer, Jackson, Mississippi, became the riders' focus. More than three hundred people were arrested there. Refusing bail, they served their time in Mississippi jails. By September 22, the federal Interstate Commerce Commission had issued a regulation prohibiting separate facilities for Blacks and whites.

The Freedom Rides posed a problem for the Kennedy administration. John Kennedy had campaigned supporting civil rights. But siding with the protesters would alienate the old guard white Democrats in the South. Robert Kennedy, the president's brother and U.S. attorney general, tried to cool things down among Freedom Riders and to divert their energies into voting registration. But violence against Blacks—captured in images broadcast widely—forced the administration to protect the protesters by sending federal marshals.

Through Freedom Rides, the young demonstrators discovered their power to provoke a national crisis that called attention to everyday segregation. Their work became the basis of the public accommodations section of the Civil Rights Act of 1964.

## What was SNCC, and why was it so important?

Student Nonviolent Coordinating Committee (SNCC) was formed in October of 1960, after sit-ins spread across the South. Its original mentor was a unique woman, Ella Baker, executive secretary of the Southern Christian Leadership Conference (SCLC), who had brought student leaders together at Shaw

University in April 1960 after their early sit-ins. This meeting, later known as the "Raleigh Conference," was the founding moment for SNCC.

SNCC drew on the ideas of nonviolence and pacifism pioneered by the Congress of Racial Equality (CORE), the Quaker Fellowship of Reconciliation, and SCLC. Its earliest statements said, "We affirm the philosophical or religious ideal of nonviolence as the foundation of our purpose." The reason for such allegiance was "nonviolence . . . seeks a social order of justice permeated by love."

SNCC captured the imagination of many in the civil rights movement. Young people dominated the organization, bringing to it their daring idealism, their seemingly boundless energy, and varied organizing techniques. In time, the organization would be admired as defiant and even revolutionary.

Under the guidance of Ella Baker, SNCC was shaped into a movement dedicated to building grass roots campaigns in the rural South. Thus, it relied on community people to provide the leadership and ideas necessary for breaking down segregation and other barriers to Black independence. "Local people" were preferred. SNCC was also willing to use whites in its organization and projects.

Within the first year of its organization, SNCC began to confront the worst kinds of segregated societies in the South's Black Belt counties, where Blacks were numerous and often poverty-stricken. In these places, the shadow of slavery still fell across race relations. White police forces openly used their power against Blacks. Vigilante organizations, the White Citizens Councils, and the Ku Klux Klan threatened anyone who stepped out of line. The federal government offered little or no protection to Blacks.

In these desperate communities, as vigilante and police violence focused on them, SNCC workers struggled to start a local movement. The most ardent work came in organizing Mississippi, and McComb was one of the first places. Though SNCC's voter registration drive there suffered setbacks, the group learned more about organizing, acquired full-time future workers, and became more cohesive. From there, members fanned out across the state, especially into the tough delta region. Later, they worked with the NAACP in Jackson. SNCC was becoming an active stimulus for Black mass movements and more than just a coordinating committee.

With soothing and helpful Bob Moses (born 1935) as its leader, SNCC became "a home" and "a family"; it "had a place for everyone," especially the young, with a special brand of dialogue, support, a sense of service, and moral clarity. All of these factors prepared the way for its next major step: Mississippi's Freedom Summer of 1964.

This was an important period in SNCC history. Together with the Congress of Racial Equality and other groups, SNCC brought more than a thousand Northern students, Black and white, to Mississippi to aid voter registration. These few summer months were the most violent in civil rights history. James Chaney, a twenty-year-old local Black from Meridian; Cornell graduate Michael Schwerner; and Andrew Goodman, a twenty-year-old Queens College student, were killed on Sunday, June 21, in gangland-style executions in Neshoba County. But this was not all: thirty-five shooting incidents occurred, almost seventy homes and churches were bombed, a thousand organizers were arrested, and many were beaten and murdered.

Nevertheless, the Freedom Summer was a success: it got more Blacks to vote, vitalized local Black leadership, and the violence told the nation of the true character behind Black oppression. Television and newsreels spread the summer's story in an unprecedented way.

After 1964, differences of opinion racked SNCC. Black staffers asked whether Southern Blacks could be free if they were dependent on outside help and on whites. Organization women felt that SNCC was too much controlled by men. In 1966, SNCC, under the leadership of Stokely Carmichael (born 1941), rejected nonviolence as the way to win rights and turned to the idea of Black Power. Bitter disputes raged on, while chapters in the South closed. SNCC's division made its future problematic. H. Rap Brown, Carmichael's successor in 1967, proved popular with Black youth, but was unable to revive the organization and give it new purpose, principally because of his constant confrontations with the law.

### "Nonviolent resistance" was a phrase often invoked during the civil rights movement. What did it mean?

James Lawson, a Black divinity student at Nashville's Vanderbilt University who had been active in the sit-ins, used to tell demonstrators, "Remember the teachings of Jesus, Gandhi, and Martin Luther King." He was telling demonstrators to stay committed to the idea of nonviolent resistance.

The idea of nonviolent resistance in the civil rights movement came from a combination of Christianity, the writings of Henry David Thoreau, India's Mahatma Gandhi, and pacifist A. J. Muste.

In tactical terms, the idea expressed itself as mass marches, sit-ins, boycotts, selective-buying campaigns, and mass rallies featuring prayers, songs, and chants. From 1955 until the rise of the Black Power movement in the mid 1960s, nonviolent tactics were highly successful in winning material support and psychological gains for the civil rights movement. Most Blacks embraced these tactics, and many white supporters were drawn into the movement because of its nonviolent character.

Although Dr. Martin Luther King Jr. was not the only person to use nonviolent tactics nor was he their inventor, he was their most popular advocate in 1960s America. King began his "pilgrimage to nonviolence," as he labeled it in his book *Stride Toward Freedom* (1958), during his years studying for the ministry at Crozer Theological Seminary. He continued to develop the idea until his death.

In writings and speeches, King outlined several basic tenets of nonviolent resistance. Nonviolence was "not a method for cowards." Discipline and great restraint were necessary. Nonviolence did not seek to "defeat or humiliate the opponent, but to win his friendship and cooperation." Ultimately, the goal of nonviolence was the harmonious community—what King called "the beloved community."

King told demonstrators "to accept suffering without retaliation, to accept blows from the opponent without striking back." But he went farther and urged demonstrators to rid themselves of any "internal violence of spirit." The nonviolent resister "not only refuses to shoot his opponent but also refuses to hate him." Ultimately, his great faith in nonviolent resistance was based on the conviction that the universe—God's creation—was "on the side of justice."

King's nonviolence put enormous power within the reach of ordinary people. Individuals who had never participated in demonstrations and who had rarely confronted the racist power structure discovered the inner resources needed to be civil rights warriors.

In Birmingham, Alabama, in April and May of 1963, the nation had a chance to see nonviolent resistance in action. Civil rights groups, led by King, started a major drive against bias in the city. Police Commissioner Eugene "Bull" Connor went all out to defeat the movement, using police attacks, high-powered water hoses, and fierce dogs against demonstrators. The city became a tinderbox. Across the nation, television, newspapers, and newsreels carried the images of Connor's assault against a dignified nonviolent army.

The brutality captured in the images aroused public opinion, especially in the North and West. The new public awareness was useful to President Kennedy in passing congressional civil rights proposals submitted in March. Birmingham continued to be the center of conflict until May 10, when an agreement was signed to desegregate public facilities.

Nonviolent resistance had its critics, those who felt it was asking too much of demonstrators to withstand attacks, those who thought that white racists could not be redeemed by Black suffering, and those young turks who felt that Blacks should be responding in kind to white violence. Soon, critics would drown out King's philosophy. But until the mid 1960s, nonviolent campaigns held center stage in the civil rights movement.

### What was the influence of Black women in the civil rights movement?

From the movement's beginning with the Montgomery bus boycott of 1955, Black women were a powerful resource for the civil rights movement. They were active in every phase of the movement. In fact, women such as Modjeska Simpkins of South Carolina (1899–1992), a major NAACP worker devoted to helping rural Blacks, were creating the movement before it even had a name. Early on, she saw the need for community building among Black country poor. Her personal motto was: "I cannot be bought and I will not be sold."

Once the movement began, women did it all: committee work, voter registration, front-line activism, political education, and national campaigning. They suffered along with the men, too, whenever civil rights workers were harassed, beaten, and jailed. At the 1963 Selma march, a woman was so badly beaten that she "sank to one knee in the mud, blood pouring between her fingers from her mouth." Police made one tiny exception: they did not search women. Ann Moody tells how, in the 1963 Jackson, Mississippi, protests, she had "hidden a small transistor radio in her bra" in order to get the news in jail.

Women helped develop the philosophy of the movement. Septima Clark (1898–1987) was one of the movement's vital traveling teachers who instructed Southern Blacks in the meaning of citizenship. Her classes empowered Blacks by demonstrating that the Constitution upheld their status as citizens. Even in the Free Southern Theater, a theater company operating for a time in Mississippi, spreading the movement's ideas, women were essential to its success.

Despite women's place in the movement, they have not been given adequate historical recognition. The early film, *From Montgomery to Memphis* (1970), highlighted Martin Luther King but failed to give a rightful place to Rosa Parks, a catalyst of his early campaigns. High-profile men have the spotlight in movement history.

At least the central contributions made by three Black women should be included in every civil rights history.

The Women's Political Council (WPC) lay the groundwork for the Montgomery bus boycott of 1955, which Martin Luther King Jr. successfully spearheaded. Jo Ann Gibson, an English professor at Alabama State College, founded the organization in 1949 after being mistreated by a bus driver. She and the WPC had been planning a boycott for months before Rosa Parks's arrest in December. When Parks took her stand, they used the incident to put their plan into action.

After the arrest, the women stayed up all night mimeographing thousands of leaflets announcing a bus boycott. They also contacted E. D. Nixon, a former NAACP president, and told him of their plan. The well-connected Nixon drew the Black community's traditional leaders—the ministers—into the plan. In this way, King came to lead a movement with roots in the work of the Women's Political Council.

Ella Baker's work was another brilliant contribution to civil rights history. Born in Norfolk, Virginia, in 1903, she attended North Carolina's Shaw University and in 1927 moved to New York City. There she worked in several political groups, learned political organizing, and fought for worker issues. In the 1940s, she became an NAACP field secretary in the South. In two years, she traveled more than sixteen thousand miles to nearly four hundred meetings. Later she left the NAACP because of its "top-down" structure and its focus on court cases as the key to destroying segregation.

In 1958, she became a coordinator of SCLC's voting rights campaign. When the sit-in movement began in 1960, she saw the opportunity for a new organization to harness this youthful energy. Through the Student Nonviolent Coordinating Committee (SNCC), she helped build a more radical, grass roots, bottom-up organization. Her ideal was people power. Baker evolved into the spiritual mentor of the movement: she practiced a selfless, low-profile leadership. Yet she criticized the movement's subordination of women.

Fannie Lou Hamer (1917–77) made a similar contribution. She came up the hard Delta, Mississippi, way: the youngest child of twenty in a sharecropping family, she lived all her years in poor Sunflower County and was a sharecropper on the B. D. Marlowe plantation for eighteen years. When she tried registering to vote, Marlowe evicted her. She was in her forties when she became an SNCC field secretary in 1963, attracted to the organization because "they [the students] treated us like we were special and we loved 'em. . . . We didn't feel uneasy about our language. We just felt like we could talk to 'em."

As an SNCC worker, she was smart, down to earth, a great speaker of sharp logic, a Scripture-reciting Christian, a singer, and a galvanizing field organizer. In 1964, as a founder of the Mississippi Freedom Democratic party (MFDP), she came to Atlantic City to challenge the all-white state delegation at the Democratic National Convention. In 1968, she was a delegate. Afterward, she created the Freedom Farms Corporation (FFC), a nonprofit group to help families raise food and livestock.

Ella Baker was right when she said, "The movement of the fifties and sixties was carried largely by women."

### Why is the civil rights movement often called "the singing revolution"?

In Andrew Young's *A Way Out of No Way* (1994) he remembers a 1963 Easter Sunday march of five thousand people in Birmingham. As they approached a frightening police phalanx, they began to sing an old spiritual: "I want Jesus to walk with me. All along my pilgrim journey, Lord, I want Jesus to walk with me." The marchers passed without an incident. One elderly Black woman shouted, "God done parted the Red Sea one mo' time!"

After Herbert Lee, a Mississippi Black farmer and father of nine children, was killed in September of 1961, Bertha Gober wrote, "We'll Never Turn Back" to honor him. "We have hung our head and cried/For those like Lee who died/Died for you and died for me/Died for the cause of equality." This grew into the haunting anthem of the Mississippi movement.

When the funeral march for Medgar Evers, in Jackson, Mississippi, in July 1963, was over and the crowd was milling about, it is reported that a soft soprano voice rose above the sounds of the throng with the words, "Oh, Oh, Freedom! Before I be a slave, I'll be buried in my grave and go home to my Lord and be free." The song was coming from a little slip of a girl. As the volume of the song rose, the circle of singers enlarged until hundreds had joined the chorus.

Time after time, song came to the aid of the civil rights demonstrators. At moments of crisis, someone would "raise" a song, and the singing would ease the tension. At street rallies, church meetings, in jails and prisons, on dangerous marches, song helped groups to center themselves on the moment.

When there were group disputes, as among Freedom Summer planners in 1964, the evening's discussions would end with singing. As SNCC organizer John Lewis has said, "though poor, we had voices."

No movement in American history has granted a greater role to song and to mass singing. By 1961, this was a habit, and it was sustained well into 1966. Often these songs were called "freedom songs" or "movement songs." As a result, the repertoire of civil rights songs was extensive. It was expanded by the many variations, as these songs were sung by different people and crafted to fit local situations all across the South.

Movement leaders were often great singers. Sometimes the leaders had favorite tunes. Fannie Lou Hamer's signature song was "This Little Light of Mine." She opened mass meetings with this song and others in order to bring the congregation closer together.

The movement's mass singing had its roots in Black traditions of music and in general American protest songs. Since slavery days, Black communities had employed song to communicate messages, to spend leisure time, to elaborate a worldview, to sing praises to God, and to ease their sorrow.

Freedom songs drew on this rich heritage, especially that of spirituals and gospel music. Movement songwriters revised many songs by inserting new political lyrics. At Highlander Folk School in Tennessee, singers taught labor songs to movement members. Zilphia Horton, Myles Horton, and Pete Seeger transmitted this music.

"We Shall Overcome" is easily the best-known movement song. It was originally a Black spiritual. Pete Seeger and Zilphia Horton rearranged it. During a mill strike by Black workers in 1934–35, it was sung for the first time in a protest. After strikers were shot, Horton got the workers to sing "We Shall Not Be Moved."

Freedom songs served the civil rights movement well. But they have also helped the world through moments of profound change. No one would have dreamed that one day the Berlin wall would come down with Germans singing, "We Shall Overcome."

### Did the coming of independence to African states energize the 1960s Black movement in the United States?

In 1957, the West African state of the Gold Coast became independent. Ghana was its new name, after the ancient gold-trading West African kingdom. Broad stripes of green, yellow, and red made up its flag, in whose center was a large black star, a symbol borrowed from Marcus Garvey's Black Star Line, which was to take American Blacks back to Africa.

At midnight, when the British flag was lowered over a stadium in Ghana's capital of Accra and the new flag was raised, the crowd yelled at the top of their lungs, "Ghana, Ghana, Ghana!"

Black Americans cheered the new country too. They felt a special kinship with Ghana. Kwame Nkrumah (1909–72), prime minister of the new country, had been educated at the Black Lincoln University in Pennsylvania in the 1930s and after that had lived in Harlem. Prominent Blacks, including Eslanda Robeson (1896–1965), had visited Ghana before its independence. Some Blacks claimed ancestry, back through the slave trade, to the peoples of Ghana. Ghana preached an appealing Pan-Africanism, the belief that all Blacks around the globe should seek unity. Nkrumah was honored with a huge, high-spirited Harlem rally in 1960.

It was not long before a small colony of Black Americans settled in Ghana, among them the novelist Julian Mayfield, the scholar Aphaeus Hunton, and the famous intellectual W. E. B. Du Bois and his writer wife, Shirley Graham Du Bois. Maya Angelou, in *All God's Children Need Traveling Shoes* (1986), told the story of this colony, of which she was a member. "So finally, I had come home . . . the prodigal child," she explained. Ghana evolved into a symbol of the future.

However, news from South Africa shocked Black America. On March 21, 1960, at Sharpeville, South African police opened fire on peaceful African demonstrators, killing sixty-nine and wounding almost two hundred. Most were shot in the back as they fled. With this occurring while sit-ins were still going on, many American protesters immediately began to carry signs about the incident: "Remember Our Brothers and Sisters in Sharpeville" and "Evil was at Sharpeville."

The rest of 1960 was "the Year of Africa." State after state became independent, totaling eighteen in all and including nearly all of French Africa and the major states of the Congo, Nigeria, and Senegal. Africa was declared "an awakening giant."

American Blacks were uplifted by Africa's ascendancy. As the new African states' delegations came to the United Nations, wearing their resplendent robes, the message to their Black American kin was clear: Blacks are returning to the world stage. American Black organizations with ties to African nationalists, such as the Council on African Affairs, were overjoyed.

The long struggle Africans had waged for freedom had paid off. This was a reminder that Blacks needed in the early years of the civil rights movement. Independence in Africa foreshadowed new possibilities for Blacks. When the new states suffered setbacks, as did the Congo in 1961 when its leader Patrice Lumumba was killed, American Blacks quickly organized newsmaking protests, a famous one being held at the United Nations.

The freedom of the many new Black states underscored the fact that Black Americans were still not treated as citizens in their own country. A favorite joke of the time ran, "All of Africa will be free before Black Americans can get a cup of coffee in Woolworth's."

Hours before the March on Washington on August 28, 1963, W. E. B. Du Bois died in Ghana. His death reminded Black Americans—at a shining moment in civil rights history—that they were sons and daughters of Africa.

### What was the "Black Power" movement?

"Black Power" resounded through America in 1966. Only a two-word phrase, it possessed a terrific potency. The idea of Blacks needing power was not new. It had been etched into Black political philosophy during the nineteenth century. Still, when the phrase "Black Power" arrived on the national scene, it stirred up a firestorm.

One reason for the interest was the Madison Avenue, sound-bite appeal of the phrase: succinct, aggressive, good for chanting, and perfectly suitable for broadcasting. "Black Power" was good political propaganda.

Yet the phrase stood for serious ideas and raised important issues. In Black America, the slogan soon turned into a debate about future directions and about the best method to build strong Black communities. That debate shattered the reigning consensus within the civil rights movement. Until Black Power, the assumption was that a coalition between Blacks and whites was the wave of the future. For many across the American racial spectrum, the coming of "Black Power" signaled the final days of nonviolent resistance.

All of this was set in motion by James Meredith's June 1966 one-man "pilgrimage against fear" from Memphis, Tennessee, to Jackson, Mississippi. Mississippi-born Meredith (1933– ) had already earned his place in history by taking the dangerous step of attempting to enter the University of Mississippi in 1962 and becoming its first known Black student. His effort provoked campus riots and eventually President John Kennedy had to send federal agents and troops to protect him, as well as to calm the campus. His new one-man pilgrimage was stamped with his earlier tenacity. Meredith began the two-hundred-mile march alone. After he was shot by a white segregationist, civil rights leaders Martin Luther King Jr. and SNCC chair Stokely Carmichael, along with others, joined the march.

"We want Black Power" was chanted by Stokely Carmichael when the march stopped at Greenwood, Mississippi. After being released from jail, he chanted the phrase while speaking to a group of six hundred people. The reception was immediate and enthusiastic.

From that point on, Carmichael was the slogan's disseminator. News agencies covering the march seized on his use of the phrase, and his youthful supporters chanted it too. King was thrown on the defensive by Carmichael and by the questions of reporters, who dramatized the sloganeering as a new division in the movement.

From the march until the mid 1970s, "Black Power" came to refer to a form of Black nationalism that promoted political and social separatism from whites. Just as clearly, it stood for pride in African and Black cultural and intellectual achievements.

Several organizations supported the idea: SNCC, the Black Panther Party, and religious groups such as the Nation of Islam. A host of activists spoke its language and shaped its ideals: H. Rap Brown of SNCC; Angela Davis of the Communist party; Kathleen Cleaver, Ericka Huggins, and Huey Newton of the Black Panther party; and artists LeRoi Jones and Nikki Giovanni. The book *Black Power: The Politics of Liberation* (1967), by political scientist Charles Hamilton and Stokely Carmichael, outlined the movement's issues.

Black Power was the result of disillusionment with decades of civil rights struggle and the slowness of change. Nonviolence seemed obsolete. Violence against the movement and the deaths of organizers had worn down supporters. Integration seemed just an invitation to be whitened culturally. Blacks were tired of extending the right hand of fellowship only to have it shot off. Blacks needed to build their own sense of community and strive for independence from white society.

Most Blacks took only a few ideas from the Black Power movement. One of its most magnetic aspects was the stress on Black cultural pride, which after two decades of integrationist symbolism, found a ready welcome in Black communities.

"Black Is Beautiful" was the new motto. Pop diva Nina Simone sang, "To be young, gifted, and Black" as if it were a utopian condition. The Afro hairstyle was worn as a crown. Dashikis, bubas, djellabas, caftans, head wraps, and skullcaps became essential fashion statements. Students fought for the establishment of Black studies departments at colleges and universities.

One of the most active fronts in this era was within the Black Arts movement, a broad assault on what were deemed the tired aesthetic values of the past. Consciously, these cultural workers aimed to overturn "the white aesthetic," "the Euro-American tradition," and "the Caucasian idolatry in the arts." Groups such as The Last Poets, whose spoken poetry was both critical of America and Black complacency, led the movement. Writers such as Amiri Baraka (LeRoi Jones), Sonia Sanchez, Don L. Lee, Ronald L. Fair, Ed Bullins, Nikki Giovanni, brought a host of new ideas and innovative literary forms to Black writing. The writing of this period is slowly gaining serious attention, no longer being considered as just political writing.

This fertility was echoed in street art (murals, posters, fabric designs), neighborhood performance groups (theater companies, dance ensembles), and in a number of exciting new presses (Third World in Chicago, Broadside in Detroit, Jihad in Newark). And pamphleteering became an intellectual enterprise of some magnitude. Out of this also came serious journals—the famous *Black World, Black Dialogue,* and the *Journal of Black Poetry.*

By the mid 1970s, Blacks had drifted away from Black Power. Integration into America's educational and economic structure was still valued. Many did not agree with racial separatism. "Integrative diversity," to use Charles Banner-Haley's phrase, began to reassert itself. Plus, leaders of the movement had been unable to translate their ideology into concrete legislative or economic gains for Blacks. "Black Power" was no longer much heard. Its history, though, flowed into the continuing saga of Black nationalism.

## What was Soul Music?

First off, depending on which aficionado, performer, or producer you ask, soul music is or is not distinguishably different from rhythm and blues (R & B).

Most frequently, as music scholars William Barlow and Cheryl Finley point out in *From Swing to Soul* (1994), soul music is regarded as "the high watermark in the development of rhythm and blues." It was the fullest, most dynamic phase of rhythm and blues. Robert Pruter, in *Soul Recordings* (1993), says that "what made . . . early soul records was mainly the employment of a more gospelized approach" than early rhythm and blues. Samuel A.

Floyd Jr., in *The Power of Black Music* (1995), argues the soul music "label was quickly appropriated by the record industry and became associated with black popular music of R & B derivation."

So, what was soul music? Soul music, like blues, gospel, jazz, and rock and roll was a Southern-grown musical idiom. It was a secular form of musical expression that used gospel conventions to plumb the depths of its primary subject: love. Rising as it did in the mid–1950s, soul music would become the musical accompaniment to the Black thrust for justice, civil rights, and Black power. Sam Cooke's 1965 "A Change is Gonna Come," The Impressions' 1964 "Keep on Pushing" and their 1965 "People Get Ready," Aretha Franklin's 1967 "Respect": these were the new anthems.

Soul music as a term did not come into being until the 1950s. But, the roots of soul were since the 1940s sometimes called rhythm and blues, a black synthesis of jump-blues, upbeat big band dance numbers, swing, and early rock and roll. This early rhythm and blues was one influence on soul music. To this must be added another less well-known influence. Tens of thousands of doo-wop groups—a term for 1950s Black singing groups—that dotted Black neighborhoods from coast-to-coast would help soul music claim its place at the forefront of youth music.

If the "soul explosion" began with anyone it was Ray Charles. Born in Albany, Georgia, in 1932, blind by the age of six from glaucoma, he grew up in Greenville, Florida. When his mother died he quit school at fifteen and started his career in music. In 1952, Atlantic Records bought his contract from Down Beat of Seattle, where Charles had relocated after stints with Count Basie-like big bands and smaller juke-joint combos. Since 1948, he had cut regional rhythm and blues hits that presaged soul music. But it was in the mid–1950s with Atlantic that he converted gospel standards into secular stompers with all the pageantry of a Sunday church service: 1954's "I Got A Woman" and "Hallelujah, I Love Her So" were cataclysms. In a world where gospel music was church and blues the devil's music, Charles had braved the divide: 1959's "What'd I Say" is the high-water mark of blending the call and response of gospel with blues' sexual exhilaration.

The legion of doo-wop groups that performed or cut at least one single in the 1950s is said to be about fifteen thousand. They were secularized versions of the gospel quartets like the Soul Stirrers, The Five Blind Boys of Alabama, and the Pilgrim Travelers. Of all the gospel quartets, the Soul Stirrers with Sam Cooke (1931–1964) was the group most emulated and loved by the young. Nevertheless, in 1957 when Sam Cooke made the leap from gospel to soul singer it was a defection more controversial than Bob Dylan's first using an electric guitar at Newport in 1965. When Cooke's "You Send Me" went to No. 1 in 1957, he was not only an icon of youthful Black cool and sweet success but, in his inimitable way, had taken a legacy of music born in slavery and once reserved only for the church and soulfully had placed it high for all America to see. From 1958–1963, Cooke had fifteen top rhythm and blues hits. Neither the church nor the radio nor the stage would be the same afterward.

If Ray Charles and Sam Cooke stylistically and commercially consolidated soul music, a legion of forerunners and peers contributed to the making of soul. In the late 1940s vocal groups like the Orioles, Inkspots, and Clyde McPhatter's Dominoes sculpted a distinctive pre-soul group sound. Many outstanding records were also issued. For example, there was "Little Fever" by Little Willie John (1937– ) in 1957—one of the hottest love songs put to wax. Mickey and Sylvia's "Love Is Strange" in 1957 had a melody good for dancing and lyrics that were comic plays on love talk. Jackie Wilson (1934–1984) lent his special singing style, decorated with falsetto, to the making of soul. Two of his 1958 records voiced love's passions and disappointments—"To Be Loved" and "Lonely Teardrops." The Drifters' 1959 "There Goes My Baby" used orchestral strings to support Ben E. King's wrenching song of love loss. Robert Townsend's film *Five Heartbeats* (1991) did an honest job of depicting the highs and lows of soul singing groups.

Like all of twentieth-century Black music, soul music is a story of like minds in a few key cities, in commerce, and on the road. In soul music those cities were Memphis (the site of Stax/Volt labels) and Muscle Shoals (the site of the Fame recording studio). Interestingly enough, the white businessmen and Black performers who sprung soul music on the world from these cities had a unique relationship. In most cases, these relations were true partnerships between Black performers and eccentric white soul enthusiasts and producers, be they the Ertegun brothers and Jerry Wexler at Atlantic or white backing musicians like Steve Cropper and Duck Dunn. Emphasis was on making the best music possible, not compromised because the audience was mainly Black.

Ironically, it is Berry Gordy's black-run dynasty of Motown, in Detroit, that in the 1960s engineered the then seemingly impossible: selling soul music directly to the white middle class teen, and producing one of America's most successful Black businesses. The Motown story has been often told. Recently, Berry Gordy's autobiography *To Be Loved* (1994) has told this story in personal terms, complete with small revelations. But, it is still remarkable how many great hits came from Motown's studios. Just think of a mere three examples: Martha and the Vandellas' 1965 "Nowhere to Run," the Four Tops' 1966 "Reach Out I'll Be There," Marvin Gaye's 1968 "I Heard It Through the Grapevine." It is also remarkable that so many talented women performed for the Motown labels—Martha Reeves, Gladys Knight, Mary Wells, Brenda Holloway, Diana Ross, Mary Wilson, Florence Ballard, Tammi Terrell, Kim Weston. For more than a decade now, at least since the film *The Big Chill* (1983), Motown's music has been kept alive in the nation's subconscious through film and ad soundtracks.

On the road soul music was at its most expressive. As with gospel music, the relation between soul singer and listener was best maintained in live performances. In soul music's early days, performers traveled the "chitlin' circuit," a cross-country network of clubs and ballrooms. Later, they traveled by package tours or "soul revues," featuring several performers. Night after night, soul performers toured the country trying to top each other: Wilson Pickett, Rufus Thomas, Otis Redding, Aretha Franklin, Ruth Brown, Jackie Wilson, Solomon Burke. The tour rosters are like a list of the greatest pop vocalists of the second half of the twentieth century.

Of all soul artists it was James Brown (1928– ) who emerged as "the king of the road," one of the greatest performers in modern entertainment history. With his band, he crafted the most solid canon in modern pop music history serving as the template of funk and rap. Brown fashioned one of the greatest live shows in modern performance history, where his astounding stage moves, the exhaustive emotionalism, the acrobatic dancing, and the stage props—the cape and the suitcase—are legend. The names of his hits are a familiar litany of soul history—"Please, Please, Please," "Papa's Got a Brand New Bag," "Out of Sight," "I Feel Good (I Got You)," "Sex Machine," and "Living in America." For many in the 1960s, he was a manifestation of Black power—strong, direct, authentically Black, and naturally uplifting. Today, it is nearly impossible to turn on a radio and escape his influence.

With the rise of funk and disco in the 1970s most true soul fans would say the music died, was watered down and turned into just another type of pop music. Today many of the great performers continue to tour but the soul impulse as a statement of Black communal genesis is past.

### What was the Black Panther party ?

The Black Panther party (BPP) was the brainchild of Huey P. Newton (1942–89) and Bobby Seale (born 1937), who met as members of the African-American Student Association while at Merritt Junior College in Oakland, California. Its full name was the "Black Panther Party for Self-Defense."

Typical of many Black Californians, Newton was born in Louisiana, Seale in Texas, and when they were young, their families moved to Oakland, a city to which many Blacks migrated during the Second World War.

The party was started on October 15, 1966, and lasted until the early 1980s. Its name, partly borrowed from a SNCC Mississippi voter registration campaign, suggested vigilance, quickness, muscularity—a far cry from the long, staid names of the older civil rights groups.

In its first meetings, a ten-point program was outlined in a document entitled "What We Want, What We Believe." It was an omnibus program, with a variety of demands, ranging from the "forty acres and two mules!" derived from the nineteenth-century Reconstruction era, to better housing and education, to a United Nations–supervised election to be held throughout American Black communities ("the Black colony") to determine the national destiny of Black people.

Of the ten demands, the most important was the seventh: "We want an immediate end to POLICE BRUTALITY and MURDER of Black people." This demand applied to many police forces but particularly to Oakland's. In 1966, the Oakland police department had only nineteen Black officers out of six hundred and a history of ferocious treatment of Blacks.

Within a year of its founding, Huey Newton was arrested for killing white Oakland policeman John Frey, who had accumulated a twenty-three-year record as a vicious cop. Newton's trial and imprisonment on voluntary manslaughter charges transformed the Panthers into a party whose goal was to "Free Huey," as its battle cry proclaimed until his 1970 release.

The year 1969 was a pivotal one for the BPP. "Serve the people" was the theme for the party's programs in the Oakland community, where free breakfast programs were started, highlighting the contradiction between an affluent America and hungry children. Health clinics were also established. The BPP attempted to organize beyond its Oakland base and to counter police violence across the country.

Although top-heavy with male leadership in the beginning, women began to play influential roles. The most prominent were Elaine Brown, Kathleen Cleaver, Ericka Huggins, and Assata Shakur, who were responsible for the party's most successful programs: the breakfast and schools programs, in which young Blacks were provided extra basic schooling and fed, both of which were well received by local communities. During the twenty years of the party's existence, only one woman was ever chair: Elaine Brown (1943– ) from 1974 to 1977.

In the 1970s, the group was locked in a death struggle with the FBI and the police. J. Edgar Hoover claimed the BPP was "the greatest threat to the internal security of the country." The

FBI mounted COINTELPRO, a covert operation to destroy the group. At one point, Ericka Huggins commemorated the party's twenty-eight "fallen comrades," most of whom were killed in altercations with the law. The Senate Select Committee, when investigating the Panther-FBI feud, concluded, "J. Edgar Hoover and the FBI engaged in lawless tactics against the Black Panther party."

Today, the reputation of the BPP is being attacked. Alice Walker has called them "punks" who were "running on empty." There are charges that the party was severely sexist, controlling and abusing its women members. Most serious is the charge that the 1970s Panthers were little more than a violent gang, extorting payment from Oakland citizens, beating members and nonmembers mercilessly, and even killing people. Party officers have said the charges are part of a police-FBI plan to discredit them. Answering these charges is Mario Van Peebles (born 1957) and his movie *Panther,* which seeks to rehabilitate the BPP. Based on a novel written by his father, Melvin Van Peebles (born 1932), the film turns a favorable light on the Panthers. Peebles called the reaction to the Panthers an "All-American Rorschach Test."

The BPP also made an unmistakable impact on 1960s popular culture. When the Panthers made public appearances—wearing black polo shirts, black leather jackets, dark glasses, and black berets—they created an aura of danger. In the spring of 1967, when Panthers entered the California statehouse in Sacramento armed with guns, their stock as daring revolutionaries shot up among the left. Later, the famous poster of Huey Newton—seated in a thronelike wicker chair like an African king, with a spear in one hand and a gun in the other—was hung in radical meeting places across the country.

The rise of the BPP was a pure 1960s phenomenon. But its reputation is still undergoing a variety of postmortems. Meanwhile, many former members continue to make a name for themselves: Bobby Rush as a congressional representative from Chicago, Kathleen Cleaver as an Emory University law professor, and Paul Coates as the brain behind Baltimore's Black Classic Press.

### How did Black America react to the Vietnam War and the peace movement?

The Vietnam War greatly escalated in 1964, causing hundreds of thousands of Americans to be sent into combat. Of this number, fifty thousand were Black.

Usually wars had been for Blacks a time to seek advancement as individuals and as a race. In wartime, Blacks were also able to prove their patriotism. World War II and the Korean War played this historic role. But with Vietnam Blacks faced a more complex fate.

Black combatants served well in Vietnam, earning numerous medals and citations. Repeatedly throughout the war, the Black soldier was praised. In one war correspondent's words, he had "won a Black badge of courage that his nation must forever honor."

At home, though, the Vietnam War produced great dissension in Black communities. Just as federal domestic programs were making a dent in poverty, the war ate up many of the dollars earmarked for clinics, preschools, and job training. Dr. Martin Luther King Jr. called the Vietnam War a "demoniacal destructive suction pump." The battlefront competed with the home front.

Blacks were angry that their young men were being inducted into the military at a higher rate than whites. More than half of the Blacks examined for the military were taken, whereas only a third of whites were taken. Many infantry units consisted of as many as 20 percent Black troops. Some claim that Blacks made up 40 to 50 percent of some units. Blacks were channeled into the infantry and then kept for prolonged stays on the front lines.

From 1961 to 1966, Black deaths often ran as high as 23 percent of those killed. It has been said that "the war was a white man's war, a Black man's fight" and that Blacks were little more than "human fodder." Many Blacks carried the main firepower—the M-60 machine gun—for their units, which exposed them to great danger.

Militant Black groups joined the white left in relentlessly attacking the war. Their position was that stripped of all flag-waving, the war was a form of "color imperialism." Whites were once again dictating to the world's people of color how to live.

Moreover, Blacks should not be fighting another people of color but rather seeking unity with them. The famous line "No Vietnamese ever called me 'nigger'" expressed Blacks' skepticism about their role. Militants said that as long as Blacks were dying in order to vote, they had no reason to fight in Vietnam. Middle-class Black students supported the militants and attacked military recruiters on Black college campuses.

By the late 1960s, Black troops in Vietnam had changed. They now came from the urban riot zones, had read Malcolm X's words, were touched by the Panthers' ideas, had listened to the empowering lyrics of Curtis Mayfield and Aretha Franklin. They called themselves "bloods," signifying racial fraternity even deeper than skin color. Between these racially aware Black troops and whites, tensions were open. Confederate flags were flown by whites, cross burnings occurred, and racial fights broke out. "I wasn't fighting the enemy. I was fighting the white man," one soldier said. Bases were divided into racial niches. Black troops started all-Black clubs.

In April of 1967, Martin Luther King Jr. came out against the war and tapped into the disillusionment that Blacks felt about it. King came to this position from a background in pacifist principles. But Black Power antiwar groups, SNCC, and the Black Panther party did not have strong ties to the peace movement. The Panthers even advocated arms for Black self-defense in American cities. As time went on, the antiwar drive overshadowed the Black movement. Still, many Blacks supported the war in Vietnam.

The war is still on Blacks' minds, as Albert Hughes and Allen Hughes's excellent film *Dead Presidents* (1995) demonstrates. They set the story of a robbery against the backdrop of Black soldiers returning from Vietnam, trying to repair their battered psyches and make it in American society.

## What was the historical significance of Malcolm X?

The Americans who had heard of Malcolm X probably thought of him in a one-dimensional way: he was only the in-your-face Malcolm X, haranguing white America, threatening violence. He was in danger of becoming a cartoon.

Thanks to Spike Lee's film *Malcolm X* (1993), Thulani Davis's *Malcolm X: The Great Photographs* (1993), and William Strickland's fabulously complete documentary *Make It Plain* (1993), the real, complex Malcolm X is reemerging.

Malcolm X was a severe judge of America's treatment of Black people. But he was also reflective and philosophical, intellectually curious, and a master teacher who loved Black history. He was fascinated by words, liked quoting Hamlet, prized good company and friendship, was a tender father and husband, and was humble in enlightened company. His anger was never wild or uncontrolled like a volcano erupting; he had disciplined it.

Repeatedly, Malcolm X told the story of his journey from being born Malcolm Little in Omaha, Nebraska, in 1925 to becoming the principal minister for the Harlem Black Muslims in the 1950s.

Few prominent Americans of that era had a more compelling history. His family, settled in Lansing, Michigan, was vilified as "smart niggers." When Malcolm was four, their home was burned. Earl Little, his father—a "race man"—was killed by a white supremacist group. Louise Little, his mother, struggled to keep the family together but collapsed and was sent to a state mental institution. (Jan Carew's 1994 *Ghosts in Our Blood* revises totally the history of his mother.) Foster homes became his existence. Smart as a whip in school, he was still discouraged by a teacher.

At eighteen, he drifted into a hustler's life, peddled drugs and bootleg whiskey, and picked up the nickname "Detroit Red." For a few years, he cut a dashing figure, wearing zoot suits, dancing the lindy hop, going to bebop joints. This life came to a stop in 1946, when he was arrested in a burglary and sent to a Massachusetts prison. Behind bars, he began a radical self-appraisal, educated himself, and joined Elijah Muhammad's Nation of Islam. There, behind bars, he transformed himself.

In August of 1952, he was paroled and for the first time heard Elijah Muhammad speak in Chicago. In September, the Nation of Islam gave him the surname of "X." A year later, he became a minister and speaker for the organization. He made the most of the opportunity.

During the 1950s, few Americans had heard of the Nation of Islam. Started in Detroit in the 1930s by Elijah Muhammad (born Elijah Poole in Georgia), it was a religious group based loosely on Islamic religious ideals and Black nationalism. Muhammad had a stern message for Blacks: their beloved Christianity kept them submissive, whites were a race of deceivers, and Blacks must restore their past glory. Urging them to an austere life, he denounced seedy lifestyles, rampant buying, alcohol, tobacco, and eating pork.

From 1953 to 1963, while the civil rights movements were growing enormously, Malcolm was making a name for himself by traveling around the country, organizing temples in different cities, and giving sharp speeches. He worked mostly within his religious group until a Mike Wallace TV show on "The Hate That Hate Produced" brought him national recognition in July 1959. By 1963, the *New York Times* said he was the second most sought-after speaker—following Barry Goldwater—on the nation's college campuses. Malcolm X spoke, though, mostly to Blacks and about Black problems.

Four messages dominated his teaching. First, Blacks had to know their proud history. Without a knowledge of their past, Blacks would be duped into thinking their only destiny was to be slaves and the oppressed. Second, Blacks had to think about the structure of American society and how that structure held them down. Third, Blacks had to build positive, creative communities that could overcome the ravages of past injustice. Finally, American Blacks belonged to a wider world of issues, particularly African issues. It was the height of provincialism to think only of their role in America.

Tensions between Elijah Muhammad and Malcolm X erupted in the spring of 1963. After Malcolm X's remark on December 1, 1963, that John F. Kennedy's assassination was "chickens coming home to roost," he was suspended as a minister. In March 1964, he announced his break from the organization and Elijah Muhammad.

What followed was a tremendous period of growth, with him

reaching out to the civil rights leaders, traveling on a pilgrimage to Mecca, and announcing the formation of the Organization of Afro-American Unity. He undertook a vast schedule of African travels and American appearances, a few in support of SNCC.

In July 1964, four knife-carrying men confronted him at his home. Threats against his life grew. The FBI and CIA kept constant tabs on him. Near dawn on February 14, 1965, his home was firebombed. One week later, just as he began speaking at New York's Audubon Ballroom, he was gunned down.

More than fifteen hundred people attended his funeral on February 27, where actor Ossie Davis delivered one of the greatest funeral orations in American history. Malcolm X was buried in Hartsdale, New York, with his pilgrimage name of "El-Hajj Malik El-Shabazz" on his gravestone.

**Why is Muhammad Ali being called "one of the greatest protagonists in the ring of public history?"**

In *Muhammad Ali, The People's Champ* (1995), edited by Elliott J. Gorn, it is claimed that Ali did more than achieve boxing fame. He entered into the realm of the public's imagination. He became a personality in the collective thought of the nation, even the world. The book looks at "Ali the boxer, Ali the Black Muslim, Ali the cultural icon, Ali the telecelebrity, Ali the narcissist . . . "

Muhammad Ali became a major multifaceted public personality during the 1960s–70s. In fact, he excited people like few athletes and public figures. Was he the most popular person in the world? Many people thought so. Writers, such as Ken Kesey, Marianne Moore, and Norman Mailer, were mesmerized by him. A rising sportcaster, such as Howard Cosell, nearly became his publicist. Photographs of him—showing his boxing skills, greeting crowds, mugging, and later at political rallies—were used all the time by papers and magazines. *Sports Illustrated* in 1964 featured a cover of him sitting in a vault behind a stack of dollar bills, with the caption "I'll take my million and run."

In popular culture, dolls, cartoons, and even comics made

him the favorite of children. His imitators—satirists mimicking his speech and improvised verse—spread his fame. Of course, television took to him, for his fights were widely viewed. Actually, he did not need help to get publicity. After all, he was in his element before a crowd of admirers, as quick with words and boastings as he was in the ring. He was not called "the mouth" for nothing.

Born in Louisville, Kentucky, in 1942, he was named by his mother Cassius (Marcellus) Clay, Jr.—after Cassius Marcellus Clay (1810–1903), the Kentucky politician, a believer in gradual and compensated emancipation of Black slaves. Through the support of his family and his astonishing physical ability, Clay entered amateur boxing during 1954–60 (fighting over one hundred matches).

When he was sixteen to seventeen, a Golden Gloves contender, he was already an awesome boxer. He had speedy footwork, sophisticated technique and good depth perception. In the 1960 Rome Olympics, he became the light heavyweight champion. His victory might have been eclipsed, though, by all the other medals won by Black athletes. In particular, his victory was probably overshadowed by "the Tennessee Tornado," Wilma Rudolph (1940-1994) who won three gold medals in track, a first for an American woman in a single Olympics, and by the decathlon record set by Rafer Johnson (1935– ). Still, Clay returned from Rome as an American hero. He then became a professional boxer.

Sports historian Jeremy Sammons has written, "When Cassius Clay first appeared on the (professional) boxing scene he was a breath of fresh air, someone who might restore boxing's tainted image and sagging fortunes." On February 25, 1964, Clay defeated Sonny Liston for the heavyweight championship, a victory that assumed almost mythic proportions in the public mind. Liston (1917?–1970) was regarded—in James Baldwin's phrase—as a "dark spirit" in boxing. He had served two long prison terms. He had defeated morally upright Floyd Patterson (1935– ) in a contest that the press portrayed as a battle between a saint and a devil. Regardless, Liston was a big, ultrapowerful puncher. Clay's challenge to Liston was not taken seriously by many boxing pundits. But, that evening in Lewiston, Maine, he was triumphant.

Ken Kesey wrote of the fight: "Fancy pants Cassius hit sullen Sonny about twenty-seven times while the old bear was still studying over which paw to swing with—and that's just in the first few seconds of the first round. For six rounds this goes on. The challenger's speed is so remarkable, so unprecedented so . . . so lucid, that it carries even over the radio." No sports photo is more dramatic than the one of Muhammad Ali standing over Liston after his right punch had floored him. Every chest muscle was accentuated. Ali's mouth was stretched wide. He taunted Liston, daring him to rise.

While this story in Clay's life was unfolding in public, another was developing in private—away from the cameras. In 1963, Clay had met Malcolm X in Detroit, in a mosque belonging to the Nation of Islam (NOI). Malcolm X convinced Clay of his usefulness to the Muslim cause, and to the Black cause by fighting Christian influence among Blacks. According to the Nation of Islam's preachings, Christianity had caused Blacks to accept their second-class status. Islam would give Blacks their manhood and womanhood. Malcolm X visited Clay's training camp before the Liston match. It was not long before Clay announced his affiliation and his newly assumed name. By this time, many Americans had a negative view of NOI. They saw it as an organization seeking revenge against whites for injustices imposed on Blacks. Many did not like the connection, and in March 1964, the World Boxing Association temporarily took his heavyweight title.

When in 1967 Muhammad Ali was inducted into the Army, he refused on religious grounds. He claimed his status as a minister. It was a matter of his faith. A federal court ruled that he was guilty of draft evasion and gave him five years in prison. The New York State Athletic Commission stripped him of his title and banned him from the ring. Other boxing commissions did the same.

Generally, Ali's decision reflected the increasing doubts of Blacks about the war. Specifically, it reflected a new mood among Black athletes. With Black power ideology advancing since 1965, Black athletes were forced to examine their status and politics. Many Black athletes supported Ali. Bill Russell, the Boston Celtics basketball star, was one of his earliest supporters. The next year, in

1968, Black athletes' assertion really came out when Tommie Smith and John Carlos—Olympic track winners—on the winners' stand, raised their clenched fists to protest Black subordination.

Ali fought his sentencing by taking his case to the Supreme Court. The court dragged its feet for two court terms, refusing to consider it. In 1971, his conviction was overturned by a lopsided vote. By 1974, he had regained his title by defeating George Foreman in eight rounds. Later, he lost it and then regained it again. He won the heavyweight title three times—an unprecedented achievement.

It is narrow in evaluating his life to love Cassius Clay and fail to appreciate Muhammad Ali. Cassius Clay was the young lightning fast boxer and talker. Two of his lines are legendary: "I am the greatest" and "Float like a butterfly, sting like a bee." Muhammad Ali was the older, more seasoned, politically conscious man, deeply committed to religion and family, poised to take a stand for his beliefs. The totality of his life can only be explained by appreciating both identities.

### What did the riots of the 1960s mean to America?

One hot Los Angeles night, August 11, 1965, a white patrolman stopped and then arrested a young Black man, allegedly a drunk driver, for speeding. The scene of the arrest was the neighborhood of Watts, a mostly Black area southwest of downtown, with one hundred thousand inhabitants.

For many Los Angelenos, Watts was famous for seven towers of steel wire, cement, seashells, tile pieces, and broken colored glass. A life's work, Italian immigrant Simon Rodia had built them over thirty years to thank the United States for affording immigrants a good life. Postcards featured this classic of American folk art, and even tour buses visited.

That August night, Watts would become famous for something else. As the arrest was being made, a crowd gathered, and additional police were called to the scene, one of whom struck a Black bystander.

After the police left, members of the crowd began throwing

rocks at passing cars, beating white motorists, and overturning cars. Police were slow to respond, and within thirty hours, looting and firebombing had begun. The riot had started.

By August 15, Governor Pat Brown had called out the National Guard to quell the disturbances. When they were over, Watts was in ashes, many businesses had been destroyed, thirty-four people were dead, hundreds were injured, and nearly four thousand people arrested. The bill for this outbreak was forty million dollars.

The country was shocked. In 1964, President Lyndon Johnson had launched "an unconditional war on poverty" that would eliminate poverty from the nation. The Economic Opportunity Act, which created the Office of Economic Opportunity (OEO), was to fund a variety of local anti-poverty programs, such as the Job Corps and the Model Cities program for urban development. By 1966, these programs would grow into Johnson's Great Society vision. But his "war on poverty" did not put a brake on discontent.

Watts began a pattern of summer urban riots, centered in Black communities, lasting to the end of the decade. The year 1967 saw the biggest disorders. In July, Newark was the scene of burning shops, shoot-outs, and harsh police assaults on Blacks. At the end of July, Detroit's riot resulted in over seven thousand arrests, forty-three dead (thirty-three Blacks and ten whites), and another forty million dollars in destroyed property. Other cities also had explosions.

The National Advisory Commission on Civil Disorders—popularly known as the Kerner Commission after its chair Otto Kerner, governor of Illinois—was President Lyndon Johnson's response to the fast-breaking events. The commission, made up of prominent citizens, sought to find the deeper causes of the riots.

Their conclusion identified a wide array of problems underlying the unrest: unemployment, poor housing, bad schools, oppressive treatment by police, and a sense of being isolated, segregated, within cities. The report implicated white society in these disadvantages. The report was strongly colored by pessimism, for it foresaw an America becoming "two nations," one Black and one white.

Since the 1967 riots happened mostly outside the South, they shattered the illusion of Black progress in the North and West. Newspapers said that Black leaders did not have control over Blacks. The riots focused national attention on the economics of Black problems. They stigmatized Black areas as unsafe to enter and prone to violence. The Black Panther party and SNCC proclaimed the riots as a step toward revolutionary times.

Little more than a month after the Kerner report, Rev. Martin Luther King Jr. was assassinated, setting off a cycle of unprecedented violence, with riots in 126 cities. Seventy-five thousand federal troops and national guardsmen were called out to restore order.

One positive result of the riots was the election of Black mayors for the first time in America's great cities, beginning with Carl Stokes in Cleveland in 1967. Blacks and some white voters assumed that Black mayors were the key to a stable urban future.

## Why was the Afro hairstyle a major development of the 1960s?

Ever since the 1920s, a few Black writers had inveighed against hair straightening. In 1922, Carter G. Woodson—the founder of Black history week—had urged, "We must cease trying to straighten our hair and bleach our faces, and be Negroes—and be good ones." Almost half a century had to pass before Woodson got his wish.

By the early 1960s, the Afro—or "'fro" to the hip—was gaining popularity among Blacks. If a hairstyle can be said to have an ideology, this one did: the Afro was a conscious repudiation of the long history of hair straightening. "The natural," as it was also called, was Black hair in all its original, bushy, free glory. The movement rallied supporters with this cry: wearers of the Afro, unite; you have nothing to lose but your pomades and straightening irons!

The Afro had a deeper significance than freedom from Black hair technology. Black hair specialist Willie Murrow argues that it was "a fundamental psychic necessity." It was an act of emancipation for Blacks from fake frills, waves, and curls, all of which had been endured to meet European standards of beauty. It wasn't just the hair that was being liberated.

Black political changes assisted in this hair emancipation. Once the young militancy had sprung up and integration began to be questioned, Blacks harkened back to African traditions, and the Afro became a pivotal element in the campaign for reclaiming Blackness.

The Afro came none too soon. Black magazines of the 1950s were chock full of advertisements for greases, oils, pomades, and relaxers. "Your hair will look marvelously smooth and soft and lustrous," promised a Dixie Peach Hair Pomade ad. Two drawings of women accompanied the ad—both of them with perfectly straight hair. Silky-Strate could straighten hair with "no hot combs, no risk, no danger, no lye." Men could rely on SuperGroom for straightening with "No Water or Hot Towels Needed."

Madam C. J. Walker (1867–1919), the doyenne of Black hair straightening, had dreamed of restoring beauty to Black women after the trauma of slavery. But greasing and straightening did not have the same magic in the 1960s as it had had at the turn of the century. Now it was not viewed as beautification but as suppression. Even Lorraine Hansberry's *A Raisin in the Sun* (1959) ventured an incisive critique of straightened hair. In 1965, Malcolm X went much further. *The Autobiography of Malcolm X* contained the most devastating critique of straightened hair—of the conk—that had ever been written. By retelling the story of his feverish desire to have a conk and then ridiculing his own racial disloyalty, he provided the paradigm for others' personal change. Many report using Malcolm's self-critique as their own guide.

The Afro spread slowly at first. Street folk were the first to wear the style. Students at Oakland City College, UCLA, and Atlanta University came next. The Rangers, a gang in Chicago, sported an Afro they called "Ranger bush." By the fall of 1966, the homecoming queen of Howard University—at the heart of middle-class Black education—wore an Afro. Sly Stone wore a towering Afro on his first American tour. In 1968, James Brown screamed, "Say it loud, I'm Black and I'm proud" and got rid of his elaborate conk to prove it.

By this time, the Afro had been adopted by Black radicals. The high-rise Afro appeared, a towering creation trained to grow upward and outward into a rounded shape. Angela Davis, Communist and ally of the Black Panther party, became a charged symbol wearing this hairstyle.

As the Afro gained converts, business and media realized that the new Black hairstyle opened a market. The same magazines that had carried ads for relaxers now advertised hair sprays (the leader being Afro-Sheen), special Afro combs (the "pick"), and that weird fusion of artificiality and naturalness, Afro wigs. The Afro was now fashion, detached from a political message.

It was also now becoming an object of satire. By 1970, George Clinton—a musical genius who loved zany performances—appeared wearing a five-foot-high blond Afro wig. By the late 1980s, the now-famous debate was created in George C. Wolfe's play *The Colored Museum* (1988) between an Afro wig and a long flowing wig over which one should grace its owner's head.

### What was Angela Davis's role in the 1960s?

It is often asked how Angela Davis became one of America's leading 1960s radicals after being born into a middle-class family, graduating as a French major from Brandeis University, and going to Germany to study philosophy. She seemed destined for an ivory-tower existence at a great university.

Born in 1944 in Birmingham, Alabama, she experienced fully from a child's perspective the Black condition—the dilapidated Black school with no gym and "never enough textbooks to go around"; when it rained outside, it rained inside. Her poor friends hung around the lunchroom door because they had no money for food. "Most Southern Black children," she remembered, "learned how to read the words 'Colored' and 'White' long before they learned 'Look, Dick, look.'" Bombs ripped Black houses apart.

At the same time, young Angela was given the best training Black Birmingham could offer. Despite poor facilities, solid teachers gave her good instruction and held her to high standards. The schools taught her to identify with Black people. "The

Star-Spangled Banner" was sung in school, but so was the "Negro National Anthem" by James Weldon Johnson, honoring the Black struggle. Students learned about Frederick Douglass, Sojourner Truth, and Harriet Tubman (but not slave rebels Denmark Vesey and Nat Turner!). Black history week was the school year's highlight.

By the time Davis left Birmingham for a New York private school, she was mature. Most important, she had seen through the racist claim that Blacks were treated as inferior because they *were* inferior. At fifteen, she read Karl Marx's *Communist Manifesto,* a document that "cut cataracts" from her eyes. Marx painted a future world without the exploited and exploiters, so unlike Birmingham.

Davis loved the world of ideas. This propelled her to top academic achievement, and in 1965 she went overseas to study philosophy. But the Black American uprising lured her back home. Ever since Davis had seen photos of Freedom Riders arriving in Birmingham in 1961, she had wanted to join the fight. She seized the time.

Returning to America in 1967, Davis combined her studies with working with local militant groups. She enrolled at the University of California, San Diego, as a student of the famous philosopher Herbert Marcuse. On her free days, she would drive to Los Angeles to work on radical projects. The year 1968 was a momentous one for Davis: she helped launch the Los Angeles chapter of SNCC, and she joined the Che Guevara–Patrice Lumumba chapter of the Communist party. Her initial dues were fifty cents.

Between 1968 and 1972, Angela Davis was catapulted into fame. In 1969, she was hired to teach philosophy at UCLA. When her Communist party membership was revealed, then-governor Ronald Reagan and the university regents fired her, citing a 1949 rule prohibiting party members from teaching at state universities. She challenged them in court and won.

Davis was drawn increasingly to the plight of prison inmates, whom she thought were hostages of class warfare. She became a well-known public supporter of George Jackson and W. L. Nolen, two Black Marxist inmates at Soledad Prison. This finally cost her the UCLA job.

What happened next is still surprising. George Jackson's brother Jonathan tried to rescue a defendant from a trial in Marin County, north of San Francisco. Using guns from Davis's home, he took hostages and tried to escape, but died in the attempt, as did two prisoners and a judge. It was a disaster. Because the guns were Davis's, a warrant was issued for her arrest. She went underground.

In August 1970 Davis was placed on the FBI's ten-most-wanted list. An FBI poster read, "Wanted by the FBI. Interstate Flight—Murder, Kidnapping. Angela Yvonne Davis." It carried two photos of Davis—one somber, one with dark glasses, both with her famous defiant Afro hairstyle. In fact, she became the cultural epitome of the Afro style for millions; she was a revolutionary and her hair a political fashion statement.

Two months later, in October 1970, Davis was captured in New York and sent back to California to stand trial. After one of the most high-profile trials of the 1970s, she was acquitted of all charges. She has continued some political activity. Mostly, she has lectured extensively and written several important books, in particular *Women, Race, and Class* (1983) and *Women, Culture and Politics* (1989). Today, she is a professor in the department of ethnic studies at San Francisco State University.

## RESOURCES

Aptheker, Herbert. *A Documentary History of the Negro People in the United States.* Vols. 6 (1993) and 7 (1994). New York: Citadel Press, 1994.

Branch, Taylor. *Parting the Waters: America in the King Years, 1954–1963.* New York: Simon & Schuster, 1988.

Brown, Elaine. *A Taste of Power: A Black Woman's Story.* New York: Pantheon Books, 1992.

Carson, Clayborne, David J. Garrow, Gerald Gill, Vincent Harding, and Darlene Clark Hine, eds. *Eyes on the Prize: Civil Rights Reader.* New York: Penguin Books, 1991.

Carson, Clayborne. *In Struggle: SNCC and the Black Awakening of the 1960s.* Cambridge: Harvard University Press, 1981.

Cobbs, Elizabeth H., and Petric J. Smith. *Long Time Coming.* Birmingham: Crescent Hill Publishing, 1994.

Crawford, Vicki, Anne Rouse, and Barbara Woods, eds. *Women in the Civil Rights Movement: Trailblazers and Torchbearers, 1941–1965.* Bloomington: Indiana University Press, 1990.

Cruise, Harold. *The Crisis of the Negro Intellectual: A Historical Analysis of the Failure of Black Leadership.* New York: Quill Press, 1984.

Davis, Thulani. *Malcolm X: The Great Photographs.* New York: Steward, Tabori, and Chang, 1992. Photographs by Howard Chapnick.

Dittmer, John. *Local People: The Struggle for Civil Rights in Mississippi.* Urbana: University of Illinois Press, 1994.

Floyd Jr., Samuel A. *The Power of Black Music: Interpreting Its History from Africa to the United States.* New York: Oxford University Press, 1995.

Garrow, David. *Bearing the Cross: Martin Luther King Jr. and the Southern Christian Leadership Conference.* New York: William Morrow, 1986.

Grant, Joanne. *Black Protest: History, Documents, and Analyses.* New York: Fawcett Press, 1968.

Harris, William J., ed. *The LeRoi Jones/Amiri Baraka Reader.* New York: Thunder's Mouth Press, 1991.

Mayer, Michael S. *Diary of a Sit-In.* 2nd ed. Urbana: University of Illinois Press, 1990.

*The Motown Album.* New York: St. Martin's Press, 1990.

Null, Gary. *Black Hollywood: From 1920 to Today.* New York: Citadel Press, 1993.

Payne, Charles M. *I've Got the Light of Freedom: The Organizing Tradition and the Mississippi Freedom Struggle.* Berkeley: University of California Press, 1995.

Pearson, Hugh. *The Shadow of the Panther: Huey Newton and the Price of Black Power in America.* New York: Addison-Wesley, 1994.

Sitkoff, Harvard. *The Struggle for Black Equality, 1954–1980.* New York: Hill and Wang, 1981.

Strickland, William, and Cheryll Y. Greene. *Malcolm X: Make It Plain.* New York: Viking, 1994. Text with photos.

Taulbert, Clifton L. *The Last Train North.* Tulsa, Okla.: Council Oak Books, 1989.

Van Deburg, William. *New Day in Babylon: The Black Movement and American Culture, 1965–1975.* Chicago: University of Chicago Press, 1992.

Vollers, Maryanne. *Ghosts of Mississippi: The Murder of Medgar Evers.* Boston: Little, Brown, 1995.

Wolff, Daniel. *You Send Me: The Life and Times of Sam Cooke.* New York: William Morrow, 1995.

TEXTS FOR YOUTH

Dolan, Sean. *Pursuing the Dream: 1965–1971.* New York: Chelsea House Publishing, 1995.

Duncan, Alice Faye. *Everyday People.* New York: Bridgewater Books, 1995. Text with photographs by J. Gerard Smith.

Stine, Megan. *The Story of Malcolm X: Civil Rights Leader.* New York: Dell Publishing, 1994.

# 9

# The Conflicted Time

## The 1970s–1980s

1971: The Congressional Black Caucus is consolidated, after starting in 1969 with nine members

1971: At New York's Attica Correctional facility, Black and Puerto Rican convicts riot

1972: A Black political convention is held in Gary, Indiana, attracting three thousand delegates and five thousand observers from across the nation

1972: Representative Shirley Chisholm (D–New York) runs for president

1973: Black mayors are elected in several cities: Thomas Bradley in Los Angeles, Maynard Jackson in Atlanta, Coleman Young in Detroit

1974: Atlanta Braves' Henry "Hank" Aaron breaks Babe Ruth's record for the most home runs

1975: Arthur Ashe wins the Wimbledon singles championship

1976: Barbara Jordan is the first Black woman to give a Democratic National Convention keynote address

1977: Jimmy Carter is inaugurated as president, appoints Andrew Young as United Nations ambassador, Patricia Roberts Harris as HUD secretary

1977: TransAfrica, a lobbying organization on African affairs, is founded by Randall Robinson

1981: Ronald Reagan is inaugurated as president

1981: Urban League focuses on the ills of Black families

1981: Lena Horne opens her one-woman show, *Lena Horne: The Lady and Her Music,* on Broadway

1983: President Reagan appoints Clarence Thomas to chair the Equal Employment Opportunity Commission

1983: Jesse Jackson launches a bid to become Democratic nominee for president

1983: Dr. Martin Luther King Jr.'s birthday, January 15, is declared a national holiday

1980s: Black conservatives, led by economist Thomas Sowell, gain adherents

1984: *The Cosby Show* is launched and has an extended run as America's most popular TV program

1985– : Rap music is generating significant artists and audiences

1985: United States initiates sanctions against South Africa

1987: Molefi Kete-Asante publishes *The Afrocentric Idea*

1989: George Bush is inaugurated as president, pledging a "kinder and gentler" nation

1989: Army four-star general Colin L. Powell is named chairman of the Joint Chiefs of Staff, the nation's highest military post

1989: David Dinkins is elected New York City's mayor

By 1970, Stevie Wonder was becoming Black America's singing poet, the bard from Motown. Just twenty, he was already famous for such high-energy, joyous hits as "Fingertips," "Uptight: Everything's Alright," and "I Was Made to Love Her." As he left his teenage years, he was no longer "Little Stevie

Wonder"; his voice deepened and his repertoire broadened.

As a bard, Wonder spoke the thoughts of 1970s Black America as it encountered conflicting conditions—many positive, many negative, and many in between. Into the 1980s, his songs continued to mirror Black aspirations.

Stevie Wonder's music had dual messages. In one set of songs, he was the chief celebrant of love—personal, interracial, and global—keeping faith with the early 1960s credo of harmony among all peoples. In his own words, his mission was to "spread love mentalism." That he did. His albums, *Talking Book* and *Fulfillingness' First Finale,* were crammed with love odes.

In a second set of songs (such as "Living for the City" and "You Haven't Done Nothing"), Wonder's lyrics said that America was betraying the 1960s humane vision. "Village Ghetto Land" told the story of a Black urban world that had not changed: "Now some folks say that we should be/Glad for what we have./Tell me, would you be happy in Village Ghetto Land?"

Sightless, Wonder absorbed and spoke what other Blacks could see happening in the seventies and eighties. With the 1960s momentum behind them, Blacks—twenty-two million strong (11.1 percent of Americans)—arrived in the 1970s anxious to complete the past decade's agenda.

Despite tragedies like Martin Luther King's assassination in 1968, Blacks pressed on. During the next two decades, progress was significant. But it was not as great as expected, nor was it uniform. Important successes came to Blacks, especially in pursuing higher education and elected office. However, Blacks also faced setbacks and new adversaries. By the end of the 1980s, some Black goals were faded dreams.

### How were desegregation and busing derailed in the 1970s?

On that spring day, May 17, 1954, when the Supreme Court outlawed public school segregation, it looked like the worst of the battles for desegregation were over. In a triumphant gesture, the NAACP lawyers, dressed in pinstripe suits and wearing broad smiles, walked down the Court steps, posed for photos, and answered questions from the press corps. Had they been asked how long the decision's implementation would take, none would have answered, "Decades." They never imagined that nearly twenty years later, desegregation would still be doggedly resisted in many communities across America.

Yet by the 1970s, that was exactly what was happening. The so-called "white backlash" had set in. More than two hundred school districts in the South were still holding on to segregation. Each fall, as schools desegregated, racial tensions peaked, resulting in violence in places as widespread as Michigan, New Jersey, and South Carolina. In 1971, the governors of two Southern states pledged to do everything in their power to thwart desegregation. The nation's federal courts were clogged with suits and countersuits over plans to desegregate. Desegregation was showing no sign of becoming a less explosive issue.

When the twentieth anniversary of the Brown case arrived, the mood among Blacks was joyless. Blacks recognized the strides that had been made, as in the South where more than 40 percent of Black students attended integrated schools. But most probably agreed with noted Black columnist Carl Rowan: "Some of the litigants in that 1954 decision never saw a day of desegregated education. They saw evasion, circumvention, massive resistance, and a generation of litigation."

Many forces blocked the path to desegregation. The 1954 Supreme Court decision did not say exactly when schools had to change. In 1955, in the so-called Brown II decision, the Court said only that desegregation should proceed with "all deliberate speed." This vagueness invited circumvention.

For a decade after Brown, some of the white South's best legal minds schemed to get around desegregation—designing private segregated academies, freedom-of-choice plans, elaborate assignment schemes, and illegal ways to fund private schools. Prince Edward County, Virginia, closed its public schools from 1959 to 1964 to avoid integration.

But blame does not lie just at the South's door. Northerners were also adept at fighting desegregation. New York, for example, enacted a statute that made it illegal for school boards to assign students to schools in order to achieve racial balance without first asking their parents. The statute was later ruled unconstitutional.

The greatest resistance to desegregation came with the widespread hostility to busing. Busing was a method chosen by the federal courts in the 1970s to achieve racial mixing. Students from a white neighborhood could be bused to school in a Black neighborhood or Black students could be taken into a white community to school. The goal was racial balance.

Busing became the hot issue of the 1970s. Whites condemned it as destroying their neighborhood schools, as too exhausting for the students and too expensive for the taxpayers, as a means of sending white students to inferior schools and into dangerous neighborhoods. Fewer Blacks criticized busing, but those who did were no less vocal.

In the 1970s, Blacks had come to doubt the benefits of desegregation. Across the South, thousands of Black teachers had been dismissed and principals demoted. Black schools, once community cultural centers, had been closed. Blacks missed their caring teachers and tough standards. Now busing was taking Black students on long journeys to schools where they were treated indifferently or worse.

Busing's chief critic was President Richard M. Nixon, who directed Health, Education, and Welfare Secretary Elliot Richardson in 1971 to "hold busing to the minimum required by law." When a federal court in 1972 issued a Detroit desegregation plan using busing, Nixon wasted no time and minced no words in condemning it as "the most flagrant example" of busing decisions. As usual, he had his eye on the upcoming presidential elections.

With so much opposition, the pressing question was this: What was accomplished by all the noble work for school desegregation? Several gains came from this work. Desegregation proceeded smoothly in many districts. Black businesspeople, physicians, engineers, bankers, and lawyers produced by the first integrated schools were soon to make their mark. Segregation also fell in colleges, prompting over one hundred thousand Blacks to enroll. Most important, school boards were taught that it was illegal to place Black students in segregated schools intentionally.

### Why were blaxploitation films so popular?

"Have you seen the three S's yet?" This was the way many Blacks referred to that trio of Black action films that packed the theaters in 1971 and 1972: *Sweet Sweetback's Baadasssss Song, Shaft,* and *Superfly.*

In the late 1960s, Hollywood was still cranking out the Black assimilationist film, with the proper-speaking male hero who was as asexual as a nun, when suddenly the cinema capital was ambushed by a militant, gritty, raunchy Black film. *Guess Who's Coming to Dinner?*, the story of interracial courting and marriage with mild-mannered Sidney Poitier in the lead, was not a satisfying film diet for Blacks who were living in an atmosphere of urban riots and H. Rap Brown militancy. Hollywood should have known something was up when Black audiences cheered as Raymond St. Jacques beat up a white man in *If He Hollers, Let Him Go.*

Melvin Van Peebles's *Sweet Sweetback* was the film that assaulted the old formula. Made on a low budget, shot in nineteen days, and held together by a loose narrative, Van Peebles created an outrageous Black film that is radical even by today's standards. It opened with the saying, "This film is dedicated to all the Brothers and Sisters who have had enough of The Man." That alone resonated with Black viewers.

Sweet Sweetback (played by Van Peebles) was a raw character, holding nothing back. He was sexually alive at twelve, later a ghetto pimp, and still later a man who beat a cop to death because the cop had brutalized a brother. Sweetback becomes a fugitive and manages to stay just one step ahead of the police. Finally he escapes into Mexico, vowing to return for revenge. Van Peebles's film gave audiences the feeling that they had entered an authentic Black life, one that had been hidden from view. It was a cinema vérité of Black oppression.

When the film made almost five million dollars and even drew thousands of whites to see it, Hollywood followed immediately with two Black action films. Radio advertisers had already discovered that they could sell their products by appealing to Black political ideals. Now, it was the movies' turn.

The first film, *Shaft*, directed by photographer Gordon Parks Sr., was filled with sepia beauty and had the driving music of Isaac Hayes's Oscar-winning theme song. Actor Richard Roundtree made detective John Shaft into a hip, trash-talking (even to whites), nervy "Black private dick that's a sex machine with all the chicks." Audiences were entertained both by the film's relentless action and by the idea that Shaft was his own man, free from the control of cops, crime lords, and Black Power revolutionaries. With aplomb, Shaft won against his urban enemies.

*Superfly*, directed by Gordon Parks Jr., glamorized the life of a Black drug dealer, to a soundtrack studded with Curtis Mayfield's hits, like "Pusherman." Youngblood Priest, Ron O'Neal's character, was a Harlem kingpin of dope and a fancy dresser (long conked hair, wide-brimmed hats, long coats, gold cross necklace). Like a cutthroat CEO, Priest ran a vast cocaine trade, lived well, had a flashy Cadillac, and two women—one Black and one white—to serve his needs. He was living an American dream. Priest, though, wanted out—he wanted one final score and then retirement.

Both films cleaned up at the box office. They were not high art, nor were they intended to be, but they were good Saturday night recreation. They also succeeded because they offered a vision—to some, warped—of Black power, Black autonomy, and Black resistance. It did not seem to matter that Superfly was an amoral tin god who was running on a low-octane spirituality.

Unfortunately, the success of these films launched a deluge of even cheaper and more shallow imitations, all highlighting drugs, violence, and sex—with titles like *Black Gun, Slaughter, Black Mama, White Mama,* and *Black Caesar.*

One of these films' features was to have a Black woman justice-warrior posing as the cinematic centerfold. Pam Grier became this genre's leading action heroine. *Coffy*, for example, narrated her double life: nurse during the day and nighttime vigilante who attacks the drug cartels. *Foxy Brown* allowed Grier to do in a female drug lord named "Miss Katherine." Gutsy, determined, butt kicking, Grier was a cult heroine.

Blaxploitation films were so named because they exploited

common stereotypes—often, the meanest ones—of Black life. In 1972, civil rights groups formed the Coalition Against Blaxploitation, protesting the films' images. They claimed, "The transformation from the stereotype 'Stepin Fetchit' to 'Super-Nigger' on the screen is just another form of cultural genocide." By 1973, Black action films were no longer surefire box office.

It is too soon, though, to write this film genre's epitaph. As film critic Chuck Stevens has observed, paint *Shaft* pink and you get Clint Eastwood in *Dirty Harry* and Charles Bronson in *Death Wish. Boyz in the 'Hood* and *Sugar Hill* are offspring of the 1970s films. Today's "cult of the 'hood" film is *Sweetback*'s progeny.

### Why did a strong bond exist between Blacks and Southerner Jimmy Carter?

As the 1976 presidential elections approached, Americans were looking for a person to restore respect to the presidency after Richard Nixon had resigned over his role in the Watergate scandal. Black Americans shared this interest, as personified by Congresswoman Barbara Jordan's (D–Texas) outstanding performance during the House Judiciary Committee's impeachment proceedings. But Blacks were also anxious about more practical issues, such as the new president's view of the country's racial problems.

Nixon and Blacks were never on the same wavelength. In his first term, starting in 1968, he appointed only three high-ranking Black officials, nominated two Southern conservatives to the Supreme Court, was indifferent to desegregation issues, and opposed busing. To his credit, he supported Black business development and honored Duke Ellington.

In 1972, more Blacks (21 percent) voted for Nixon than in 1968, but overwhelmingly they gave their votes (79 percent) to Democratic candidate George McGovern, a cold fish for many Blacks because he had shown so little knowledge of them and their politics. Blacks voted almost the exact opposite of whites. After Nixon left office, Blacks grew somewhat closer to his successor, Gerald Ford.

When Jimmy Carter, former governor of Georgia, emerged in 1976 as a leading Democratic presidential candidate, it was not love at first sight for Blacks. They were suspicious of him as a white Southerner, and Georgia was well known for its past bigotry and for brandishing Confederate symbols in its flag. Carter also had been a loose talker about the value of "ethnic purity." But Carter gained Black support before the July convention in New York City.

Southern Black leaders such as Martin Luther "Daddy" King Sr., Coretta Scott King, Andrew Young, and John Lewis shifted their allegiance to him. Daddy King sent a telegram to a Black group saying that Carter stood for equal justice, that he had spoken for integration of his Plains, Georgia, church, appointed Blacks to judgeships while governor, passed the state's first fair-housing law, and honored Martin Luther King Jr. in the state capital. While governor, Carter had defused a tense racial situation in a town over a hundred miles from Atlanta by flying there and meeting with Blacks and whites. These deeds were persuasive.

At the convention that nominated Carter, Blacks were highly visible. Barbara Jordan gave the keynote address, a first for a Black woman, in which she urged a "great healing of the national heart and soul." At the final session, Daddy King gave the benediction, and the King family joined the Carters in singing "We Shall Overcome." It seemed like the old civil rights spirit was returning through Carter's presidential quest.

Blacks put forth the maximum effort for Carter. They were poised to deliver. After years of campaigning for the vote, thousands of Blacks were registered, particularly in the South. With sixteen congressional Democrats (forming the Congressional Black Caucus in 1971) and thousands of local Democratic officials, Carter had a sizable army of Black supporters.

Carter beat Ford, with 50.1 percent of the popular vote and 297 electoral votes. He owed his victory in crucial states to the more than five million Black votes he received.

Blacks looked to the Carter administration with great hope. In forming his cabinet, he named Andrew Young, formerly in the Georgia state legislature, to be the United States ambassador to the United Nations and Patricia Roberts Harris, a prominent lawyer and former ambassador, as secretary of housing and urban development. During his administration, Carter moved swiftly to appoint numerous Black federal judges, more than all of his predecessors put together.

In any close relationship, moments of division arise. From the start, Blacks complained that more high-level appointments were needed. Congressional Blacks were angry that Carter's budget spent too much on the military and not enough on social problems. Carter was criticized because he did not clearly endorse diversity in university admissions, as in the famous Bakke case in 1978, when a white applicant to the University of California contested affirmative action.

Perhaps Black expectations for Carter were too high and were bound to be disappointed. Although he had used his office for Blacks more than any president since Lyndon Johnson, Blacks failed to support him as much in his 1980 bid for reelection. He was defeated by a conservative landslide.

### Why did Alex Haley's *Roots* take the country by storm?

Sugar Hill was usually a lively San Francisco tavern, a watering hole for a middle-class Black clientele. Yet every night for the week of January 23 to 30, 1977, at 10:00, it was totally quiet except for the tinkling of glasses.

All eyes were fixed on the television adaptation of Alex Haley's best-selling book. Three days after Jimmy Carter's inauguration, the patrons were immersed in a Black family's story.

Across town at the Caffe Trieste in North Beach, business was unusually slow. Regulars were at home watching *Roots.* City officials caught the fever and proclaimed a "*Roots* Week." Schools used the television show in classroom discussions. B. Dalton's bookstores reported that Haley's book, already selling well since its publication in October 1976, was now flying out the door at a rate of thirty copies a day.

Nationally, the reaction was no different. Barely halfway through the series, *Roots* had become one of the most widely viewed programs in television history. Each of the eight episodes produced a record. The third episode ranked behind only *Gone With the Wind* as the top-rated television show. Thirty million homes, or 68 percent of total viewers, were tuned in. In all, some 130 million Americans watched a part of the series.

Alex Haley was hardly an unknown before *Roots.* Born in 1921, Haley became an accomplished journalist, earning the post of chief journalist in the U.S. Coast Guard and inventing the interview feature for *Playboy.* In 1965, he assisted in the writing of *The Autobiography of Malcolm X,* one of the enduring classics of American memoirs.

Nevertheless, the journey to the *Roots* success was a long one for Haley. As a child in Tennessee, he had heard from his maternal grandmother, called "Miss Sis," while she was "just sittin'," a story about "the African" who had been captured and brought to America. This man was the Haleys' historic link to Africa. At first, all that Haley had to go on was this slender story.

Eventually, his search focused on the history of just one person—Kunta Kinte, Haley's great-great-great-great-grandfather, "the African" his grandmother had spoken of. Haley went to a small village, Jufurre, in Gambia, West Africa, to discover the full story of Kunta Kinte.

There, a local oral historian—a griot—told Haley the story of a Kunta who had been captured at sixteen and taken into slavery. The year was 1750. Haley was overjoyed. Miss Sis's tale was true. Once this riddle had been solved, Haley researched his family's odyssey from West Africa into slavery and finally into freedom after the Civil War.

Why did *Roots* succeed so phenomenally? The settings were well chosen, particularly the lush, verdant Savannah, Georgia, coastline used for the West African episodes. A cavalcade of stars and celebrities—John Amos, Maya Angelou, Lloyd Bridges, Lorne Greene, Burl Ives, O. J. Simpson, Ben Vereen, Robert Reed, Cicely Tyson—drew viewers. But the real draw was the uncomplicated story. "Middlebrow" was the word critics used, and they were right. But that was precisely what put the viewers at ease. Nor did the series preach at the viewer. Sixties rhetoric was nowhere to be found.

Viewers also identified with *Roots* because it was told as an American story. Oppressive slavery was presented. But the story always stressed the power of the underdog and his or her struggle for freedom—traditional American ideals. Another point of identification was the family as a source of support and hope. For many viewers, *Roots* succeeded because it told the Black story in a way similar to other American immigrant stories.

Even with thirteen hours of television time, *Roots* did not deeply analyze American race relations, nor did it usher in a new era of good feeling. Later Haley was attacked as having been duped by the wily West African griot who told him of Kunta Kinte. It is true that his methods were slightly naive and that his idea of "faction" (facts plus fiction) allowed him to take interpretative flights. But none of this robbed him of the achievement that captivated a nation. His enormous vision gave birth to the most popular Black history narrative of all time. For a week, America's race problems were suspended.

### What was Ronald Reagan's relation to Black America?

In the mid 1980s, there was a joke about a Black college professor who told a largely Black audience that he was planning to write a book on "Ronald Reagan and Black America." The idea was greeted with silence, until the master of ceremonies said, "Well, professor, it could be either one page or a thousand pages, but the result would be the same." In other words, the conclusion would condemn Reagan.

Why did Black Americans come to see Ronald Reagan and his presidency—a presidency about which the rest of America was enthusiastic—in such a negative light? Polls show that Blacks wanted change in 1980, when Reagan first ran. Like other Americans, Blacks were sick of the Iran hostage crisis, sick of double-digit inflation, and sick of a listless economy. Reagan promised that his policies could fix these.

Yet in the Jimmy Carter–Ronald Reagan contest, less than one in ten Blacks voted for Reagan. Despite the fact that civil rights leaders Ralph Abernathy and Hosea Williams had endorsed Reagan, this was one of the lowest Black votes ever for a Republican presidential candidate. What happened?

Some Republicans suggested that Black voters had been on the "liberal plantation" so long that they could not escape their Democrat politician-masters and vote their true interests. Others blamed the party for not welcoming Blacks or telling them how Reaganomics would help them. Neither explanation works.

Blacks were suspicious of Reagan as a candidate and as president because of his record. They judged him by his words and deeds.

It was during the 1980 presidential campaign that Blacks first became seriously worried by Reagan. (Nationally, Blacks did not know his record as California governor.) He criticized policies they favored while wooing white voters with messages and language that offended Blacks. He campaigned as the leader of a white American tribe.

Early in his campaign, he stated his opposition to affirmative action, a policy many Blacks felt was essential for advancement. He was also against job training programs and busing to achieve desegregation. He repeatedly attacked welfare as if Blacks were its only recipients.

When invited to speak at the NAACP's annual convention, he turned down the invitation. Later that summer, Reagan promoted "states' rights" at the Mississippi State Fair, which took place in the area where civil rights workers Chaney, Goodman, and Schwerner had been killed in 1964. To Blacks, "states' rights" belonged to the era when whites dominated the Deep South, an era Blacks had fought and died to end.

After Reagan won the election and began forming his administration, Blacks looked for signs that he would be more inclusive than he had been during the campaign. His inaugural address was generous in spirit. "This administration's objective will be a healthy, vigorous, growing economy for all Americans," he said, "with no barriers born of bigotry or discrimination."

During his first term, Reagan named only one Black to a high post—the outstanding lawyer Samuel Riley Pierce Jr. as secretary of housing and urban development. He attacked the United States Civil Rights Commission, fired its chair, and installed a Black, Clarence M. Pendleton Jr., who lacked political savvy. He appointed only two Blacks to federal judgeships.

In 1983 Reagan did—reluctantly—sign the bill that established a federal holiday honoring slain civil rights leader Martin Luther King Jr. In a press conference, he hinted that future FBI document disclosures might prove that King was a Communist. In foreign affairs, the Reagan administration made overtures to racist South Africa.

The lowest point came when the Reagan administration tried to give tax-exempt status to two South Carolina private schools that discriminated against Blacks. This was against Internal Revenue Service (IRS) policy. The White House announced its decision on a Friday afternoon so it would not get a lot of media coverage. Later, the Supreme Court upheld the IRS.

Black participation in the 1984 Walter Mondale–Ronald Reagan election was the greatest since the 1964 presidential campaign. Part of this upsurge was because of Jesse Jackson's inspiring presidential candidacy in the Democratic party primaries. But many Blacks registered to vote in order to get rid of Reagan. Even so more Blacks voted for Reagan than had in 1980; but they were less than 20 percent of the Black vote. Some of Reagan's support came from affluent Blacks who had profited from the economic recovery and tax cuts. And Reagan had shown public goodwill to Blacks, in visits to families and to predominantly Black schools. In the end, though, Blacks never gave him more than a 40 percent approval rating in polls.

In the last days of his administration, Reagan defended his record on Black America and attacked Black leaders as having a stake in distorting it. Blacks answered by pointing to the increased poverty in their communities. At best, the Reagan-Black experience was a stalemate.

### What has been Jesse Jackson's contribution to Black America and American history?

Jesse Jackson has created as many political organizations and movements as any figure in twentieth-century America. Unfortunately, Jackson's achievements have been underestimated. It is often said, and not only by Blacks, that if Jackson were white, his true historical stature would be more appreciated.

Jackson was eighteen, a student at North Carolina's Agricultural and Technical College, when the sit-ins started in early 1960. He became a protest leader, learning organizing strategies and discovering that he liked being on history's stage, at the forefront of change.

In the summer of 1966, Jackson was propelled to the center of civil rights action by his participation in Martin Luther King Jr.'s Chicago Freedom Movement—the first entrance by King into the North. Although the fight for open housing failed, Jackson became indispensable to King. His energy, resilience, and ability to assess tricky situations caught King's attention. From this sprang a close relationship, which caused him to take training at Chicago Theological Seminary and led to his ordination as a Baptist minister in 1968.

When King left Chicago for the South, Jackson remained, building a new group, Operation Breadbasket (1967–71), which focused on economic goals, getting jobs for Blacks, and promoting products made by Blacks.

The darkest day in Jesse Jackson's young life—he was twenty-eight—came on April 4, 1968, when an assassin's bullet struck down King at the Lorraine Motel in Memphis, Tennessee. Jackson lost his leader, mentor, and friend and had been a witness to the hateful act. Surviving a time of deep sorrow and introspection, he emerged more passionately dedicated to King's ideals.

Throughout the 1970s and 1980s, he became both a premier racial politician and a first-class coalition builder. His strategy was to consolidate first his base among Blacks and then to reach out audaciously to other groups.

In 1971, Jackson designed Chicago's People United to Save Humanity (PUSH), an aggressive Black economic and political pressure group. This was his launching pad, and Jackson became even more prominent as PUSH went national. As early as 1974, he advocated a movement of "Blacks in cooperation with Mexican-Americans, Puerto Ricans, American Indians, Asian-Americans—all the diverse segments of the population." In 1980, he urged, "We must find a new focus. We must do it as a coalition."

Only four years later, in 1984, Jackson decided to seek the Democratic party's nomination for president. Two main goals drove him. First, the Reagan presidency had seriously damaged Black America, and its policies needed to be opposed. He also wanted to shake up the Democratic party for failing to reward Black loyalty and votes adequately.

The 1984 campaign was a good time for a Black presidential candidate. The number of Blacks voters increased in the 1970s and then leaped upward in the 1980s, making for a dramatic rise in the number of Black officeholders. And many Blacks wanted to defeat Reagan in his reelection bid.

Jackson now had formed the Rainbow Coalition, another group advertising his candidacy. His oratory was known nationally. And his success in winning the release of Lieutenant Robert O. Goodman—a Black Navy bombardier-navigator whose jet had been shot down by Syrian artillery—from a Syrian prison camp catapulted Jackson into the national news. This drama unfolded in January 1984, just weeks before the primaries.

Just as Jackson's candidacy gathered steam, he was exposed in the *Washington Post* as having referred to Jewish Americans as "hymies" and New York City as "Hymietown" in private conversation. This was morally reprehensible. At first, Jackson stonewalled, denying the remarks. That was a mistake. Later, he appeared at a Jewish synagogue in Manchester, New Hampshire, and apologized, but the damage had been done.

Jackson won more than three million votes in the primaries, translating into 384 delegates at the San Francisco convention. The convention nominated Walter Mondale.

Jackson decided to try again in 1988. His second campaign focused on broader issues, more appeals to labor and farmers, seasoned campaign managers, and a better campaign to attract white voters. It paid off, with nearly seven million votes in the primaries; 92 percent of his support came from Blacks and 13 percent from whites.

When he went to the Atlanta convention, Jackson had 1,122 delegates, almost one-third of the total. With so many delegates, Jackson was able to influence the platform on South Africa and minority contracts. Although Massachusetts Governor Michael

Dukakis was nominated, Jackson's crowning moment came when he addressed the packed convention hall on the theme of "Keep Hope Alive." That night, he was "the spellbinder of spellbinders."

In 1990, Jesse Jackson, a resident now of Washington, D.C., won the new post of "statehood senator," who will lobby for the district's statehood. Recently, he has hosted CNN-TV's *Both Sides,* a show debating major issues.

Jesse Jackson opened up America's politics. His two presidential challenges produced thousands of new voters, launched new political careers, and made the country think about hard issues. For example, the voter base he created in New York City helped elect David Dinkins mayor in 1989. He revitalized Black politics in the 1980s. Even conservative commentators who loathed his politics have conceded that he is a great American populist.

### Where did the 1980s Black conservatives come from, and what were their ideas?

"Black conservatives" seems an oxymoron, a contradiction in terms. It is not thought natural for Blacks to be conservative. It is almost like talking about a unicorn.

Most Blacks vote for the Democratic party and elect liberals to represent them. Whereas recent polls show many Blacks with conservative views on family and crime issues, Blacks still think of conservatives as being mostly white and Republican. So does the rest of the country.

Once upon a time, Blacks could be conservative without causing any consternation. In Booker T. Washington's heyday after 1895, many worked on making money, spurned politics, warned against involvement with racial issues, and did not question white supremacy.

Also once upon a time, Blacks were heavily Republican. From the Civil War's end in 1865 to Franklin D. Roosevelt's second race in 1936, Blacks had voted for Republicans. As in a ritual, Blacks thanked Republicans every four years for Lincoln's freeing of the slaves and the Reconstruction's gains by boycotting bigoted Democrats.

So being conservative and being Black, being Republican and being Black were not such abnormalities.

George Schuyler (1895–1977), for many years the leading columnist for the Black-owned *Pittsburgh Courier,* was good at being both Black and conservative. In his autobiography *Black and Conservative* (1966), he announced, "I learned very early in life that I was colored, but from the beginning this fact of life did not distress, restrain, or overburden me." To him, the Black conservative stood for middle-class industry and gradual change, was patriotic and deeply anti-Communist, and rejected Black firebrands.

As American conservatism grew in the 1970s, it sprouted a new branch of Black conservatives. Ronald Reagan's election in 1980 focused the spotlight on this group and its ideas.

Leading this group was Thomas Sowell (born 1930), a distinguished economist, well versed in conservative theory. He was a prolific author and opinion columnist. In 1972, Sowell wrote the first of many books on racial issues, *Black Education: Myths and Tragedies,* which criticized a host of liberal approaches to Black education. During the 1980s, Sowell was one of the most read and quoted intellectuals in America. His enormous knowledge, machete-like logic on complex issues, and wit earned him a vast readership.

Prominent also were Walter Williams (born 1936) and Glenn C. Loury (born 1946), both economists like Sowell. Reagan appointees were also conservatives—Clarence M. Pendleton Jr. (1930–88), chair of the U.S. Civil Rights Commission, and Alan Keyes (born 1950), an assistant secretary of state. Marva Collins (born 1936), Chicago educator and founder of the successful private Westside Preparatory School in 1975, was often placed in this group.

Common ideas ran through the writings and speeches of the Black conservatives:

- Blacks should not fear capitalism. The capitalist marketplace will reward Blacks who are willing to compete.

- Blacks should be self-reliant. This is the only path to true freedom.

- White racism is not responsible for the major Black problems. Instead, these problems stem from fundamental failures in Black society.

- The poorer parts of Black America are sinking deeper into poverty. New, daring policies—such as a subminimum wage for the hard-to-employ—are needed to correct this situation.

- Public schools have failed Black parents and students. Vouchers, or money grants from the state, would allow parents to send their children to better private schools.

- Affirmative action or other special policies that promote Blacks and other minorities actually undermine Black progress by creating doubt about Black merit. They also create tensions between Blacks and other groups.

- Blacks deserve fair opportunities in college education but not special deals, such as preferential programs of admission and financial support.

- Welfare policies have given birth to generations of dependent Blacks, mostly in cities. The cycle of dependency must be broken.

- A return to traditional values of moral responsibility is necessary to turn the tide of family fragmentation, single-parent households, and teenage pregnancy.

Shelby Steele's best-selling *The Content of Our Character*, published in 1990, continued the Black conservative thrust. Steele (born 1946) argued that racism had declined, too many Blacks acted as victims, it was wrong to look for redress, and self-reliance was the answer.

This minority within a minority was much criticized. The term "Oreos"—Black on the outside, white on the inside—was used as one attack. Another was that Black conservatives were outsiders to Black communities. Others said their arguments against affirmative action were really off center. Plus, the Reagan-Bush administrations never translated any of their ideas into reality.

Black conservatives have caused changes in Black opinion. It is unlikely, though, that they will get credit for this shift. Thus, to many of them, Justice Clarence Thomas (born 1948) being seated on the Supreme Court in 1992 to fill Justice Thurgood Marshall's seat was their greatest tangible public triumph. That is, until the 104th Congress began. This Congress has two Black Republican members—Gary Franks from Connecticut and J. C. Watts Jr. from Oklahoma. Many other Blacks ran for Congress in 1994 on the Republican ticket. This has prompted Black conservative authors—such as talk-show host Armstrong Williams, writers for the conservative Black magazine *Destiny,* and economist Glenn Loury—to seize the time as they attempt to capitalize on this momentum.

Maybe these efforts are beginning to pay off. In the *New York Times* article "Under the Press of Social Ills, Some Blacks Are Turning to the GOP to Find Answers" (October 8, 1995), Karen DeWitt argued that there "was evidence during the 1994 campaign that blacks were developing an increasing interest in what has become a Republican message of conservative family values." She was reporting on the second annual National Leadership Conference in Washington, D.C., a gathering organized by *National Minority Politics* magazine, a nine-year-old Black conservative publication. Obviously some Black political loyalties are now in flux.

### What did the dazzling 1970s and 1980s Black women writers accomplish?

Until the 1970s, whenever Black literature was discussed, male writers such as Langston Hughes, Countee Cullen, Richard Wright, Ralph Ellison, James Baldwin, and Amiri Baraka were most often mentioned. Yet the writing of tremendously talented Black women like Zora Neale Hurston, Ann Petry, Paule Marshall, Gwendolyn Brooks, Kristin Hunter, and Lorraine Hansberry was clearly the equal of men's.

The 1970s got off to a fantastic start for Black women writers with two autobiographies, Maya Angelou's *I Know Why the Caged Bird Sings* (1970) and Nikki Giovanni's *Gemini* (1971). Angelou's was a tender recounting of her life, from the time when she arrived at age three to live with her grandmother in Stamps, Arkansas, to when she lived in San Francisco at sixteen and gave birth out of wedlock to her only child, a boy. Giovanni, at only twenty-eight, retold selected moments from her life in a memoir that addresses 1970s issues of Black history, politics, male-female relations, and art's purpose.

Both works were masterpieces, Angelou's because it was so lyrical and honest (telling of her rape by an uncle) and Giovanni's because of her engagement and audacious wordplay (coining such words as "huemanity").

As if these were not sumptuous enough offerings, there was Toni Morrison's compelling first novel *The Bluest Eye* (1970). Pecola Breedlove, its central character, is a frail eleven-year-old Black girl who thinks that if she only had blue eyes, she could escape loneliness and lovelessness. She wants white features. The community and her peers abuse Pecola, who is raped by her father and goes mad. By the novel's end, Morrison has forced the reader to look hard at the vicious scapegoating of a person. The novel foreshadowed Morrison's enormous literary talents (prose invention, uncommon characters, multilayered plots) and her intellectual toughness.

In the mid 1970s, new stars appeared: Alice Walker in her second novel *Meridian* (1976) and Ntozake Shange in the play *For Colored Girls Who Have Considered Suicide/When the Rainbow Is Enuf* (1976). Walker presented a fully realized female character in Meridian Hill, who rebels against stale family traditions, racism, and sexism, joins the civil rights movement, and seeks a higher spirituality. Ultimately, Meridian becomes her whole, true self.

Shange's play was an eye-opener. Seven women characters, each dressed in a color of the rainbow (plus brown), addressed directly the pain that Black women suffer at the hands of arrogant Black men and a mean world. Each woman recites a poem—or choreopoem—telling of her life. At the play's end, the Lady in Red asserts, "I found god in myself & I loved her . . . fiercely."

As the decade progressed, Black women writers were generating incredible excitement. Unexpected riches appeared, as in Gayle Jones's first novel *Corregidora* (1975), the story of a young blues singer, Ursa Corregidora, who expresses her despair through her singing and overcomes her pain. Toni Morrison's *Sula* (1975) and *Song of Solomon* (1977) were thunderclaps over the literary terrain. *Sula* continued Morrison's fascination with complex women characters, Sula being a woman who is open to feeling pain and pleasure and who gives both to others. *Song of Solomon* was the story of Macon Dead III (Milkman) as he moves beyond his comfortable family to find his past. The novel is a suspenseful journey through Black culture and history, and it put Morrison in the front rank of American writers.

The 1980s witnessed more major writers coming to the foreground. Toni Cade Bambara's *The Salt Eaters* (1980) shattered many storytelling rules by mingling past, present, and future and by having characters live close to the miraculous. Gloria Naylor's first novel *The Women of Brewster Place* (1982) made a great impact because of its lively retelling of seven women's stories and their life in a walled-up slum, Brewster Place. Naylor followed with *Linden Hills* (1985) and *Mama Day* (1988). Both novelists showed their women reflecting, laughing, and healing themselves.

No chronicle of Black women's writing would be complete without spotlighting Audre Lorde's strikingly feminist, lesbian, and human poetry; Lucille Clifton's poetic pursuit of memory and personal definitions; Alice Walker's highly popular *The Color Purple* (1982), a fervent hymn of appreciation for the woman as underdog; and Toni Morrison's *Beloved* (1987), a historical novel based on the true story of Margaret Garner, a runaway slave who killed her daughter rather than let her be returned to slavery by patrollers. For these works, Walker and Morrison won the Pulitzer Prize.

Black women packed a lot of accomplishment into two decades. At least twenty-five major novels were produced. They left a legacy of writing experimentation and entered new places in women's thinking and feeling. They made readers see women and men differently. They gave us a host of memorable characters—Lady in Red, Celie, Shug, Sethe, Beloved, Baby Suggs,

Mattie, Willie Mason, Minnie Ransom. And for many, they restored the pleasures of reading.

## Increasingly, Blacks say, "We do Kwanza." What does that mean?

Kwanza is an African-American holiday that is celebrated for seven days, from December 26 to January 1.

In 1966, a small assembly of people gathered in Los Angeles to hold the first Kwanza celebration to honor Black culture around the globe and uphold Black togetherness and the Black family. "Kwanza" means "the first fruits of harvest" in Swahili, the lingua franca of East Africa. Many African villagers celebrate the first reaping from the harvest: nature is praised and the gods are thanked; singing and dancing honor the gods' universe. Kwanza broadened this fundamental Africanism.

Today, Kwanza is celebrated by perhaps as many as six million Black Americans. Kwanza's spread has transformed it into a historical event. But the loyal adherence of tiny 1970s communities made this possible.

Cultural nationalist Dr. Maulana Karenga was the originator of Kwanza. A native of southern California and now chairman of Black studies at California State University, Long Beach, Karenga was during the 1960s a serious student of African languages, rapidly achieving fluency in both Swahili and Zulu. His controversial 1960s political organization, Unified Slaves (US), was based on African ideas.

Nguzo Saba—or the "seven principles" to live by—establish the overall framework for Kwanza. Each day, a different principle is honored. The seven principles, in Swahili and English, are Umoja (unity), Kujichagulia (self-determination), Ujima (collective work and responsibility), Ujamaa (cooperation), Nia (purpose), Kuumba (creativity), and Imani (faith). In small family celebrations, a family member might lead a discussion of the principles, whereas in large community celebrations, speakers, plays, and dancing are often used to illustrate them.

Seven symbols are used in the ceremony: a straw place mat *(mkeka)*, symbolizing the history of Black Americans; a seven-branched candleholder *(kinara),* representing Africa and the African ancestors; seven candles *(mishumaa saba),* three green, one black, and three red; fruits and vegetables *(mazao),* representing the fruits of collective work; ears of corn *(muhindi),* symbolizing Black children; a communal unity cup *(kikombe cha umoja),* to be passed among celebrants; and gifts *(zawadi),* shared on the seventh day (January 1) as a reward for principles upheld.

Kwanza filled a collective need among Black Americans. In the past, no general holiday allowed Blacks to recognize their strength and triumphs as a people. No holiday existed where Blacks could rededicate themselves to their ideals for the future. Blacks needed a moment, a psychological place where stock-taking and rededication could enter. Including African culture made the holiday even more appealing, especially in the 1970s climate of American Pan-Africanism.

The holiday is not a substitute for Christmas, and often people celebrate both. However, Kwanza's original attractiveness was that it was an alternative to the Christmas season, regarded by some Blacks as a European holiday and as far too commercialized; commercialization had robbed it of positive content, especially as a way of teaching values to children.

In recent years, Kwanza and pre-Kwanza festivities have been present in every major American city. Increasingly, markets and fairs precede the week of celebration—featuring handicrafts, hair braiding, Black history quizzes, food, children's book sales, artwork, dance, musical performances, workshops in capoeira (a form of martial arts developed by Brazilian slaves), tie-dying, and food preparation. Most popular are the storyteller performances for children. Libraries and museums now hold Kwanza days. Kwanza information telephone lines exist in many cities, as do Kwanza clearinghouses.

In the midst of all its success, Kwanza is threatened now by too much gift giving, marketing, and trendiness. Years ago, the items for the Kwanza celebration were often gathered by the children in a family. Today, "Kwanza kits" are carried by supermarket chains.

Kwanza has changed the Black American calendar and infused a sense of high purpose as each year closes and a new one begins. Its success is deserved.

**What is the "Kente phenomenon"?**

These days, a colorful fabric has been turning heads as a new element in Black American fashion. One sees it on the street, in offices, at college commencements, even at black-tie functions. The members of Sweet Honey in the Rock, a Black women's singing group, have bedecked themselves in hats and scarves made of Kente. Marion Barry, in his recent comeback, rarely appeared before Blacks without Kente accessories—a tie, a cap, a suit with a lapel border. In mid-September, the O. J. Simpson defense team wore ties with Kente designs—"for demonstrating our team unity," they said.

Kente is a striking West African golden-colored cloth into which are woven complex designs and stripes of red, blue, black, and sometimes green. Real Kente shines, drawing and reflecting light. Kente has been so eye-catching that several national publications—*Emerge, Essence,* the *New York Times,* the *Wall Street Journal*—have had to explain to their readers what the bright-colored cloth was all about.

For centuries, Kente cloth has been woven in Ghana, originating from the Ashanti region, homeland of the Asante people. Originally, Kente was made for Asante leaders to wear at festive occasions. It was reserved for the elites. As times changed, Kente began to be worn by all types of people, even though it was expensive. Men wore it in togalike robes. Women wore it in three smaller cloths (as a long wraparound skirt, a covering for the upper body, and a shoulder cloth).

Strip weaving is the technique used to make Kente. The cloth is first woven into strips four and a half inches wide by 144 inches long; then the strips are sewn together to make a broad cloth. No training school creates professional Kente weavers. The art of weaving Kente is passed down along family lines. The weavers build their own looms and other equipment, or inherit them from their uncle or father, and they do all the work on the cloth from beginning to end. Generally, the cloth is produced in households and backyards.

Many designs make up the panoply of Kente. Designs are given names, and legends are attached to them. The design called *Adjwini-Asa* means "the end of design, the ultimate design." According to legend, the weaver was asked by a king to weave the best Kente design ever. The king's cloth had to be unique at any gathering. *Kyemia* was a design for the most prominent royal women.

Kente made its first major American public appearance when independent Ghana's leaders appeared in the 1960s at New York's United Nations, in Harlem, and at Black gatherings wearing their Kente togas. Black Americans visiting West Africa brought back swatches of the fabric.

By the late 1960s, the cloth was stylish among a few Black Americans in the form of a stole or scarf worn around the neck over a dress or a coat. Simultaneously, it came to symbolize a renewed pride in African heritage. The many urban African novelty shops opened in the 1970s further spread the fame of Kente. African street merchants, such as those on 125th Street in Harlem, did the rest.

During the 1980s, Kente was accepted in the highest reaches of Black American society. Black celebrities wore it. Pastors wore it over their Sunday robes. Kente cummerbunds and bow-ties appeared at formal parties.

Today, Kente has changed a lot to meet the large demand. To many Kente traditionalists, these changes degrade the high standards of the cloth. Now Kente is made out of synthetic fabrics and is machine-woven. Even worse, from the traditionalists' viewpoint, are products like Kente baseball caps, duffel bags, shirts, jeans, socks, curtains, and bedspreads. A rumor about a BMW with Kente upholstery has gained currency.

Corporations have been quick to choose Kente patterns in order to target Black consumers, using it on book covers, greeting cards, dolls' clothes, bank checks. Coca-Cola has recently launched ads with its logo against a Kente background. Even Ralph Lauren, not a friend of ethnic designs, has run ads featuring Kente (a young Black woman, with braids, dressed in Ralph Lauren plaids, holds a Kente-covered book—a contradictory set of images). And JCPenney, Macy's, and Bloomingdale's have entered the market.

Has the dramatic spread of Kente designs cheapened the glorious history of the cloth? As Kente has met the Godzilla of American capitalism, it has been torn from its roots. This was inevitable and sad. The positive side is that many thousands of Americans have shared in a brilliant artifact from West Africa.

### Did Black America develop into "two nations" in the 1970s and 1980s?

If Black America had been a nation-state in the 1970s and 1980s, it would have placed well on the World Bank's chart of developing small countries. By any measure, Black America's economic progress since the 1960s was great. Overall, the numbers looked good.

During these two decades, total Black income grew. Black America was a market of billions of dollars. The top one hundred Black businesses, as identified by *Black Enterprise* magazine, became firmly established as midsize enterprises with total sales reaching nearly eight billion dollars as the 1980s ended. Black youth were entering the nation's colleges in record numbers, some of them passing through the portals of the most prestigious universities. Blacks worked at the largest variety of jobs—in business, media, technology, science, medicine, law, public policy, factories, offices, and homes—that they had ever held.

Yet also during the 1980s, a serious division in the internal structure of Black America was emerging. There was increasing class stratification.

Upper-middle-class, middle-class, and working-class Blacks grew in numbers and in economic clout. They were moving ahead (although sometimes slowly) toward prosperity. But many other Blacks began to fall behind, and some got lost altogether. Black America seemed to be splitting into two nations, the haves and have-nots.

The Black middle class had been particularly successful in the 1960s and 1970s. Thousands of households did well in these years. Federal laws were ensuring access to jobs, and the economy was expanding. From 1960 to 1970, the Black middle class doubled in size. Part of this success came from barrier-breaking educated women, who went after new job opportunities aggressively. In 1976, just after the recession, surveys showed that prosperous Blacks were still confident about their future.

In the 1980s, Black middle-class progress slowed. Job opportunities were curtailed as affirmative action hiring and contracts were undermined in the face of a severe recession. Even with the downturn, however, one in three Black middle-class families earned more than the average white family.

You would have thought this Black middle-class success would have been plastered across the front page of every newspaper! What is America about if not economic success? However, by 1985, most attention was being given to the Black "underclass." William Julius Wilson, a University of Chicago sociology professor, was the proposer of this term. Wilson had discovered that Black poverty had increased dramatically in the nation's fifty largest cities between 1970 and 1980.

High-poverty Black areas had many strikes against them: high unemployment rates; long-term joblessness; few middle-class and working-class people living there; many single-parent households, usually headed by a woman; and high crime rates. Most disturbing, children were greatly at risk. In sum, these were socially isolated communities, where Blacks were cut off from the information, ideas, jobs, and schools that advanced the rest of American society.

The underclass concept was barely off the drawing board before its critics came forth with some good arguments. It was a simplification to look at Black society as two large economic blocs—the middle class plus the skilled working class versus the underclass. It was also a distortion to talk about the underclass without looking behind the label to see deeper problems. Many inner-city poor were women with children.

Julianne Malveaux, an economist and social commentator, doubted the whole idea of the underclass. "The concept of the 'underclass' has moved," she wrote, "from nebulous to factual without a hint of data, from a whisper to a shout without a definition." But she did agree that Blacks were experiencing a new deep poverty. The bottom-line issue was, what could be done about this class before it became a permanent drag on Black American progress?

The 1980s did not resolve the increasing class stratification in Black America. In the future, there would be two Black economic trajectories. This was something new for a people used to struggling as one mighty force against economic injustice. How these two entities could communicate with each other was to be a constant theme of the early 1990s.

### What is the meaning of Afrocentrism?

In a way, Afrocentrism has made it: it appears in the latest edition of *Webster's College Dictionary.* "Afrocentric" means "centered on Africa or on African-derived cultures such as those of Brazil, Cuba, and Haiti." This is right, as far as it goes.

Afrocentrism as a word and a philosophy appeared in the 1970s, fertilized by the 1960s Black consciousness movement. The idea was created for Blacks and by Blacks. It did not take long for it to spread among some Black thinkers. Considerably later, the word made its crossover debut in the national media. The first national newspaper articles appeared, in the *Washington Post* and the *Los Angeles Times,* on Afrocentric-oriented programs, such as a drug treatment project.

The main contention of Afrocentricism is that African culture and traditions should be at the center of a Black person's growth. Blacks need to know African history and culture to understand who they are today. According to the concepts, Blacks scattered all over the globe—including American Blacks—still belong to Africa. The Afrocentric thinker sees the continent's flavor in Black language, logic, in personal psychology, spiritual ideas, and notions of time and space.

Afrocentrism is a bookish movement. Several books are its classics: W. E. B. Du Bois's *The Souls of Black Folk* (1903), George G. M. James's *Stolen Legacy* (reissued in 1976), Carter G. Woodson's *Mis-education of the Negro* (1933), Cheikh Anta Diop's *African Origin of Civilization* (1974), Ivan Van Sertima's *They Came Before Columbus* (1976), and Chancellor Williams's *The Destruction of Black Civilization* (1974). These books have one trait in common: they tell of a Black vision of culture.

The movement's recent thinkers are psychologists Na'im Akbar, Wade Nobles, and educator Jawanza Kunjufu, whose books on educating Black boys have gone through many editions.

To this list must be added Molefi Kete-Asante (born 1942), now a professor in Temple University's African-American Studies Department, who is a remarkably energetic exponent of Afrocentric ideas. Since 1980, he has written more than ten books for the movement, which show the differences between Black and European ways of expression or thinking.

Dressed in West African garb, Kete-Asante has become a popular lecturer. One of his favorite stories concerns how, during a talk before Black college students, he asked them to name various people who made up America. Easily they named the Germans, French, Swedes, Italians, and the English. But no one could name a single African group, such as the Yoruba, Wolof, or Ewe, that had peopled the country—proof that today's educators had not taught them about their heritage.

For this reason, promoters of Afrocentrism want to topple the Eurocentric worldview from its pedestal. They say that Black Americans are bombarded daily with images—from television, movies, writing—coming from European or Euro-American traditions. Black youth are especially vulnerable to this barrage at a time when they are forming their self-images. A world of white images sends them a clear negative message: whatever you are, it will not be acknowledged, and if acknowledged, certainly not honored.

Naturally, Afrocentrism's advocates target schools as the places that must change. They want many subjects taught to young Blacks: ancient Egyptian history; West African kingdom history; African ideas of family and religion; African-American stories; Black scientific and technological achievements; and Black wisdom. Training students in this knowledge is supposed to light a fire under them.

In fact, many Black independent schools (such as Los Angeles's Marcus Garvey School) already use Black heritage to get students interested in traditional subjects. To teach math, they tell of Black contributions to mathematics. As one teacher put it, "we explain that math is in their genes, in their DNA." To teach reading, they use stories with Black contemporary names. The newly fashionable and very controversial all-male Black schools, as in Milwaukee, are doing the same.

Critics are passionate that Afrocentricism is filled with racial illusions because it exaggerates Black achievements and separates Blacks from whites. They say the Afrocentric thinkers are "new pharaohs," tying Blacks to outdated notions of culture. Afrocentrics argue that their mission is to lift up Black culture for all to share. Beyond the arguments lies the reality that several Afrocentric schools are producing smart kids with mastery in difficult subjects, something public schools have often failed to do. Afrocentrism is here to stay. But its critics are durable too. In a recent massive essay entitled "Understanding Afrocentrism: Why Blacks Dream of a World Without Whites" (*Civilization,* July–August 1995), Gerald Early concludes, "Today, Afrocentrism is not a mature political movement but rather a cultural style and moral stance."

### What was the impact of *The Cosby Show*?

"Tonto" was the name a few cynics fixed on Bill Cosby in his costarring role in NBC-TV's *I Spy* in 1965, a series about two undercover agents posing as a tennis champ and his trainer. Just as Tonto had been for the Lone Ranger, Cosby was seen as Robert Culp's sidekick, almost a silent partner. It was predicted that years in the future, Trivial Pursuit buffs would strain their memories to remember him.

Cosby certainly made the cynics into buffoons. He quickly established his dramatic talent and used it as a springboard to a 1970s career of television shows (*The Bill Cosby Show,* 1969–71, *Fat Albert and the Cosby Kids,* 1972–77), comedy records, and stand-up comedy appearances. By the 1980s, Cosby had finely honed a broad repertoire of characters, imitations, storytelling, and a way of inserting wisdom in comedy. All of this came to fruition in the 1980s *The Cosby Show.*

Appropriately, *The Cosby Show* started in 1984, the only year in the 1980s when the Black middle class grew. It centered almost exclusively on an upper-middle-class Black family's life. In itself, an upscale Black family was not new TV territory: *The Jeffersons* (1975–85), with its bombastic father, was already there.

What distinguished *The Cosby Show* was the high quality of its scripts: they dealt sensitively with family issues and human behavior rather than just milking situations for laughs. A lot of instruction, coated with light humor, was transmitted by the best episodes. Television audiences immediately bonded with the Huxtables, defying media prophets' predictions that viewers were tired of Black family sit-coms and would watch only shows where Black and white characters were mixed, such as *Diff'rent Strokes.*

Americans of all races were able to see a new class of Blacks depicted on television. The seven-person Huxtable household, presided over by obstetrician Cliff Huxtable (played by Bill Cosby) and his attorney wife Claire (played by Phylicia Rashad), offered a collective education for the country. Television stereotypes of Blacks from the 1970s—stock humor, drab parent characters, dysfunctional family members—were gone. The 1970s *Good Times* stereotypes had swamped the excellent Esther Rolle and John Amos.

This family was nested in a well-furnished New York City brownstone. Cliff and Claire were modern parents: knowledgeable, clever, and forward-looking. The oldest daughter was in her sophomore year at Princeton. Each family member was on his or her way to success. Suddenly, America faced a Black family every bit the equal of any American family—and not only that but a family whose dilemmas and questions could be their own.

Parenting was the great theme of *The Cosby Show.* Cliff could be opinionated and obstinate, but he often changed his mind and admitted mistakes. He was more of a diplomat, a negotiator, a reasoner. Claire was not a second-string parent but was often instrumental in resolving family issues, and she did not shy away from criticizing—albeit fairly—her husband.

The nation was able to watch the Huxtable children grow under Cliff and Claire's unique style of caring. In one episode viewers could see Cliff closely counseling his son on dating or romance and in the next episode dealing with his daughter leaving the nest for college. Claire disciplined her daughters' suitors, reminded her son how preparation paid off, and helped her youngest to not feel overshadowed by her older siblings.

Critics pounded at Cosby's creation. Cliff was too noble a Black. It was a Black *Father Knows Best.* The Huxtables were not representative of Black people: how many lived in such a fancy house and wore such stylish clothes? Cliff was still a patriarch. Even in the 1990s, the series is still controversial.

Critics failed to grasp the tremendous change *The Cosby Show* ushered in: it made obsolete latter-day minstrelsy and all those 1970s "Mammy" figures. What is more, nearly every week offered "Black Culture Appreciation"—the scripts showcased entertainment celebrities, Black colleges, Black celebrations, and Black artists (such as Varnette Honeywood).

*The Cosby Show*'s success silenced the doubters of Black ingenuity in TV-land.

### What are some landmarks in the history of rap music and hip-hop culture?

First, it is necessary to point out that rap music is only one facet of hip-hop culture. The four facets of the original hip-hop culture are rapping, DJing, break dancing, and graffiti writing.

Rapping is the verbal component of rap music, and a rapper is called an MC. Hence, another term for rapping is MCing. MCing uses the conventions of the Black American colloquial game "the dozens," verbal contests in which boasting and sarcasm figure heavily. It is also indebted to the 1960s art of toasting, entertaining games of one-upmanship, and streetwise showmanship. DJing is the art of using records, two turntables, and a mixer to create a dynamic sound backing for rappers and the essential beat for dancers.

Rap and hip-hop culture as we know it really began in the mid 1970s in the South Bronx in New York City. From that point, the major events in its growth were these:

- Kool DJ Herc (Clive Campbell), inspired by the huge sound systems of Caribbean carnivals and street festivals, creates his own mammoth speaker setup and hooks it up to power lines throughout city parks and school yards in the Bronx. His record collection—containing Blaxploitation soundtracks, classic soul grooves, Caribbean dance hits (both reggae and calypso), and novelty records—form the sound foundation of hip-hop.

- In the 1970s, Afrika Bambaata from the Bronx takes his voluminous musical curiosity and percolating vision of a vibrant flowering of African-American nationalist youth and founds the Zulu Nation in the Bronx River Community Center. From its ranks come the pioneers of break dance, the Rock Steady Crew, many of the greatest graffiti writers of the 1980s, and Bambaata's sonically groundbreaking unit, the Soul Sonic Force.

- Grandmaster Flash (Joseph Saddler), a teen obsessed by records and electronics, begins to experiment with technology and invents the concept of scratching, the abrasive sonic squall that turns the turntable into an instrument by forcing repetitions of the same beats and producing sonic static.

- In 1979 Sugar Hill Records in New Jersey releases "Rapper's Delight." This rap novelty record extolling the exploits of Wonder Mike, Big Bank Hank, and Master Gee goes to number one on the pop singles charts. It takes rap out of the "'hood" and to worldwide prominence.

  The musical track beneath it is the disco group Chic's "Good Times." The uncredited use of the tune presages all of today's battles over sampling—the use of a primary recording of someone else's as a backing track for a brand-new record with or without the original artist's consent. The general public thinks of rap as a novelty and not as real music—a popular sentiment that it will take almost another decade to eradicate.

- As rap artistry continues to evolve underground, the pop group Blondie releases "Rapture" in 1980, a tribute to rap and the first popular acknowledgment of the form by white musicians.

- In 1982, "Planet Rock" by Afrika Bambaata and the Soul Sonic Force and "The Message" by Grandmaster Flash and the Furious Five serve as the sonic blueprint of hip-hop and the political blueprint of the hip-hop nation. "The Message" remains one of the greatest singles of the decade, one of the greatest raps ever committed to vinyl, and, most of all, a harrowing documentation of inner-city life that exhausts all the basic pitfalls of the modern ghetto existence.

- In 1983 Herbie Hancock, the jazz pioneer, mates hip-hop and jazz with his record "Rockit."

- On their 1989 masterpiece album *Bring the Noise,* Public Enemy declares, "Run DMC first said a DJ could be a band/stand on your own feet/do the watusi."

- LL Cool J in 1985 releases "Radio" on Def Jam records, a new label started by Russel Simmons, brother of Run in Run DMC, and New York University student Rick Rubin.

- Run DMC continues its string of hits, and meanwhile three white kids from New York, formerly a hard-core band, the Beastie Boys, release *Licensed to Ill* on Def Jam. It sells six million copies and becomes the best-selling rap record to that time.

- Run DMC collaborates with hard-rock superstars Aerosmith, cementing the relationship between heavy metal and rap.

- Run DMC plays to a sold-out Madison Square Garden audience in 1986.

- Public Enemy's "Fight the Power" is used as the opening theme music to Spike Lee's controversial movie *Do the Right Thing.* The song contains the line "Elvis was a hero to most but he never meant shit to me."

- N.W.A. (Niggaz wit' Attitude) releases *Straight Outta Compton.* The era of "gangsta rap" has officially arrived. N.W.A., Ice Cube, Easy E, Doctor Dre, and M. C. Ren take the zeitgeist of rap with them to their home of Compton, California. Queen Latifah, M. C. Lyte, and Salt and Pepa release groundbreaking female raps. Ice Cube, having left N.W.A., releases *Amerikkka's Most Wanted,* the "gangsta rap classic." He follows it up with *Death Certificate.*

- April 30, 1992—The Los Angeles riots, the largest peacetime civil disturbance of this century, break out after the four Los Angeles Police Department officers who beat Rodney King are declared not guilty. Fifty-two people die. Rap music is credited with prophesying the "fire next time."

- *Fear of the Black Hat,* a film parodying rap and rappers, comes out in the summer of 1994, reflecting that rap is so established that its rituals can be satirized. More important, C. Delores Tucker—head of the National Political Congress of Black Women—has led a concerted campaign against the violent, sexist lyrics of gangsta rap and has managed to make it a major public issue.

## RESOURCES

Banner-Haley, Charles T. *The Fruits of Integration: Black Middle-Class Ideology and Culture, 1960–1990.* Jackson: University of Mississippi Press, 1994.

Barker, Lucius, and Mack H. Jones. *African-Americans and the American Political System.* 3rd ed. Englewood Cliffs, N.J.: Prentice-Hall, 1994.

Billingsley, Andrew, Ph.D. *Climbing Jacob's Ladder: The Enduring Legacy of African-American Families.* New York: Simon & Schuster, 1991.

Bogle, Donald. *Blacks in American Films and Television: An Encyclopedia.* New York: Simon & Schuster, 1989.

Bogle, Donald. *Toms, Coons, Mulattoes, Mammies, and Bucks: An Interpretive History of Blacks in American Films.* New York: Continuum, 1993.

Butler, John Sibley. *Entrepreneurship and Self-Help Among Black Americans: A Reconsideration of Race and Economics.* Albany: State University of New York Press, 1991.

Diggs, Anita. *The African-American Resource Guide.* New York: Barricade Books, 1994.

George, Nelson. *Buppies, B-Boys, Baps, and Bohos: Notes on Post-Soul Black Culture.* New York: HarperCollins, 1992.

Giddings, Paula. *When and Where I Enter: The Impact of Black Women on Race and Sex in America.* New York: William Morrow, 1984.

Hager, Steven. *Hip-Hop: The Illustrated History of Break-dancing, Rap Music, and Graffiti.* New York: St. Martin's Press, 1984.

Hull, Gloria T., Patricia Bell Scott, and Barbara Smith. *All the Women Are White, All the Blacks Are Men But Some of Us Are Brave: Black Women's Studies.* New York: Feminist Press, 1982.

Landry, Bart. *The New Black Middle Class.* Berkeley: University of California Press, 1987.

Majors, Richard, and Janet Mancini Billson. *Cool Pose: The Dilemmas of Black Manhood in America.* New York: Lexington Books, 1992.

Omi, Michael, and Howard Winant. *Racial Formation in the United States: From the 1960s to the 1990s.* 2nd ed. New York: Routledge, 1994.

Rose, Tricia. *Black Noise, Rap Music, and Black Culture in Contemporary America.* Middletown: Wesleyan Press.

Wallace, Michele, and Gina Dent. *Black Popular Culture.* Seattle, Wash.: Bay Press, 1992.

Wilson, William Julius. *The Truly Disadvantaged: The Inner City, the Underclass, and Public Policy.* Chicago: University of Chicago Press, 1987.

A TEXT FOR YOUTH

Kennedy, Pagan. *Platforms: A Microwaved Cultural Chronicle of the 1970s.* New York: St. Martin's Press, 1994. Texts with illustrations.

# 10

# Preparing for the Twenty-first Century

## The 1990s

1990: George Washington Carver (plant scientist) and Percy Julian (chemist noted for synthesis of cortisone) are inducted into the National Inventors Hall of Fame

1990: L. Douglas Wilder is inaugurated as Virginia's governor

1990: Oprah Winfrey begins building Harpo Productions, the media production company she purchased

1990: George Bush is cited by Black respondents to a poll as a highly successful president

1990: Shelby Steele publishes the best-selling *The Content of Our Character*

1990: Nelson Mandela, accompanied by his activist wife Winnie, begins a speaking tour of the country, is given a ticker-tape parade in New York City, praises Crispus Attucks in Boston

1991: Dr. Mae C. Jemison journeys in space shuttle *Endeavor* mission

1991: Clarence Thomas is nominated by President George Bush to the Supreme Court; his confirmation hearings are a charged affair

1991–92: Discovery of skeletons from New York City eighteenth-century "African burial ground"; dispute over their fate rages

1992: Four Los Angeles police officers who were videotaped while beating Rodney King are acquitted; the decision sparks the biggest racial eruption in U.S. history

1992: Blacks give William "Bill" Clinton crucial votes in critical states, helping him become president

1992: Congressional Black Caucus increases its membership to a record of forty, including Senator Carol Mosley Braun from Illinois

1992: Spike Lee's *Malcolm X* hits theaters

1993: Clinton's inauguration features Maya Angelou and her poem "On the Pulse of the Morning"

1993: New Clinton cabinet includes five Blacks—Hazel O'Leary (energy), Michael Espy (agriculture), Ronald Brown (commerce), Jesse Brown (veteran affairs), and Joycelyn Elders (surgeon general)

1993: Arthur Ashe dies; his *Days of Grace* is published posthumously, reminding the country of its great loss

1993: *Waiting to Exhale*, Terry McMillan's novel, appears on the best-seller lists, along with Toni Morrison's *Playing in the Dark* and Alice Walker's *Possessing the Secret of Joy*

1993: Cornel West's *Race Matters* brings new insight to an old issue

1993: Toni Morrison wins the Nobel Prize for literature

1993: The Delany Sisters publish *Having Our Say,* sharing their wisdom from more than one hundred years of living

1994: O. J. Simpson, football and advertising celebrity, takes his famous ride on a Los Angeles freeway, drawing a television audience of ninety-three million viewers; his pretrial hearing and trial, for the murder of Nicole Brown Simpson and Ronald Goldman, become a national obsession

1995: Death of Rapper Eric "Easy-E" Wright (1964–95) from AIDS sends a message to youth

1995: Supreme Court decisions on affirmative action suggest a paring of the programs

1995: Henry Louis Gates Jr. hits his stride as *New Yorker* contributing editor

1995: Madeleine Kunin, Department of Education official, notes the test scores for young Blacks have improved nationally

1995: Wynton Marsalis is credited with restoration of jazz music

1995: The posthumous release of Marlon Riggs's film "Black Is . . . Black Ain't" brings applause

1995: Dr. Henry Foster fails to become U.S. surgeon general; his advocacy on teenage pregnancy is spotlighted

1995: Black unemployment falls beneath the usual double-digit percent

1995: Honoring the United Nations' fiftieth anniversary, Maya Angelou delivers a poem—"A Brave and Startling Truth"—at San Francisco ceremony

1995: Dorothy West, the last survivor of the Harlem Renaissance, publishes *The Wedding*

1995: President Clinton honors Johnson Chesnut Whittaker, a Black West Point cadet who was driven from the school by bigoted classmates in 1880. His descendants receive his Bible and gold-plated bars, making him a lieutenant

1995: O. J. Simpson is acquitted on October 3rd in Los Angeles's "Trial of the Century" after less than four hours of jury deliberation

1995: Louis Farrakhan, head of the Nation of Islam, plans for "Million Man March" on October 16, 1995, amid controversy and criticism

1995: Sentencing Commission of Washington, D.C., issues report showing one in three Black males from twenty to thirty is in jail or under supervision of the judicial system and one-half of United States prison population is Black

1995: A new word enters the nation's vocabulary— "Powellmania"—as Gen. Colin Powell's book tour for *An American Journey* generates a huge reception. Speculation arises as to whether he plans presidential bid

On Monday night from 10 P.M. to midnight, the switchboard at San Francisco's KMEL-FM is all lit up. If the switchboard were a parking lot, it would be jammed.

Young people are calling "Street Soldiers" with their questions and comments. A boy asks, "How can I study and still be 'down with my homies'?" A girl asks, "My boyfriend keeps pressuring me to commit myself to him; but I am too young, I want to stay free. . . . You know how I mean." Another girl says, "I am only sixteen years old now, and I eventually want to get married and have kids, but it seems like by the time I'm old enough, there ain't gonna be any Black men left. All this killing for nothing . . . "

Thousands of Bay Area young people—particularly Blacks—tune in to Joe Marshall and Margaret Norris's award-winning show (with DJ Kevin Nash) and their special brand of advice. They do not preach at the young but get them to think critically about the choices and decisions they make. They talk calmly and carefully to callers.

Marshall, Norris, and Nash's program is just one sign that Black children have moved to the center of Black America's concerns. They are the theme of the Black 1990s. Every day, Black Americans confront statistics demonstrating that far too many Black children live in jeopardy—are poorly fed, badly educated, live in violent conditions, and are often headed for a life of more difficulties. These figures have set off an alarm across Black America. Today, Blacks are working like never before to save their children—the only way to have a future. In fact, the 1994 Congressional Black Caucus's annual conference was held under the banner "Embracing Our Youth for a New Tomorrow."

In 1995, Black Americans were 12.6 percent of Americans. (The percentage is expected to rise to 15.7 by 2050.) Black America in the 1990s comprises a tremendously varied population in opinions, cultures, skills, intelligence, status, geography, and possibilities. Blacks of the 1990s are participating in American and world society in ways that could hardly have been imagined even a decade ago.

Yet this progress is haunted by the sense that Blacks—particularly the eleven million Black young—are also facing a stringent testing. John Hope Franklin, the dean of Black historians, says that not since Reconstruction have Blacks faced such challenges. To secure a solid base in the twenty-first century, Blacks know that these problems' solutions must be found during this decade. This chapter's questions deal with Black America as it heads for the millennium.

### What do people mean when they say that young Black males are "an endangered species"?

"Endangered species" was a catchy concept whose currency peaked in the early nineties. The phrase was often used as a shorthand for Black males' prospects in today's America. So popular was this term that nearly every major newspaper and magazine invoked it, and Black commentators far and wide wrote about it. Ice Cube, the rap performer, created a song around the idea. Los Angeles's Theatre of the Arts presented a sketch using it.

At first, the phrase "endangered species" was part of a serious analysis and had the positive effect of focusing attention on the special problems that Black males face. But once it became overused and new, unintended, mostly derogatory meanings had been given to it, a Black backlash set in. The phrase went from being considered helpful to being denounced as hype.

It all began in 1988 when clinical psychologist Jewelle Taylor Gibbs at the University of California wrote a wide-ranging book on Black male problems, called *Young, Black, and Male in America: An Endangered Species.* Her argument was straightforward: Black males were being undermined by many negative conditions. If this continued, they would find it hard to survive. Thus arose the idea of a special group that was endangered.

Bleak statistics supported her contention. The unemployment rate among Black males had tripled since 1960. The leading cause of death for young Black males was homicide. Suicide was the third leading cause of death, and the rate was almost twice as high as for young white males. Black males made up one-third of all juvenile arrests. Police arrested Black males at a rate that was twice as high as for young white males for the same offenses.

Appearing in 1991 before a Senate committee, Gibbs said, "The Black male is threatened with physical, psychological, and social annihilation." Explaining this problem in another way, Richard Majors and Janet Billson wrote, "African-American men have defined manhood in terms familiar to white men: breadwinner, provider, procreator, protector. But they have not had . . . the same means to fulfill their dreams."

Good came initially out of the "endangered species" argument. New initiatives were tried. Scholarship funds were extablished by churches, clubs, and individuals focused on young Black males. Nearly every major Black community developed a program where Black male concerns were a priority. Elizabeth Dole, U.S. secretary of labor, made job training take account of Black male issues. Detroit opened three educational academies to benefit mainly Black males—an option criticized as too limiting and resegregating. (A judge has ruled that these schools had to include girls.) Dallas instituted special school projects for building character.

The tide turned against the "endangered species" phrase when Blacks saw it as stigmatizing. Many questions were raised. Didn't it suggest that Black males were like animals in game parks, besieged by poachers and hunters? Wasn't it hinting that Black males were a species different from the rest of humans? Was it suggesting that Black men could slam-dunk basketballs but not score high on tests? Did it make people see Black young men as lacking a future? One mother pleaded, "When you see me walking with my Black boys, please don't shake your head and say, 'It's so hard raising young Black boys during these times.'" Did "endangered species" suggest that Black men were dependent on other forces to save them?

By the mid 1990s, the concept was not much used. However, it had made its contribution by highlighting a coming crisis. Now the emphasis had slightly shifted. As writer Pearl Cleage said, "We have to claim our children, understand what racism is and that we have been declared an endangered species and fight back."

**How did four Black women, living on a novel's pages, make Black history?**

When the stories of Savannah, Bernadine, Robin, and Gloria were told by Terry McMillan in *Waiting to Exhale* (1992), Black history was being made.

McMillan's novel appeared just in time for summer reading, zoomed up the best-seller list, and stayed there for more than twenty weeks. After six weeks, her publisher sold the book's paperback rights for $2.64 million. Terry McMillan had arrived. Also arriving on the best-seller lists that summer of 1992 were two other Black women writers: Toni Morrison, for *Jazz* and *Playing in the Dark*, and Alice Walker, for *Possessing the Secret of Joy*.

McMillan's book sold well all over but especially in Black bookstores, where huge stacks of the novel covered tables and signs in the windows yelled, "It's here—*Waiting to Exhale.*" Soon, it was known simply as "*WTE.*" The book's distinctive cover picturing four Black women, dressed in primary colors and spectacularly chapeaued, could be spotted from a distance, while being read in airport and shopping malls. Her success broke open the book market for Black books and Black authors. Publishers finally were made to understand: Blacks read books, Blacks will buy books, but Blacks want to read books where they count as real.

The novel and McMillan's book signings became communal events in cities across the country. McMillan would appear in casual attire, wearing her signature long earrings and a distinctive cap. Her audiences were mainly women, mostly Black, but many Black men also showed up with their worn copies of *WTE.* A Terry McMillan reading was a cross between a positive group therapy session and a celebration of women's survival, occasionally interrupted by the audiences' shouted responses: "Tell it, girl!," "It's the truth," "You are talking to me."

*Waiting to Exhale* was a superbly told, deeply felt examination of Black women and men's relationships. Its frank dialogue was that of actual people, its encounters typical, its heartaches endured by real folks, and its victories those everyone wants. For women, the book was like having a conversation with their girlfriends. For men, it was a chance to eavesdrop on women's thoughts. Among Black men and women, the novel became a reason to have another round of discussion on what had gone wrong in their relations.

McMillan had been working in the literary field for a long time before her breakthrough. In 1987, she began her engagement with Black women characters in *Mama,* whose main character, Mildred Peacock, is a mother of five who, during the 1960s and 1970s, is determined to keep her family together. McMillan considers this her favorite book. The novel is the inside view of a Black family. *Disappearing Acts,* in 1989, an urban love story, also explored Black women's lives. Her anthology *Breaking Ice* collected contemporary Black writing.

*Waiting to Exhale* slightly shifted her interests: to Black women who are contemporaries, the West, spicier talk, and volatile relations between men and women. Each of her four women is trying to find that one man who is reliable, caring, and working hard for the future. When they do, they believe they can relax, feel secure, "exhale."

Terry McMillan has proved that popular Black literature can be a potent force, providing a vivid vision of personal issues that Black Americans are eager to address.

## Why was Nelson Mandela's 1990 American visit and 1994 presidential inauguration greeted so enthusiastically by Black America?

In June of 1990, Nelson Rolihlahla Mandela (born 1918) made an eleven-day, eight-city tour of America that turned into a very important public event for people across the political and racial spectrum. From 1944, Mandela had been active in the opposition to South Africa's apartheid policy. (The word *apartheid* means "separation" in Afrikaans—a major language in South Africa.) His valiant fight was being honored.

Just released from twenty-seven years' imprisonment in South Africa's Robben Island jail, Mandela was striding as a free man on American soil.

His first stop in America was New York. Randall Robinson, Black lobbyist and a leader of the American campaign against South Africa, remembers the city's incredible reception: "We came in from Kennedy [Airport] in this motorcade to downtown, and there was not a block—it is a long way from the airport—that was not lined with people. It was the most incredible thing that I have seen in my lifetime." In other cities, thousands came out to see him, wave at him, sing to him, applaud him, dance the toyi-toyi (the street dance of South African protests) before him, and yell, "MANDELA!"

As he stopped in Black communities—at churches, schools, even playgrounds—a love-fest developed between him and Black Americans. He came at a moment when Blacks were, according to polls, worried about the future.

Hungry for real heroes, many Blacks saw Mandela's visit as a beacon of hope and a concrete opportunity to demand freedom for South Africa's Blacks. Mandela, in turn, paid tribute to Black American giants. In his speech before a joint session of Congress, he said, "We could not have heard of and admired John Brown, Sojourner Truth, Frederick Douglass, W. E. B. Du Bois, Marcus Garvey, Martin Luther King Jr., and others and not be moved to act as they were moved to act."

He laid a wreath of chrysanthemums on the tomb of Dr. Martin Luther King Jr. Speaking in Harlem, he said, "The kinship that the ANC [African National Congress] feels for the people of Harlem goes deeper than skin color. It is the kinship of our shared historical experience. . . . To our people, Harlem symbolizes the strength and beauty in resistance." Mandela's careful weaving of Black America's and Black South Africa's histories had great impact.

The tour was significant for Black Americans for more reasons than easily meet the eye. One, it was the first global tour completely controlled by Black Americans. Ed Maddux, who did advance work for President Jimmy Carter, organized the coast-to-coast journey. Black mayors in major cities came together to sponsor Mandela's appearances. Black communities raised considerable sums of money for Mandela's political agenda.

By 1994, Mandela had steered his nationalist party, the ANC, to victory in the first democratic all-race national elections in South Africa's history. On May 10 of that year, he took the oath of office before a crowd of over one hundred thousand people. In the distinguished-guests section was an official forty-four-member American delegation, led by Vice President Al Gore and including cultural celebrities like Maya Angelou and Quincy Jones, political leaders such as Governor Lowell Weicker of Connecticut, Representative Ron Dellums of California, and Representative John Lewis of Georgia.

One ceremony held during the inauguration had no name and no place on any program; it consisted of calling the spirits, raising the names of ancestors who had made the moment possible. Held in Johannesburg's Market Theater, it was a moment of bonding between Africans and their Black visitors from around the world. Maya Angelou raised the names of W. E. B. Du Bois and Patrice Lumumba, prime minister of the Congo in 1960. Someone mentioned young Black leader Steve Biko (1946–77), who was killed by the South African police. Don Mattera, South African poet, raised the names of American poets Claude McKay and Gwendolyn Brooks, who had "given South Africans strong words and feelings to shape our dream."

The feelings of Black Americans for Mandela and his country prove, in Randall Robinson's words, that "the blood that unites us is thicker than the water that divides us."

**How did African-Americans view their participation in the 1991 Gulf War and during 1992 in Somalia?**

On August 2, 1990, Saddam Hussein, the world's newest empire builder, ordered Iraqi tanks to invade neighboring

Kuwait. Quickly, President George Bush decided to counter this with a military buildup in the region and, if this did not break Hussein's grip on Kuwait, to prepare for war. Five months later, war erupted, and with it, Blacks were thrown into a major widespread debate.

Indeed, the high percentage of Black troops being shipped to the Persian Gulf sent a wave of anxiety through Black America. Blacks numbered 104,000 in the Gulf War forces, nearly 25 percent of American troops and almost 30 percent of all army troops there. Additionally, 43 percent of all the women in the Gulf War were Black. One in eight Americans are Black, while one in four people serving in the Gulf War were Black. Numbers like these caused flashbacks of Vietnam, where Blacks served and suffered in far greater numbers than other Americans.

Naturally, these factors shaped Black opinion. The Washington *Afro-American* sent special correspondent Randi Payton to the war front so that the Black forces' story would not be lost. In polls before the war, half of Blacks wanted to try economic sanctions against Iraq rather than use military intervention, whereas whites were four to one in favor of military action.

Beyond this, Blacks were unhappy about the war. They were opposed to the idea that their young men and women would die for oil. Sending Blacks to fight for George Bush's administration, which Blacks believed was insensitive to them, was another issue. While some Black leaders claimed that the military has been more egalitarian than other areas of society and hence has offered a good opportunity for Blacks, others, like Jesse Jackson and Benjamin Hooks, said extensive Black military participation was due mainly to a lack of economic opportunity, producing what they called a "poverty draft."

All Democratic Black members of Congress voted against the measure authorizing the use of force in the Persian Gulf. That vote took place four days before the war started.

On the other hand, many Blacks supported armed action and, in particular, General Colin Powell (born 1937), the Black head of the Joint Chiefs of Staff and a leader of the intervention. Blacks took special pride in him; a Black marine sergeant voiced the sentiment of many when he called Powell "the American dream come true." Also, Black leaders were criticized by conservatives that they had thrown away a golden chance to remind Americans of the loyalty and willingness to sacrifice of Blacks—a tactic that might have a political payoff.

Once Operation Desert Storm started, on January 15, 1991, Blacks rallied behind their troops. A columnist wrote, "Let's wave the flag and tie yellow ribbons in support of our troops, aside from our misgivings about war." When the war ended, America reigned victorious, and Blacks in the military shared the glory, with Bush calling attention to their great contribution. But Blacks still questioned the venture.

The 1992 American military mission in Somalia—alongside other troops from other United Nations countries—did not produce much fervor among Black troops or Blacks at home. One of the reasons for this low-pulse response was that it was hard for Blacks to see how Somalia could be helped: the television images showed a country beyond repair, decimated by warlords. TransAfrica, the Black Washington, D.C., lobbying organization, wanted American troops to stay there to set up an effective government that would postpone future disaster. But this proposal got little attention. After the killing of American soldiers in Somalia, Americans had no taste for staying.

Black soldiers experienced mixed emotions, from kinship with the Somali people to disillusionment with their African heritage because of Somalia's anarchy. A few Black soldiers found themselves taunted by Somalis, and one marine told a reporter that a woman condemned his features (broad nose and so on), which are considered low class in Somalia. It was harder for some Black soldiers to identify with the Somali people than they had expected.

In both these cases, Black Americans got their first taste of military participation in a world where communism has collapsed and American troops are used in small regional conflicts. Haiti, in September 1994, would be the next example.

Black troops, in Operation Uphold Democracy, were part of the American forces that accompanied Bertrand Aristide back to Haiti to assume his role as the country's newly elected president.

### How did Black Americans react to Clarence Thomas's nomination to the Supreme Court?

President George Bush threw Black America a smart curveball with his nomination of Clarence Thomas to the Supreme Court.

The Black establishment was afraid to come out against the Black nominee, even if they disagreed vehemently with his political philosophy. As Melvin Oliver wrote, "President Bush scored a resounding victory by dividing Black Americans on the Clarence Thomas issue and then by winning their support for a candidate seemingly so out of step with the mainstream of Black political thought."

Blacks had strongly criticized the Bush administration for its indifference ever since his race-baiting 1988 presidential campaign. When Thurgood Marshall announced his retirement from the Court on June 21, 1991, after twenty-four years' service (1967–91), Bush was handed an opportunity to address Black concerns. He did it in his own way, by choosing a Black who was deeply conservative.

It surprised many that Black leaders came out in support of Thomas. National Urban League president John Jacobs hoped that Thomas would honor the legacy of Justice Marshall and protect minority rights, since he was of humble origins himself. Benjamin Hooks of the NAACP said that he was happy that Bush had nominated a Black for the position. Niara Sudarkasa, president of Pennsylvania's Lincoln University, urged Blacks to support Thomas, saying his experience of childhood poverty would temper his conservatism.

But Blacks opposed to him came out swinging. Representative Louis Stokes of Ohio claimed that although Thomas had humble origins, he had forgotten the poverty, adversity, and racism of his past. Representative Major Owens of New York called Thomas "a monstrous negative role model" and compared him to Benedict Arnold. His elevation to the Supreme Court would be "a cruel slap in the face of all African-Americans."

Actually, hardly any Blacks—or whites—knew Clarence Thomas. Unlike Thurgood Marshall, a major figure in civil rights litigation by the time he was nominated to the Supreme Court, Thomas had limited public exposure.

Born in 1948 in Savannah, Georgia, Thomas was educated in Catholic schools, including Holy Cross College. (While in college, he was the founder of the Black student union.) In 1974, he graduated from Yale Law School, one of the nation's finest. For several years, he worked in Missouri under John Danforth, who later as a senator won Senate confirmation for Thomas.

Thomas's career really took off when President Ronald Reagan appointed him chair (1982–89) of the Equal Employment Opportunity Commission (EEOC), which processes workplace discrimination claims. His greatest coup was his appointment to the U.S. Court of Appeals for the District of Columbia, where he served for one year. So when Bush nominated Thomas, he was more of a federal administrator than a judge. For a decade before his judgeship, he had not practiced law at all.

With the Black establishment divided, Thomas's appointment seemed likely, although no cakewalk. He still had to undergo grueling Senate confirmation hearings. This was when a lone Black woman's story about Thomas almost derailed the entire process. Anita Hill, law professor at the University of Oklahoma, accused him of sexually harassing her when she worked at the EEOC. The accusations surfaced slowly in late September but grew quickly into a roar that could not be ignored. She said Thomas attempted to date her, made explicitly sexual remarks to her, and talked of pornography.

In 1987, the country had seen acrimonious confirmation hearings when Robert Bork was nominated. But the lurid inquisition on Thomas's character and the attack on Hill were even more of a shocker, easily the year's media event.

On October 12, while Black churchpeople supporting Thomas milled outside the Senate Caucus Room, Thomas read his defense before the judiciary committee, attacking the process as a "lynching." Anita Hill followed with her testimony, which Senator Arlen Specter of Pennsylvania charged was "flat-out perjury." Both galvanized their supporters: Thomas by his emotional statement and Hill by her disciplined testimony.

Blacks had definite opinions on the hearings. Many experienced pain as the Thomas-Hill contest progressed. They felt that the hearings revived the myth of oversexed Black men. Some felt that the hearings were a "peep show" for whites, where Blacks were exposed and violated. A Black testifying against a Black was also painful. Many were upset that Thomas used images of "lynching" to get sympathy from Blacks.

Others saw the hearings as widening the gender gap between Black men and women, with men seeing advancement of Blacks as more important than sexual harassment and with women supporting Hill.

In the days following the hearings, 60 to 70 percent of Blacks supported his confirmation. These figures influenced senators, especially those from the South, to vote for Thomas. On October 15, 1991, the Senate confirmed Thomas by a vote of fifty-two to forty-eight, the most negative votes for a successful Supreme Court nominee in history.

It was over. Yet the Thomas-Hill disputes would continue to be hot items in Black America. Since then, Thomas's high approval rating has collapsed. In a late 1994 poll, only 2 percent of Blacks still admired him.

In early 1995, Armstrong Williams, a friend of Thomas's from their days at the EEOC and a radio talk-show host, arranged for mettings between Thomas and various Blacks, including a group of Black reporters. According to reports, their meetings were extremely cordial.

### What is Bill Clinton's connection to Black America?

No one paid much attention to it, but under President George Bush, Republicans conducted an aggressive effort to appeal to Blacks. The GOP resuscitated its heritage as the party of Abraham Lincoln.

Immediately after the 1988 election, the campaign for Black support unfolded. Playing to symbolism, Bush invited Jesse Jackson and Coretta Scott King to the White House. Appointments of numerous Blacks followed, the most impressive being Colin Powell as chairman of the Joint Chiefs of Staff.

Nominating Clarence Thomas to the Supreme Court was also a move intended to placate Blacks. But to most Blacks, Bush was still not living up to his 1988 convention pledge of a "kinder and gentler" conservatism.

As the 1992 election approached, Blacks were caught in recession's vise. In addition, many of their vital signs—family stability, neighborhood safety, graduation rates—had not improved or were declining.

In the primary season, less than a year before the election, Bill Clinton was an unknown to Blacks. Whenever Clinton made forays into core Black voting districts, as in the New York primary, he ran into opposition from Rev. Al Sharpton and local politicians. Clinton's membership in an all-white Little Rock country club aroused old suspicions about Southern white politicians.

Clinton conducted a strong primary campaign for the Democratic nomination, eventually swamping his rivals Paul Tsongas and Jerry Brown. Reversing the image that Democrats were just interest-group hostages, Clinton campaigned on the general theme of the necessity for change rather than the specific needs of groups.

Many Blacks understood this strategy. Said one, "As a politician, I understand why Clinton is playing down overt policy commitments to the Black community. . . . He is trying to reach white middle America. . . . I think his strategy is paying off." Another thought Clinton had "hinted at race enough to say to minorities, 'We are not going to reject you.'" However, Jesse Jackson's camp viewed this approach as sending an anti-Black signal.

A defining campaign moment for many Blacks was Clinton's attack on Sister Souljah, a Black rap artist invited to Jesse Jackson's Rainbow Coalition's "Rebuild America" conference in June 1992. Sister Souljah was invited because of her activism in Black causes. What brought Clinton's criticism was her remark, "Why not have a week and kill white people?," which he viewed as hate mongering in the same vein as comments made by David Duke, Louisiana's racist political figure.

The Clinton speech created a small tornado of condemnation. Jackson, at first quiet on it, later blasted him for a "very well-planned sneak attack," "a Machiavellian maneuver" that "exposed a character flaw." Others said the speech stank like the Willie Horton ads of the 1988 Bush campaign. Polls showed that Blacks disagreed with Clinton three to one. All agreed, though, that Clinton attracted white voters with the attack.

By the election, much of this was forgotten. Blacks gave Clinton 83 percent of their votes and voted more heavily than in any past presidential election. Georgia and Louisiana went for Clinton because of Black voters.

After the election, Clinton drew closer to Blacks, appointing four Blacks to cabinet posts (Ron Brown, commerce; Mike Espy, agriculture; Hazel O'Leary, energy; Jesse Brown, veterans affairs) and to other positions (Joycelyn Elders as surgeon general). He traveled to several Black Washington neighborhoods.

At the inauguration, Maya Angelou's poem "On the Pulse of the Morning" embellished the ceremony with a soaring sense of possibility. Clinton further endeared himself by proposing an economic plan that included a thirty-billion-dollar stimulus package, aimed at job training, more Head Start for children, and new education programs.

Was this Clinton and Blacks' finest hour, the high point of their togetherness? It seems so. To come was a most destructive episode: Clinton's nomination of University of Pennsylvania law professor Lani Guinier to head the Justice Department's civil rights division. After a vicious attack on her legal theories and idea of voting reforms, Clinton withdrew her nomination without allowing her to appear before the Senate Judiciary Committee to defend her views.

A howl followed. Marchers protested outside the White House: "Reagan backed Bork. Bush backed Thomas. Clinton backed down." Congressional Black Caucus head Kweisi Mfume said, "The president has succumbed to fear, innuendo, and a whispering campaign." In a dignified response, Guinier said simply, "My views have been distorted in the media" and "I have always believed in democracy, and nothing I have ever written is inconsistent with that." In polls, 55 percent of Black Americans disagreed with Clinton's yanking of Guinier's nomination.

Clinton's Black support rebounded slightly, especially as the economy improved. But in 1994, differences appeared between him and the Black Caucus. The caucus wanted a racial justice act, allowing racial statistics to be used to challenge death sentences, attached to Clinton's crime bill, which they threatened to sink. In the end, the Black Caucus dropped their demand, the bill passed, and Black communities received more prevention program money.

Halfway through Bill Clinton's presidency, signs showed increasing satisfaction among Blacks. His economic plans were buoying Black communities, like Harlem. Operation Uphold Democracy in Haiti got Black applause. But Joycelyn Elders's controversial departure from his administration did him no good. Clinton was operating under new conditions, with Blacks and their representatives in a Rosa Parks mood.

**What have Black gays and lesbians done to reclaim their place in Black history?**

When the video-film *Tongues Untied* appeared in 1989, it drew big appreciative audiences, and their applause made Marlon Riggs (1957–94), the film's creator, into the sweetheart of a culture. The short, tam-wearing, Bay Area filmmaker had created something unusual and needed: an ode to the life of Black gays. In the film, poetry, personal testimony, and drama were brought together into a statement affirming Black gay life. Riggs's film unflinchingly confronted derogatory accusations and judgments in American culture.

For once, Black homosexuals could see themselves as they had wanted to be seen for a long time—not as invisible and silenced but as an important culture within the larger Black and gay life, as a significant minority within two minorities, and as a culture that would not retreat because of prejudice.

PBS had planned to show Riggs's film in 1989, but some affiliates refused to air it, and conservative presidential candidate Pat Buchanan attacked it.

In the months and years that followed, Black gay and lesbian culture exploded unashamedly into the public. Isaac Julien's film *Looking for Langston* (1989), a stimulating surreal work of art, brought up the Harlem Renaissance's gay culture by looking at poet Langston Hughes. Essex Hemphill's poetry was filling books and reading halls. *Aché,* a lesbian magazine, and *BLK,* a gay magazine, began publishing. Jennie Livingston's documentary about voguing, *Paris Is Burning* (1990), was a hit and underlined how Madonna had pirated Black gay culture. Jewelle Gomez's *The Gilda Stories* (1991), which linked a vampire with a Black gay theme, provoked a lot of interest.

In performance art, a San Francisco troupe, Pomo Afro Homos (the name is an abbreviated version of "Postmodern African-American Homosexuals"), led the way with *Fierce Love* and *Dark Fruit,* short snapshots of Black gay life—some campy and playful, some darker in tone. By 1992, Riggs had produced another assertive film, *Non, Je Ne Regrette Rien,* about HIV-positive Black gays. Out of this ferment came Black drag pop star RuPaul.

As Black gays and lesbians have risen up, they have connected themselves with their history. It is not enough for them to have a presence today; they want to reclaim their forerunners. Sylvia Rhue's documentary *We Have a Legacy* makes this point by discussing past Black gays and lesbians.

Black gays and lesbians have always been around. From the 1920s forward, however, they made up a distinct culture within Black life. As Blacks migrated to the North's cities, gays and lesbians built complex cultures that sheltered and nurtured them. In a place like Harlem, the variety of Black homosexual life could be seen.

The Harlem Renaissance was at one level a gay resurgence. Two of its great figures, Alain Locke and Carl Van Vechten, were gay men who were patrons to young writers. Many leading writers were gay-oriented or bisexual: Countee Cullen, Wallace Thurman, Claude McKay, and maybe Langston Hughes. McKay's *Home to Harlem* (1928) and Thurman's *Blacker the Berry* (1929) and *Infants of the Spring* (1932) contained gay and lesbian figures. Bruce Nugent's short story "Smoke, Lilies, and Jade!" set tongues wagging with its drugs and homosexuality.

As Harlem grew, it became famous for its nightlife. Black gay men insisted that that nightlife tolerate them. At night, speakeasies flung open their doors to men who wanted to dance together and drag queens who wanted to strut in their finery. Cyril's Café and the Hot Cha on Seventh Avenue were gathering places for artists, intellectuals, gays, and lesbians. Many of Harlem's prominent blues singers were lesbians or bisexuals: Alberta Hunter, Ma Rainey, Bessie Smith, Ethel Waters. Most obvious was Gladys Bentley, who performed in tuxedo and top hat while she gave sexy interpretations of popular songs.

Spectacular annual balls were given at Harlem's Hamilton Lodge in the 1920s, featuring lesbians as male impersonators and many working-class men in drag. The ball came to be known as the Faggots Ball. Much of Harlem went to it, and thousands came to stare as the revelers entered. The *Amsterdam News* reported in 1934, "Four thousand citizens, numbering some of Harlem's best, elbowed and shoved each other aside . . . to obtain a better eyeful." Two years later, it ran the headline: "Pansies Cavort in Most Delovely Manner at That Annual Hamilton Lodge 'Bawl.'"

In other words, the current upsurge among Black gays and lesbians can claim a solid history. Not only are there classic texts like James Baldwin's *Giovanni's Room* (1956), the story of a Parisian love affair between two men, but there are also pioneers like Bayard Rustin (1910–87), the 1963 March on Washington's organizing genius, and poets like Audre Lorde (1934–92).

In 1994–1995, Black gays and lesbians continued to chalk up advances as Jacqueline Woodson's *The Autobiography of a Family Photo* and James Earl Hardy's *B-Boy Blues* met with great excitement and sold extremely well.

**What was the importance of Toni Morrison receiving the 1993 Nobel Prize for literature?**

From the predawn call on October 7, 1993, from a friend who said simply, "Did you hear?" to the Nobel Prize award ceremony in Stockholm, Sweden, on December 7, 1993, Toni Morrison's life was changed forever.

In selecting her, the Swedish Academy said, "She delves into the language itself, a language she wants to liberate from the fetters of race. And she addresses us with the luster of poetry."

That "poetry" was revealed in a stream of unusual novels, all centered within "the village of Black history." The *Bluest Eye* (1970) came first and met with a cool reception. But her stock rose with *Sula* (1973). *Song of Solomon* (1977) brought her the first real taste of near-universal acclaim. *Tar Baby* (1981) was not everyone's favorite Morrison book, but it found a niche. *Beloved* (1987) was the greatest of her achievements. And her stature climbed steadily higher with *Jazz* (1992).

Although Morrison had won many major prizes (including the Pulitzer in 1988 for *Beloved*), had collected her share of honorary degrees (fourteen, by recent count), and had been on the best-seller list (in 1992 for both *Jazz* and a book of essays, *Playing in the Dark*), she was now entering the company of literature's immortals, that rarefied world of Nobel laureates that includes William Faulkner, Ernest Hemingway, Boris Pasternak, John Steinbeck, Gabriel Garcia Marquez, and Nadine Gordimer.

It was a tremendous honor for an American author. Since 1901, when the award was established, only nine Americans had been chosen. Not since 1978 had an American writer, Isaac Bashevis Singer, won. (Singer was born in Poland in 1904, emigrated to the United States in 1935, and became a naturalized citizen in 1943.)

Of these, only one American woman had ever been voted the prize, Pearl S. Buck in 1938. (Only eight women in the world have won the award.)

Morrison saw the award as standing for more than American achievement. Only two Black writers had won before her—Nigerian writer Wole Soyinka in 1988 and Caribbean poet Derek Walcott in 1992.

In an interview months later, she said, "I felt a lot of 'we' excitement. It was as if the whole category of 'female writer' and 'Black writer' had been redeemed. I felt I represented a whole world of women who either were silenced or who had never received the imprimatur of the established literary world. I felt the way I used to feel at commencements where I'd get an honorary degree: that it was very important for young Black people to see a Black person do that."

Morrison says that she never expected the prize, did not even know that she was being considered by the Swedish Academy. "I was aware of the cautions and the caveats and the misunderstandings that seemed to lie around the criticisms of my work." So when her friend's call came, international fame and its accompanying pandemonium were already descending on her small world, New Jersey's Princeton University where she held the position of the Robert F. Goheen professor of creative writing.

Just as important as the prize itself was her stunning public lecture at Stockholm's Grand Hall about what she called "word work." Her talk was judged to be the most important Nobel literary address since Faulkner's speech at the 1949 awards ceremony. Morrison kept her audience's attention with her song of praise to the power of language. Reports say the chandelier-lit hall was packed. She spoke slowly, almost in cadences. The listeners angled their heads to pick up every word.

Morrison's lecture started with the folktale of a wise, blind Black woman who is visited by children with a question about whether the bird they are carrying is living or dead. They are taunting her because they know she cannot see the bird. Morrison takes the old woman's careful response as the start of her statement on language. "I don't know whether the bird you are holding is dead or alive, but what I do know is that it is in your hands. It is in your hands," the old woman says. The woman is turning their game back on them. She asserts that they are in control of the tiny life they hold. For Morrison, the bird is language. To Morrison, the blind woman is a writer, a person who "draws pictures with words." Therefore, the old woman is saying that she "is worried about how the language she dreams in . . . is handled, put into service . . ." This conceit was used to advance Morrison's main issue: she is deeply worried over the future of language.

To Morrison, our modern world is filled with much dead language—language that "thwarts the intellect, stalls conscience, suppresses human potential." Today, "the systematic looting of language can be recognized by the tendency of its users to forgo its nuanced, complex, midwifery properties, replacing them with menace and subjugation." Opposed to this deadening is "word work," which is "sublime . . . because it is generative; it makes meaning that secures our difference, our human difference—the way in which we are like no other life."

Critics were yelping even during Morrison's shining moment. Stanley Crouch, a harsh judge of *Beloved,* said that maybe the Nobel Prize would inspire Morrison to write better books. Others said that "The Year of the Woman" obviously overtook the Swedish Academy. But these were niggling complaints in the face of the astonishing victory that Toni Morrison had won.

### How did a box on a census form generate a debate about the meaning of "race"?

When the United States conducted its 1990 national census, it started a controversy about what is race in today's America. Census takers allowed only the traditional categories: Black, Caucasian, or Hispanic. People in the mixed-race category—such as those with only one Black parent—were forced to choose among options that were inaccurate.

What race is the son of a Jewish father and a Ugandan (East African) mother? A daughter of a Black mother and an Italian-American father? What box do they check? Mixed-race children are faced with similar bad choices when they apply to college, fill out a job application, or apply for medical benefits. They have to deny one race in order to admit another.

Deborah Thomas, born to a German-American mother and a Jamaican father, wrote a 1993 essay on her dilemmas with the box. "Sometimes, I check the box marked 'other' just to mess with people. I did it on my application to Brown University . . . and another time on a job application that listed racial classification as optional." Then she added a significant detail: "The personnel director who knew my family called and asked me why I chose not to help the company fill its affirmative action requirements." She concluded, "Mixed-race Americans are still not officially represented on demographic information surveys. Put more simply, we have no box to check."

Our application forms are not keeping pace with our social reality. These forms were created for an America where the races were forbidden to mingle and where marriage between people of different races was illegal. But the issues raised by Thomas are not just about application forms and which boxes to check.

In today's America, interracial mingling happens frequently. While still widely disapproved of, interracial marriages are no longer illegal. In fact, they are increasing, generating thousands of mixed-race children.

In 1967, in *Loving v. Virginia*, the Supreme Court ruled unanimously that a Virginia statute prohibiting marriages between Blacks and whites was unconstitutional, thus striking down similar laws in fifteen states. (Mildred Love, of Indian and Black ancestry, and her white husband, Richard, brought the case.) Between 1980 and 1990, marriages between Blacks and whites increased 51.8 percent. In 1980, there were slightly over 120,000 Black-white marriages, but in 1990, 183,000 were registered.

America has encouraged the idea of "the melting pot," but it has suppressed the reality of racial mixing. In fact, Americans have not grown more tolerant on this score. Spike Lee's film *Jungle Fever* showed how a Black man–white woman love affair caused the woman's family to throw her out of the house and the man's wife to feel doubly cheated. The film's characters almost denied the possibility of interracial love.

In real life, the country's reluctance to admit racial mixing has produced some peculiar and sad situations. At Dallas's Children's Theater, a kiss in a school play between a Black actor and a white actress had to be removed after an investment banker threatened to picket the theater and demanded that the roles be recast. Montel Williams, Black TV talk-show host, lashed out at critics who attacked him for having a white wife. At Wedowee High School in Alabama, principal Hulond Humphries threatened to cancel the prom if mixed couples attended.

More of the issues facing mixed-race Americans are revealed in the lives of actress Halle Berry and pop singer Mariah Carey. Berry's father is Black and Venezuelan; her mother is Irish. Berry had much contact with her father's family and is very aware of her Black heritage, but says, "I have a mother who is 100 percent Irish who raised me from birth and who is my best friend. If I were to say that I'm Black only, that would be negating everything she is." Mariah Carey agrees and says that while growing up, she often felt ostracized by both white kids and Black kids because she was not "100 percent like either."

Mixed-race relations, marriages, and children are issues that are likely to remain contentious. But Americans will have to abandon their notion that races are pure things and have to be kept pure at all costs. Otherwise, we will become a nation of ethnosaurs.

## How did Spike Lee's success open up Black filmmaking?

Spike Lee has been a meteoric presence in American culture since the mid 1980s. Born Shelton Jackson Lee in Atlanta in 1957, he was reared in New York, educated at the elite Morehouse College in Atlanta (1975–79), and studied film at New York University. Now he has made a special niche for himself among major American culture makers. Lee is not simply his own creator: his family, especially his jazz musician father—Morehouse graduate Billy Lee—emphasized art and Black tradition in his upbringing.

As much as any one person, Spike Lee has been responsible for educating the Black intelligentsia about the contemporary power of film. He also advanced, for the 1980s and 1990s, that there could be Black visions in film.

When Lee's first feature film *She's Gotta Have It* appeared in 1986, it stirred up moviegoers, particularly Black Americans. It defied expectations about what Black filmmakers were likely to produce. Lines stretched down the street and around the corner from the art-house theaters showing it. Audiences were attracted by the film's fresh theme, a funny exploration of the sexual double standard that American society holds for men and women. Visual techniques were also part of the film's appeal, as Lee was influenced by Japanese filmmaker Akira Kurosawa, French New Wave film, and Warren Beatty's *Reds* (1981).

Equally important were the film's fascinating characters—Nola Darling, an independent, artistic young woman; Greer Child, an impossibly vain model; Mars Blackmon, a basic wily type; and Jamie Oversteet, Mr. Sane and Sober. Egocentrically, each man wants Nola for himself, but she has her own plans—she wants all of them and on her terms. Later, Lee said, "When I wrote this script, I had the Black audience in mind." Lee was exposing that audience to a new woman and sending men an alarm signal about their sexual hypocrisy.

Lee's film had another message: it stood for his overwhelming, relentless desire to make films. His mantra: "Black film by any means necessary." He did not wait for the blessing of Hollywood. Instead, he became a "film hustler." In a diary (*Spike Lee's Gotta Have It,* 1987), Lee recreated his day-by-day quest for funds. His grandmother gave her savings, he used credit cards, and every day after shooting, Monty Ross—his sidekick since college—hit the phones to plead for funds. Lee shot the film in eleven days, and it cost $175,000—as biographer Alec Patterson put it, "less than it would take to keep Steven Spielberg in turtlenecks." *She's Gotta Have It* went on to earn eight million dollars.

Nelson George is emphatic that "from 1987 to 1990 the spark of *She's Gotta Have It* ignited little filmic brushfires all over the place." Both in Black independent filmmaking and in studio films, there was increased momentum for Blacks. Actor Robert Townsend's strong eccentric comedy *Hollywood Shuffle* appeared in 1987. Lee's musical *School Daze*—a not-so-favorable look at Black colleges, fraternities, sororities, and color consciousness—appeared in 1988, with the lead role played by Larry Fishburne. *Do the Right Thing,* in 1989, undoubtedly lifted Spike Lee into some very select company, that of world-class filmmakers. It had everything—a credible story, great acting (down to the "corner men," a brilliant ensemble), wonderful color, and adept use of background rap music. Rosie Perez's dancing at the film's start set a standard for energy and style. In 1990, Charles Burnett's *To Sleep with Anger* wove a curious story around a Los Angeles family's evil visitor, played by Danny Glover. Reggie and Warrington Hudlin's *House Party,* also a 1990 film, was a highly successful teen comedy, raking in twenty-eight million dollars.

The banner year for Blacks in film was 1991. Wesley Snipes gave a solid performance in an otherwise flat *New Jack City. The Five Heartbeats,* a period film about a rhythm and blues singing group, was a great vehicle for music nostalgia. Robin Givens was a stunning performer in *A Rage in Harlem. Jungle Fever,* dissecting interracial affairs, reminded the public that Spike Lee still had the power to provoke. *Boyz 'N' the 'Hood,* directed by John Singleton, brought to the screen a vivid, insider interpretation of life in South Central Los Angeles. Larry Fishburne's Furious Styles was a paean to the responsible Black male.

Far different from these was the imaginative *Daughters of the Dust* by Julie Dash. Set on Saint Helena Island, South Carolina, the film told the story of a tight-knit family at the turn of the century just when some of its members wanted to migrate North for better jobs and schooling. Producing the film was a miracle: Dash's funding was precarious, her crew barely escaped Hurricane Hugo, and shooting lasted an exhausting twenty-eight days. The result was a visual and anthropological treasure. The film's scenes were as visually arresting as an impressionist painting. And the film is crammed full of African customs, including the Gullah language. (A few critics have hit Dash for including this potpourri of traditions.) The film had, as writer Toni Cade Bambara noted, "an emancipatory purpose," exalting quietude, the Black body, and women's lore and culture.

Not every Black film since 1991 has been without flaw—Spike Lee's *Malcolm X,* even with Denzel Washington's acting, dragged; *Boomerang,* while funny, had a bad second half; Matty Rich's *Inkwell* was too much of a cartoon; John Singleton's *Higher Learning,* while tackling a major issue, was riddled with stereotypes. But successes have continued to appear: Albert and Allen Hughes's *Menace II Society* (for its insight into the life of violence) and Haile Gerima's *Sankofa* (for treating slavery as an issue) and *Clockers* (a great film essay on the real people living in the Black inner city) are three examples. In short, the 1990s Black film promoters pushed possibilities' limits, having been motivated by Lee's work.

### What was the huge reaction to *The Bell Curve* (1994) about?

No book in recent memory has generated such a storm of controversy as did Richard J. Herrnstein and Charles Murray's

*The Bell Curve,* with its revealing subtitle, *Intelligence and Class Structure in American Life.*

The book had been in preparation since 1989, and its authors were already two of America's most controversial thinkers—Herrnstein (1930–94) for his work on the basis of intelligence and Charles Murray (born 1943) for his writing on the welfare system. Herrnstein was a Harvard professor. Holder of a Ph.D. from MIT, Murray is a "think-tanker." He has held research posts since the 1980s at the Manhattan Institute and the American Enterprise Institute. In 1984, Murray became famous for his argument, stated in *Losing Ground,* that social programs did more harm than good, and he suggested that they be eliminated. Reporter Jason DeParle said that in the Reagan White House an aide "was brandishing" Murray's book "as a sort of Exhibit A when advocating program cuts."

Charles Murray was interviewed in 1990 about his "inquiry on race and IQ," which became *The Bell Curve.* He remarked that "there is no way to exaggerate the sensitivity of the issues." Already, he said, people were asking him, "So, Charles, are you a racist?" His guess about the sensitivity of these issues proved all too true. Since the book's publication, he has been called a "racist" and worse. Elaine Jones, the NAACP Legal Defense Fund's director, spoke of the "racist pseudoscience of Charles Murray" and said *The Bell Curve* "warms over old white-supremacist arguments of African-American inferiority." Leon Wieseltier, literary editor of the *New Republic,* was more damning: "Murray's pseudoscientific tome is hate mail to Black America cloaked in numbers and charts and borne by a man in an academic gown and mortar board instead of a sheet and hood."

When one opens *The Bell Curve,* with its 845 pages of text, graphs, statistics, and footnotes, the impression is of an immensely learned book, considerate and civil. Although the book is identified as a racial manifesto, it does not actually devote much space to race. Not until chapters thirteen and fourteen (out of twenty-two) is there a discussion of "Ethnic Differences in Cognitive Ability." The book does not look explosive.

Fewer people have read *The Bell Curve* than comment on it. In fact, the book's purpose is broader than race; it looks at a lot of inequalities and their relation to intelligence. Furthermore, the book is as much about the future as anything: the authors lay out a plan for a society where the most intelligent people are the most rewarded, forming a new cognitive elite. Essayist Gerald Early perceptively says the book proclaims a new Calvinism, with a predestined elect at the apex of America and the other less intelligent classes arrayed beneath them. Herrnstein and Murray really want a new American IQ republic, where everyone will have a preordained place. For them, this would be a utopia.

This is the best that can be said about their presentation. They do argue that a constant gap exists between Black and white IQs. Accordingly, "the average white person tests higher than about 84 percent of the population of Blacks." They go on to say, "In discussing IQ tests . . . the Black mean is commonly given as 85, the white mean as 100." Herrnstein and Murray see some possibility of Blacks' IQs catching up with whites' IQs, as revealed in measures like the Scholastic Aptitude Test (SAT), but "we could expect Black and white SAT scores to reach equality sometime in the middle of the twenty-first century" (in other words, 2500 A.D.!). Ultimately, they argue that lower IQs reside in Blacks. Although they do not say absolutely that genetic background is responsible for this, they strongly imply it. Apparently, nothing can substantially remedy this. Lastly, Blacks do badly in socioeconomic terms because they have fewer IQ resources. Unless the nation deals with these issues with candor (the authors are big on candor), they see the Black future as dim.

Good counterarguments can be made against the authors. Examining their methods reveal several weaknesses:

- Nothing new is revealed in Herrnstein and Murray's data and arguments. No dramatic new evidence of racial IQ inequality is offered. The arguments are a smooth recycling of old ideas.

- Most of the work Herrnstein and Murray rely on was collected for social purposes rather than scientific purposes. It was intended for social engineering.

- Herrnstein and Murray fix on the dubious concept that an entire race has an IQ profile, a nearly homogeneous intelligence. This is not true. Even if it were, Black Americans are not such a unified group: they come from a welter of African groups, have mixed further here, and therefore should not have a monolithic IQ profile.

- Whatever the gap in IQs might be, there is no evidence that it is genetic and, therefore, passed on.

- Herrnstein and Murray succumb to the pessimistic, deterministic side of science and paint a picture of a bleak future.

- Interventions can help the disadvantaged, and their abilities can be enhanced.

Strikingly, Herrnstein and Murray's new world resembles British writer Aldous Huxley's 1932 novel *Brave New World*, where the Alpha Plus Intellectuals ruled as a permanent caste over the underclass Epsilon Minus Morons. Can this happen in America? Avoiding it will require more extensive knowledge of what Herrnstein, Murray, and their ilk have conjured.

## Why has affirmative action become such a widespread national debate?

In 1989, Randy Pech—the white owner of Colorado's Adarand Constructors—bid to build a section of highway in the San Juan National Forest. Congress stipulates that at least 10 percent of federal money on highway projects should go to businesses owned by "disadvantaged individuals." The company receiving the contract was Gonzales Construction, owned by Hispanics. Pech was upset over this outcome because his bid was $1,700 lower. He sued the federal government, challenging the program. His case became the landmark *Adarand Constructors v. Peña.*

On June 12, 1995, Pech got what he wanted. The Supreme

Court ruled five to four in his favor. Justice Sandra Day O'Connor, the author of the Court's opinion, wrote, "All governmental action based on race should . . . ensure that the personal right to equal protection of the laws has not been infringed." Joining her were Justices William Rehnquist, Anthony Kennedy, Antonin Scalia, and Clarence Thomas—all appointed during the Reagan-Bush years. Clarence Thomas wrote, "Government cannot make us equal; it can only recognize, respect, and protect us as equal before the law." The near-consensus among political observers was that this Court decision spelled the end for many federal affirmative action programs, especially set-aside programs that annually award some ten billion dollars in federal contracts. Mary Berry, a University of Pennsylvania law professor, was one of a few who thought that the decision had narrower applications.

Today, America is witnessing an all-out assault on affirmative action. From the floor of the 104th Congress to the state capitol of California, the anti–affirmative action forces that gathered strength during the 1980s are now poised to eradicate the programs. Leading the assault is a Black writer, Shelby Steele (born 1946). His writing is most often cited in this dispute. For him, affirmative action has not uplifted poorer Blacks, only the middle class. Furthermore, he says affirmative action has stigmatized Black achievements. And it has caused dependencies to develop. Steele is not wanting for company.

With him are all of the Republican 1996 presidential candidates, in particular California's Governor Pete Wilson, already repealing his state's pro-minority regulations. Generally, polls show 60 percent of white Americans want major changes in these minority programs. On the whole, conservatives have mounted more attacks on affirmative action than liberals have mounted defenses, an indication that today even many liberals are less confident about these programs' efficacy.

*Webster's Collegiate Dictionary* defines affirmative action as "a policy to increase opportunities for women and minorities, especially in employment." This sounds noble. But today, as columnist Julianne Malveaux has observed, "affirmative action has become for some more of an expletive than a description of public policy." Now, it is often called "reverse discrimination."

Where did all this begin? In 1961, Hobart Taylor Jr. (born 1920)—a Black high-ranking federal official—used the words "affirmative action" when describing racial preferences in President John Kennedy's Executive Order 10925. The order urged federal contractors to use more minority employees. In 1964, President Johnson signed the Civil Rights Act that prohibited discrimination in the workplace. In a famous 1965 Howard University commencement address, Johnson declared, "We seek not just . . . equality as a right and a theory . . . but equality as a result." This moved beyond civil rights. After the Watts riots in 1965, Johnson signed Executive Order 11246, which required employers using federal funds to set goals for hiring minorities. Affirmative action was shifted to the Labor Department. By 1969, President Richard Nixon's administration issued the Philadelphia Plan, which compelled federal contractors to develop plans for minority hiring.

In the 1970s and 1980s, affirmative action ran into its first major opposition. In 1978, the Supreme Court, in *University of California v. Bakke,* struck down a university program that set aside sixteen out of one hundred openings for minorities. This case originated from Allan Bakke's disappointment at being turned down at the University of California at Davis's medical school in favor of minority candidates. The verdict was that race can factor in admissions, but don't use quotas.

*Memphis v. Stotts,* in 1984, allowed the Supreme Court to rule that a city could lay off newly hired Black firefighters according to its seniority system. A court decision in 1986 (*Wygant v. Jackson Board of Education*) undid an affirmative action program protecting Black teachers from layoffs, declaring it was unconstitutional. In 1986, another Court decision (*Richmond v. J. A. Croson Co.*) struck down a Richmond, Virginia, set-aside program.

Defenders of affirmative action make a strong case for its continuance. First, they claim affirmative action policy has prodded professions and businesses to integrate. It opened the door. Second, it has helped create a "level playing field" for groups victimized by exclusionary practices. To its supporters, affirmative action provided Blacks and other minorities opportunities that compensated for past inequities. Third, American corporations, universities, and companies profit from having diverse work-forces; affirmative action is good business. Third, politicians cynically lambaste affirmative action to distract from the real failure of the American economy to produce more jobs that pay well for ordinary people.

In a single week in July 1995, a new chapter was written on this issue. After a review of federal programs, President Clinton delivered a major address—on July 19—favoring the basic premises of affirmative action but urging a pruning of its excesses, such as quotas. "Mend, but don't end" was his epigram. A same-day CNN poll showed that 48 percent of Americans did not want affirmative action programs eliminated entirely, versus 46 percent who did. One day later, the University of California regents, under pressure from Governor Pete Wilson and with support of Black regent Ward Connerly, voted to end affirmative action in admissions, hiring, and contracts based on race or gender. The vote came after a heated meeting filled with testimony and demonstrations and featuring an impassioned appeal by Jesse Jackson to retain the programs.

In August 1995, Senator Dole was preparing legislation to retire all federal affirmative action programs. Obviously, affirmative action will be a "hot-button" issue in the 1996 presidential election.

### What is the status of the word *nigger* today?

In Quentin Tarantino's *Pulp Fiction* (1994), the word *nigger* was liberally sprinkled in the film's opening scenes. Jules (Samuel L. Jackson) used it six times in his conversations with Vincent (John Travolta). Later in the film, it popped up a few more times.

Computer magazine *Wired* (December 1994) told of an incident where "nigger" shattered the techno-cool of hacker subculture. On a marathon phone call linking many people, when a Black hacker came on the line with "Yo, dis is Dope Fiend from MOD," a hacker from Texas yelled, "Get that nigger off the line."

Rappers' use of "nigger" is legendary. "Niggaz wit' Attitude"

(N.W.A.) flaunted the word. Rappers have invoked the word so frequently that in suburban America, their young white fans use it among themselves. This is a crossover effect. Some even call themselves "wiggers," or white niggers—a white American who is loyal to Black culture.

Basketball star Charles Barkley, in facing down press criticism, retorted, "I'm a nineties nigger. . . . The *Daily News,* the *Inquirer* has been on my back. . . . They want their Black athletes to be Uncle Toms. I told you white boys you've never heard of a nineties nigger. We do what we want to do."

Early on, the very idea of introducing the word *nigger* into the O. J. Simpson trial triggered elaborate legal maneuvers. Prosecution witness detective Mark Fuhrman was alleged to have used the word. The prosecution feared the defense would use this to impugn Fuhrman's honesty before the mostly Black jury.

All these recent episodes in the history of the term *nigger* tell us that not only is the word still in use but its use might be increasing. Is the nation's threshold against racial epithets lowering, and slowly the word is—once again—becoming an acceptable part of everyday speech? Maybe this explains why major Black magazines—such as *Emerge* (June 1993) and *Image* (October 1994)—have published articles about "the N-word."

All of this has happened in the 1990s, just a few years after many Blacks renamed themselves "African-Americans," seeking a label that better reflects their history. Blacks had finally enshrined their dual heritage and were reaching out to Africa—at that moment, especially to struggling South African Blacks. The 1980s rise of "African-American" was the second recent revolution in Black group naming. The first revolution was a spin-off of the Black Power movement in the mid 1960s, when the time-honored "Negro" was disparaged and unseated as the reigning collective name. (Harlem author Richard Moore campaigned for change with his 1960 classic *The Name "Negro": Its Origin and Evil*). In its place came "Black"—a change writer Trey Ellis notes was "an important linguistic battle for our self-determination . . . our own Russian revolution." "Black" embraced the color boldly, acknowledged "Blackness" as a worldview, and seemed more assertive.

For the sake of history, it should be noted that "Negro" had also been ardently fought for. Once, "Negro" was thought honorable. Anthropologist Alison Hamilton has identified a 1920s NAACP campaign for "Negro" to be used as the official racial designation. In 1930, it was considered a triumph when the *New York Times* stated it would capitalize "negro." During these years, "colored" ran a close second to "Negro" as a group name. But "colored" had one crucial drawback: the South's ubiquitous Jim Crow signs saying "Colored" beckoned Blacks into separate waiting rooms, train cars, and rest rooms.

In September 1994, two Gallup polls showed 60 percent of Blacks surveyed were comfortable with either "Black" or "African-American." Both were regarded as positive terms. Both had come out of transformative group renamings. It is against this idealistic background that "nigger" must be considered. "Nigger" comes from a history of subjugation and, at best, was an imposition on Blacks. Its very utterance fixed Blacks in a certain place in society.

*Webster's Collegiate Dictionary* says clearly that "nigger" is "disparaging and offensive." Yet from the seventeenth through the nineteenth centuries, the term was used interchangeably with "slave" or "negro." The 1933 *Oxford English Dictionary* equated "nigger" with "negro." But after the 1930s, "nigger" was beginning to wear thin its public acceptance; it was more often condemned, and not just by Blacks. It did not die quickly, as playwright Charles Fuller shows in *A Soldier's Play*, set in the Deep South of 1944, where the word takes center stage.

Complicating any discussion of "nigger" is an intricate Black American cultural reality. For Blacks, the word has both an inside and an outside meaning. If a non-Black assaults a Black with "nigger," the Black person feels justified in expressing outrage, in defending himself or herself. On the other hand, "nigger" has attained a value within Black society as a term of affection, bonding, and historical kinship. Actually, as any local Black community speech specialist will say, it is not "nigger" but "nigga" that has power among Blacks. In other words, the two words are not the same: "Nigger" is the outside word. "Nigga" is the inside word.

"When used by Black people among themselves," writes novelist and poet Clarence Majors in *Juba to Jive* (1994), "it is a racial term with undertones of warmth and goodwill—reflecting aside from the irony, a tragicomic sensibility that is aware . . . of the emotional history of the race." Geneva Smitherman, in *Black Talk* (1994), makes distinctions among the uses of the word *nigga,* saying that it has positive, neutral, and negative connotations. She demonstrates some positive uses: "She my main nigga" (meaning she is a close friend) and "He real, he is a shonuff nigga" (he lives Black traditions).

Just because "nigger" is appearing more does not mean it is unopposed. Communities, under pressure, are removing offensive place names: "Niggertown Marsh" (in New York), and "Nigger Bridge Road" (in Maine). Sadly, the Maine change erased a free Black settlement's name. Parents fight to get Mark Twain's *The Adventures of Huckleberry Finn* (1884) removed from high school reading lists, because Twain used "nigger" at least 160 times in the book. The National Cathedral School and Saint Albans, in Washington, D.C., no longer require the book and use it only in eleventh- and twelfth-grade elective courses.

And miracle of miracles, Richard Pryor, in his frank *Pryor Convictions* (1995), has renounced the "N-word," after an early career of using it extravagantly. His conversion came at the end of a trip to East Africa: "After three weeks, I had no doubt that being in Africa had a profound effect on me. It seemed especially so when it came time to return to the United States. The land had been timeless, the people majestic . . . I left enlightened."

He concluded: "I left also regretting I ever uttered the word 'nigger' on a stage or off it. It was a wretched word. Its connotations weren't funny, even when people laughed. To this day, I wish I'd never said the word . . . so, I vowed never to say it again. . . . A voice said, 'You see any niggers [in Africa]?' I said, 'No.'"

## What and who defines the new Black intellectual and artistic culture?

The writer bell hooks (born 1955) is an emblem of the new Black intellectual culture. Writing since the early 1980s, when she published *Ain't I a Woman* (1981) and feminist essays, hooks has worked hard to become a leader in the burgeoning Black intellectual/art scene. (Gloria Watkins is her real name, bell hooks—in lowercase letters—is the writer's pseudonym.)

Now, everyone wants part of "bell." When she recently accepted a City University of New York professorship, arts-chic *Bomb* magazine interviewed her. Talk shows featured her. She has buzz on the lecture circuit. The *Chronicle of Higher Education* put a foot-tall color photo of her on its cover. Even hip culture and fashion magazine *Paper* has profiled her. Now her audience is vast, spanning all groups, and other writers and artists are often inspired by her work.

Hooks's writing is typical of the new Black intellectual-artist class and its products. Her work became popular in the late eighties. It is accessible, easy to understand, and close to the issues that are on people's minds. Her work critiques traditional forms (such as novels, poetry, drama), but it is equally adept at dealing with trendier popular culture. She is intellectually omnivorous. She consumes and comments on everything that shapes our collective moods. She can speak with authority on everything from the Whitney art show on the Black male, to Black Indians, to self-healing among women, to rappers. Like her peers, hooks can be deeply subjective in her work, almost confessional, interrogating herself and using personal experiences to propel her argument.

Her main goal is to educate people about society's complexes that maim full individual expression—male dominance, female self-doubt, imperialist thinking, racism, self-censorship, and ethnic mythologies. Often, Blacks disagree with her occasionally blunt style and, on first contact, with her feminism, since it critiques Black male-female relations. But hooks wins over even her critics by being a caring educator and offering a positive plan for self-transformation.

In March 1995, two national magazines headlined Black intellectuals. *Atlantic* ran Robert S. Boynton's long report on "The New Intellectuals." The cover was classic: a tight Black fist clenched a Mont Blanc pen—the Black power salute had become a symbol of Black intellectual assertion. The cover's byline proclaimed, "Sophisticated, political, morally aware, at times fiercely contentious—the breed known as 'public intellectuals' has seemed close to extinction. Suddenly, they're back. And they're Black." Boynton questioned whether today's Black public intellectuals were the successors of the earlier Jewish public intellectuals, such as Irving Howe, Philip Rahv, and Lionel Trilling. Within days, the *New Republic* followed with Leon Wieseltier's "The Decline of the Black Intellectual," a denunciation of Cornel West's ideas. Despite the two magazines contradicting each other, they were dramatic evidence that Black thinkers are being taken seriously.

The list of these thinkers is long. Henry Louis Gates Jr.'s name is not yet a household word, but it is fast becoming so. Having started in the 1980s with exciting readings of Black literature, Gates has gone on to become a Harvard professor, a much-discussed essayist (lately in the *New Yorker* where he is an editor), and a cultural impresario, as when he convinced Spike Lee to teach a film course at Harvard. Cornel West has taken philosophy—that thin-air, logic-ridden field—and applied it to the nation's social conflicts. In 1991, he teamed up with bell hooks for a dialogue on a wide range of issues, which later became *Breaking Bread.* Since *Race Matters,* his attempt to get beyond racial impasses, landed on the best-seller list in 1993, West has become a major public pedagogue, whose philosophical discourse is laced with Black church oratory and street metaphor. (A favorite is "chocolate cities," meaning Black-majority American cities. The term was first applied to Washington, D.C.)

Beyond Gates and West lies a galaxy of intellectuals and artists: Arnold Rampersad (biographer and collaborator on autobiography), Toni Morrison (historical novelist, critic), Martin Puryear (sculptor), Lorna Simpson (photographer), Ben Carson (neurosurgeon, youth motivator), August Wilson (historical playwright), Robert Woodson (self-help expert), Michael Dyson (cultural critic), Julianne Malveaux (economist, columnist), Jerome Washington (prison writer), Darryl Pinckney (novelist, biography scholar), Hortense Spillers (literary critic), Lawrence Bobo (sociologist, public opinion expert), Mary Helen Washington (literary critic), Gerald Early (essayist, cultural critic), Barbara Christian (feminist literary critic), Patricia Williams (writer of critical racial and legal studies), Rita Dove (poet), Brent Staples (autobiographer, *New York Times* editor), William Julius Wilson (sociologist), Jill Nelson (journalist, autobiographer), Thulani Davis (novelist, librettist), Darlene Clark Hine (historian), David Hammons (avant-garde conceptual artist), Nelson George (movies, sports, music commentator), Frances Trudier (cultural historian), Derrick Bell (law professor, activist, and theorist), Betty Saar (ceramist), Ronald Walters (political scientist), Kobena Mercer (social critic), Mary Berry (historian), Bernice Reagon (music curator, performer), Robin D.G. Kelley (historian), David Levering Lewis (biographer and historian), Ellis Cose (journalist, social critic), Claude Steele (social psychologist), Faith Ringgold (narrative quilt artist), Stanley Crouch (essayist, biographer), Anna Deveare Smith (playwright, dramatist), Randall Kennedy (law professor, magazine editor), Wynton Marsalis (musician and music interpreter), John Wideman (novelist, essayist), Glenn Loury (policy scientist, economist), Bill T. Jones (dance choreographer), Molefi Asante (theorist), Shelby Steele (student of race dilemmas), Manning Marable (political intellectual, social critic), Lisa Jones (essayist), Essex Hemphill (poet of gayness), Suzan-Lori Parks (playwright), George C. Wolfe (playwright, New York Public Theater director), Walter Mosley (detective fiction writer), Kwame Anthony Appiah (philosopher, race critic), Gwen Daye Richardson (editor), Orlando Patterson (sociological historian), V. Y. Mudimbe (philosopher), Idris Ackamoor (performance artist), June Jordan (essayist and poet), Trey Ellis (novelist and literary critic), Garth Fagan (dance choreographer), Rhodessa Jones (dramatist of women's lives in prison), Stephen Carter (law professor, writer on race, religion, ethics), Bobby McFerrin (musician), Carl Franklin (filmmaker), Evelyn Brooks Higginbotham (historian), Jamaica Kincaid (novelist), Charles Ogletree (law professor), Forest Whittaker (filmmaker), Leslie Harris (filmmaker), Kristal Brent Zook (feminist writier), and Sara Lawrence-Lightfoot (educator, writer).

Today, America is profiting from these cultural visionaries. In fact, difficult problems—such as Jewish-Black relations—have been helped by their interventions. Cornel West and *Tikkun* editor Michael Lerner have been traveling the country discussing the issues of Black and Jewish identity in connection with their book *Jews and Blacks: Let the Healing Begin* (1995). Beyond this, the main problem these intellectuals and artists face is that their popularity might cause them to lose their edge. A certain tension has been essential to them, an at-odds spirit, a willingness to criticize—for that matter, to criticize even Blacks. Historically, kudos and acclaim have been the enemies of intellectuals. Their story's next chapter will tell us whether this will be true for these intellectuals as well.

### What does the future hold for Black America?

Today, across Black America, a historic conversation is in progress. To sample it, search late-night cable TV. Northern California's Channel 41 *Visions* program is typical: four guests and a moderator, talking freely about Black issues and taking remarkably candid call-ins. Usually, the program is serious and instructive. Smaller channels carry similar programs throughout the country.

Added to these are other conversation sites—political clubs, churches, youth centers, radio talk shows, men's and women's groups, and increasingly, bookstores. Then add the many new pamphlets, sold almost exclusively in Black-owned shops, with titles like "Whither Black America?" and "Can We Save Ourselves?" The result is a gigantic palaver, like the communal dialogues of old Africa's villages. But this palaver is taking place in America, as a gigantic coast-to-coast discussion.

The century is coming to an end. Understandably, a big question hanging over this national discussion is, what will the future bring? Will Black America be prepared to handle what the future delivers? What trends are likely? What are the best strategies for coping with these trends? Questions about the future press in on Black Americans as they stand at a truly historic crossroads. At this moment, many Blacks are torn between being optimistic and being pessimistic about the future. Is the future to be feared or welcomed?

On the one hand, Blacks are still scaling the heights of achievement. This year, Dr. Bernard Harris became the first

Black astronaut to walk in space. Recently, Isabel Wilkerson won a Pulitzer prize for news features. Carlton Guthrie has become a successful auto-parts manufacturer. Chicago high school students are state chess champions. A Black was elected to be sheriff in Jacksonville, Florida, a district that is over 70 percent white. When sociologist Andrew Beveridge analyzed 1990 census data, he found that the Black middle class had grown steadily over the last generation, despite some setbacks. In fact, in ten locations across the country of populations over fifty thousand, he even found Black median household income was greater than white median household income. Blacks continue to make their mark.

On the other hand, every day brings news of difficulties Blacks face. Unemployment still remains high for too many Blacks. For example, Katherine Newman—a Columbia University anthropologist—has shown that Harlem Blacks who diligently pursue jobs, who have proper credentials, and who work hard to keep jobs still fail to secure lasting employment. Added to unemployment, the twin epidemics of AIDS and crack cocaine have taken a disproportionate toll on Blacks. And then there are the factors of Black family instability, illiteracy among Black elementary schoolers . . . The list seems to lengthen daily. Writer Ishmael Reed is right when he tells us, in *Airing Dirty Laundry* (1993), that Black America is not—as the media often conclude—just a problem-ridden subnation. Moreover, Reed says the media is silent on "white pathology." Regardless, on their journey to the future, Black Americans will confront problems.

Today, Black America's condition is one of upsides and downsides. Analyzing this condition allows some predictions to be made of trends that are likely in the future of Black America:

- *Education:* Education will continue to be the crux of Black advancement. It will be the key to Blacks' jobs future and earning power.

  Changes for the better are already evident. A National Science Foundation study reports steady gains by Black pupils in math and science scores. College-entrance exam scores show gradual improvements. In 1994, Skylar Byrd of Washington, D.C.'s Banneker High got a perfect SAT score.

Recently, Blacks' test scores in New York City's schools rose. Many mentoring projects are improving Black school performance. For example, Claude Mayberry publishes *Science Weekly* to stimulate children's appreciation of science and math.

Expect much more activity in this field, especially in high school job skills training.

- *Business and Money:* Opening businesses and accumulating money will be the new campaign of Black America.

  In 1995, the one hundred top Black-owned firms posted a second consecutive postrecession year of solid growth. Their income rose by an astonishing 14 percent, to twelve billion dollars. New businesses, like Karl Kani sportswear, had exciting growth. A *Harvard Business Review* (May-June 1995) study claims inner-city areas offer competitive advantages for future business.

  Expect the proliferation of small Black businesses, a relentless search for start-up capital, and an aggressive entrepreneurial class. Already, the direct sales industry has 13 percent Black salespeople.

- *Political Leadership:* The Black leadership class will undergo great changes.

  After the recent Supreme Court decision (*Miller v. Johnson,* June 29, 1995), more gains in Congress might be stymied. In their decision, the court cast doubt on the constitutionality of the new Black-majority congressional districts in Georgia, North Carolina, and Louisiana. Another factor affecting Black political leadership is the aging of the civil rights pioneers. And, some Blacks are finding that they can be successful candidates in districts with a variety of populations.

  Expect Black elected officials to increase in numbers. But the political class will consist increasingly of new faces. Leadership from the civil rights movement will gradually shrink in numbers, even though its issues will remain strong. And Black political leaders will share the spotlight with non-politicians—entrepreneurs, local activists, intellectuals, artists, and entertainment and sports figures.

- *The Definition of Black America:* Future visions of Black America will be more inclusive.

  Emphasis will be on tolerance for internal diversity. Monolithic Black America will be a dinosaur. Cultural critic Henry Louis Gates Jr. has urged, "Black America . . . needs a discourse of race that is not centrally concerned with racial unanimity." Some call for a return to 1955–65 Black America, a time of variety in ideas.

  Expect many more Black voices to be valued. Also, immigrant voices—from Africa and the Caribbean—will be embraced, as will those of people with interracial backgrounds. Black America will mature into a "big-tent" society.

- *New Black Cultural Icons:* Major Black personalities, with universal appeal, will emerge.

  Oprah Winfrey (born 1954) and Colin Powell (born 1937) are this trend's current examples. In the 1980s, Winfrey created a hugely successful television empire. *Essence*'s twenty-fifth anniversary issue (May 1995) rewarded her accomplishments with an exquisite cover photo. Colin Powell, with an unparalleled military record, is now promoted by Citizens for Colin Powell for the presidency and is headed for best-sellerdom with *American Journey* (1995), his autobiography.

  Expect more such figures to appear who, as journalist Lee Daniels says, are committed both to "cosmopolitanism" and "to the race."

- *Health and Wellness:* New grass roots movements promoting health will become popular.

  Recent crises—AIDS, exploding rates of hypertension and diabetes, high infant mortality—have forced the creation of this movement. But basic worries over diet and nutrition will assist it. Already, there are health guides (mostly for Black women), *heart & soul* magazine for Black health themes, cookbooks for "lite" Black cuisine, and Afrocentric exercise tapes. Jenny Craig appeals regularly to the Black women's market.

Expect an era of health improvement for Blacks. This new focus will be touted as self-investment, a form of wealth.

- *Computer Literacy:* Joining the computer age will be a priority among Black Americans.

  Lena Williams, in "Computer Gap Worries Blacks" (*New York Times,* May 25, 1995), showed that Blacks are not abreast of the computer revolution. Few families—only 11 percent (versus 39 percent of white families)—have bought computers. Hence, too few youth are being exposed to the new technology. Responses to this problem are forthcoming. Los Angeles's Black churches are sponsoring computer workshops. Netnoir, a San Francisco Black company, is on the Internet with commentary and news on Black America.

  Expect a push to bring computer literacy to Black communities.

- *Nonviolence in the Future:* Nonviolence will be taught among Black Americans.

  Homicide rates among young Blacks are staggering, particularly among Black men. Marita Golden's *Saving Our Sons: Raising Black Children in a Turbulent World* (1995) is a moving plea for a nonviolent Black America. "Death, like some medieval plague, is now with us," she writes.

  Expect Dr. Martin Luther King Jr.'s message on nonviolence to be updated and workshops on peaceful resolutions to disputes to be specially designed for Blacks.

- *Black Places of Worship:* Black places of worship will increasingly become centers of community uplift.

  Rev. Johnny Ray Youngblood's Saint Paul's Community Baptist Church in New York City—which has built housing, helped family stores into existence, counseled drug abusers, youth, and men—offers a model of things to come. Already, Black churches have outreach programs. Islamic mosques are energized by such projects too. Other sects, such as Buddhist communities, are also engaged in Black outreach.

Expect religious groups to accelerate their social, community-building commitment.

- *Achievement and Spirit:* Individual and group achievement will greatly surpass expectations. Coming decades will witness yet another major leap forward by Black Americans.

  Spiritual renewal, so essential to a positive future, is flowering. Analysts say the worst Black problems are slowly bottoming out. Even events viewed as setbacks, like attacks on affirmative action, act as a "wake-up call." Black history is now being used to show discipline, intelligence, and self-respect, not just past glories. The Urban League's Hugh Price has urged on Black audiences new disciplines such as each middle-class Black nurturing a Black youth from a less-advantaged background. He has found ready listeners for this and other proposals.

  Expect a fireworks of talent and accomplishment that will continue well into the next century. Historical optimism is in order.

## RESOURCES

Appiah, Kwame Anthony. *In My Father's House: Africans in the Philosophy of Culture.* New York: Oxford University Press, 1992.

Berman, Paul, ed. *Blacks and Jews: Attitudes and Arguments.* New York: Delacorte Press, 1994.

Boyd, Herb, and Robert Allen, eds. *Brotherman: The Odyssey of Black Men in America—An Anthology.* New York: Ballantine Books, 1995.

Chauncey, George. *Gay New York: Gender, Urban Culture and the Making of the Gay Male World, 1890–1940.* New York: Basic Books, 1994.

Cosby, Camille O. *Television's Imageable Influences: The Self-Perception of Young African-Americans.* New York: University Press of America, 1994.

Dash, Julie. *Daughters of the Dust: The Making of a an African-American Woman's Film.* New York: New Press, 1992.

Dates, Jannette L., and William Barlow. *Split Image: African-Americans in the Mass Media.* Washington, D.C.: Howard University Press, 1993.

Dawson, Michael C. *Behind the Mule: Race and Class in African-American Politics.* Princeton, N.J.: Princeton University Press, 1994.

Early, Gerald. *Lure and Loathing: Essays on Race, Identity, and the Achievement of Assimilation.* New York: Penguin Books, 1993.

Fraser, Steven. *The Bell Curve Wars: Race, Intelligence, and the Future of America.* New York: HarperCollins, 1995.

George, Nelson. *Blackface: Reflections on African-Americans and the Movies.* New York: HarperCollins, 1994.

Goings, Kenneth W. *Mammy and Uncle Mose: Black Collectibles and American Stereotyping.* Bloomington: Indiana University Press, 1994.

hooks, bell. *Art on My Mind: Visual Politics.* New York: New Press, 1995.

Mercer, Kobena. *Welcome to the Jungle: New Positions in Black Cultural Studies.* New York: Routledge, 1994.

Turner, Patricia A. *Ceramic Uncles and Celluloid Mammies: Black Images and Their Influence on Culture.* New York: Anchor Books, 1994.

Watkins, Mel. *On the Real Side.* New York: Simon & Schuster, 1994.

West, Cornel, and Michael Lerner. *Jews and Blacks: Let the Healing Begin.* New York: Putnam and Sons, 1995.

TEXTS FOR YOUTH

Blue, Rose, and Corinne J. Naden. *Whoopi Goldberg: Entertainer.* New York: Chelsea Press, 1995.

Carrington, Mellonee. *Carol Moseley-Braun: Breaking Barriers.* Chicago: Children's Press, 1993. Text with photos.

Century, Douglass. *Toni Morrison: Author.* New York: Chelsea Press, 1994.

Dolan, Sean. *Michael Jordan: Basketball Great.* New York: Chelsea House, 1994.

Hunter, Latoya. *The Diary of Latoya Hunter: My First Year in Junior High.* New York: Vintage Books, 1992.

Moutoussamy-Ashe, Jeanne. *Daddy and Me: A Photo Story of Arthur Ashe and His Daughter Cameron.* New York: Alfred A. Knopf, 1993.

Nicholson, Lois P. *Oprah Winfrey: Entertainer.* New York: Chelsea Press, 1994.

Stefoff, Rebecca. *Nelson Mandela: Hero for Democracy.* New York: Fawcett Books, 1994.

# Index

## About the Author

KENNELL JACKSON is an associate professor of history at Stanford University, where he teaches courses in both African and Black American history. He was director of the Stanford program in Afro-American Studies for nearly a decade in the 1980s. In his Stanford teaching career, he has won the Dinkelspiel Award and the Allan V. Cox Medal, both honoring his work in undergraduate education. He relishes lecturing and writing op-eds on Black history.